DIFFRACTION 2002:
Interpretation of the New Diffractive Phenomena in Quantum Chromodynamics and in the S-Matrix Theory

NATO Science Series

A Series presenting the results of scientific meetings supported under the NATO Science Programme.

The Series is published by IOS Press, Amsterdam, and Kluwer Academic Publishers in conjunction with the NATO Scientific Affairs Division

Sub-Series

I. Life and Behavioural Sciences	IOS Press
II. Mathematics, Physics and Chemistry	Kluwer Academic Publishers
III. Computer and Systems Science	IOS Press
IV. Earth and Environmental Sciences	Kluwer Academic Publishers
V. Science and Technology Policy	IOS Press

The NATO Science Series continues the series of books published formerly as the NATO ASI Series.

The NATO Science Programme offers support for collaboration in civil science between scientists of countries of the Euro-Atlantic Partnership Council. The types of scientific meeting generally supported are "Advanced Study Institutes" and "Advanced Research Workshops", although other types of meeting are supported from time to time. The NATO Science Series collects together the results of these meetings. The meetings are co-organized bij scientists from NATO countries and scientists from NATO's Partner countries – countries of the CIS and Central and Eastern Europe.

Advanced Study Institutes are high-level tutorial courses offering in-depth study of latest advances in a field.
Advanced Research Workshops are expert meetings aimed at critical assessment of a field, and identification of directions for future action.

As a consequence of the restructuring of the NATO Science Programme in 1999, the NATO Science Series has been re-organised and there are currently Five Sub-series as noted above. Please consult the following web sites for information on previous volumes published in the Series, as well as details of earlier Sub-series.

http://www.nato.int/science
http://www.wkap.nl
http://www.iospress.nl
http://www.wtv-books.de/nato-pco.htm

Series II: Mathematics, Physics and Chemistry – Vol. 101

DIFFRACTION 2002: Interpretation of the New Diffractive Phenomena in Quantum Chromodynamics and in the S-Matrix Theory

edited by

R. Fiore

Dipartimento di Fisica,
Università degli Studi della Calabria, Cosenza, Italy

L.L. Jenkovszky

Bogolyubov Institute for Theoretical Physics,
Academy of Sciences of Ukraine, Kiev, Ukraine

M.I. Kotsky

Budker Institute for Nuclear Physics,
Novosibirsk, Russia

V.K. Magas

CFIF, Physics Department, Instituto Superior Technico,
Lisbon, Portugal

and

A. Papa

Dipartimento di Fisica,
Università degli Studi della Calabria, Cosenza, Italy

Kluwer Academic Publishers

Dordrecht / Boston / London

Published in cooperation with NATO Scientific Affairs Division

Proceedings of the NATO Advanced Research Workshop on
Interpretation of the New Diffractive Phenomena in Quantum Chromodynamics and in the S-Matrix Theory
Alushta, Crimea, Ukraine
31 August–6 September 2002

A C.I.P. Catalogue record for this book is available from the Library of Congress.

ISBN 1-4020-1306-X (HB)
ISBN 1-4020-1307-8 (PB)

Published by Kluwer Academic Publishers,
P.O. Box 17, 3300 AA Dordrecht, The Netherlands.

Sold and distributed in North, Central and South America
by Kluwer Academic Publishers,
101 Philip Drive, Norwell, MA 02061, U.S.A.

In all other countries, sold and distributed
by Kluwer Academic Publishers,
P.O. Box 322, 3300 AH Dordrecht, The Netherlands.

Printed on acid-free paper

Printed in the Netherlands.

TABLE OF CONTENTS

PREFACE

Diffraction 2002, International Workshop on Diffraction in High-Energy Physics and NATO Advanced Research Workshop, was held in Alushta (Crimea, Ukraine) from August 31 to September 5, 2002, in a beautiful resort "Dubna" near the Black Sea. The Workshop was the second of the series "Diffraction" started two years earlier in Cetraro, Italy.

The Workshop was organized by an International Committee including the organizers of Diffraction 2000 coming from Università della Calabria (Italy) and "local" organizers from Dubna, Kiev and Novosibirsk research institutions. There were 46 participants, coming from 14 countries.

The purpose of the Workshop was to review the experimental and theoretical aspects of Diffraction in high-energy physics and to discuss the new developments. There were talks devoted to Diffraction in hadron-hadron collisions, in lepton-hadron collisions and in Quantum Chromodynamics.

This volume contains the written version of 29 talks, that arrived before the deadline and ordered, somewhat arbitrarily, as experimental, phenomenological and theoretical ones. We thank all the speakers and attendees for their contribution to the scientific success of the Workshop.

The Secretariat of the Workshop was admirably held by Alla Borisenko, Tania Korzhinskay, Elena Rusakovich, Olga Ugrumova and Zoya Vakhnenko, whom we thank for their invaluable assistance.

We also thank the stuff of the Joint Institute for Nuclear Physics and in particular Prof. P.N. Bogolyubov for his professional performance in organizing and managing the computer service.

The invaluable financial support of NATO, grant ARW 977335, is gratefully acknowledged.

The next workshop from this series, "Diffraction 2004", is planned to be held in September 2004 in Dubrovnik, Croatia.

DIFFRACTIVE PHYSICS IN CDF - RUN 2

M.G.ALBROW
Fermilab
MS 122, Wilson Road
Batavia, IL 60510, USA

Abstract. After a brief review of the diffractive physics done by the Collider Detector at Fermilab (CDF) in Run 1 (ending 1996), I describe the improved detectors of Run 2 (starting 2001) and some of the new studies they make possible. These range from low mass exclusive central systems produced by double pomeron exchange, to high mass $b\bar{b}$ dijets by $D\mathbb{P}E$. I comment on the possibility of a Higgs signal in the latter channel at the Tevatron (unlikely) or at the LHC (less unlikely). Some related processes which could constrain the theoretical estimates are discussed.

1. CDF Diffractive Physics in Run 1

There have been three phases of diffractive physics in CDF which, although a "fringe" activity compared to high mass physics (e.g. top quark physics and supersymmetry searches) have nevertheless resulted in over a dozen publications. Many of these diffractive studies are in the high p_T realm, such as the diffractive production of W's and high E_T jets.

Phase I, prior to 1993, used tracking with silicon strip detectors in Roman Pots on both downstream arms (E for p's and W for $\bar{p}$'s). Elastic scattering at small $|t|$ (below 0.3 GeV^2) was measured at $\sqrt{s}$ = 546 GeV and 1800 GeV, [1] together with σ_T [2], and single diffractive dissociation inclusive measurements were made, $\bar{p} + p \rightarrow \bar{p} + X$ [3].

In phase II a series of measurements were made using the large rapidity gap signature. It was found[4][5] that events with a high p_T jet at large +ve rapidity and one at large -ve rapidity have a rapidity gap between them more frequently than would be expected by downward fluctuations of the particle density, at the 1% level. Such events had been predicted by Bjorken[6]. If we associate pomeron exchange with large rapidity gaps, this "pomeron"

R. Fiore et al. (eds.), Diffraction 2002, 1–12.

has extremely large $|t| \approx E_T^2 \approx 1000$ GeV^2. A more realistic picture is that a parton-parton scatter occurs at high Q^2 (on a very short time scale) and soft gluon(s) are exchanged on a much longer time scale to create the rapidity gap. At no one time is an "isolated" color singlet entity present. This is the Soft Color Interaction (SCI) model[7]. We can call these events Jet-Gap-Jet or JGJ events. Most work in phase II was on **hard** diffraction, with a large gap in one direction and a cluster containing high E_T jets[8], high E_T b-quarks[9], J/ψ[10] or W [11]. All these processes have similar but not identical ratios $\frac{gap\ events}{all\ events}$ at the level of 1%. From the differences one can derive an approximate ratio $\frac{gluon}{gluon+quark}$ in the exchange ("pomeron"). These (GJJ, GW etc), like the JGJ events, may be best visualised as a hard gluon exchange with soft exchange(s) on different time-scales.

For Phase III we added new Roman Pot detectors about 56 m downstream (for the $\bar{p}$) of the interaction point. Diffractively scattered antiprotons, with $\xi = 1 - x_F \approx 0.04 - 0.10$, pass through 3 superconducting quadrupoles, an electrostatic beam separator and 3 s/c dipoles before reaching the pots. There are 3 pots (moveable vacuum vessels that allow one to put detectors close to the beams) spaced over 3 m, each containing a 2 cm $\times$ 2 cm scintillator for triggering and an (x, y) fiber hodoscope for tracking. These gave a resolution $\sigma_{x,y} \approx 100\mu$m ($\sigma_t \approx 0.07$ GeV^2, $\sigma_\xi \approx 0.001$). The pots provided a diffractive trigger to read out all of the CDF detectors, both prescaled and in coincidence with high E_T jets. These were only installed for the last two months of Run 1 but allowed several physics studies to be done. We measured[12] SD dijets ($PGJJ$) where P means (anti)proton at $\sqrt{s} = 630(1800)$ GeV. Diffractive events were measured with jets up to $E_T \approx 20(35)$ GeV. This allowed us to evaluate the diffractive structure function of the antiproton and compare it to expectations from diffractive DIS at HERA. We then looked in these events for rapidity gaps in the proton direction [$PGJJG$] as candidates for $DI\!PE$ dijet production, yielding[13] a very significant (although with very modest statistics) signal (132 events on a background of 14 events). Note that this data was taken in a very short time at low luminosity with a trigger not optimised for $DI\!PE$. In Run 2 we can get such a sample in < 30 minutes. Comparing the $DI\!PE$, SD and non-diffractive (ND) dijet rates we find a breakdown of factorization: The $DI\!PE$/SD ratio is about 5 $\times$ the SD/ND ratio. A third study was done[14] with the $\bar{p}$ triggered data of additional gaps in soft SD events: [$PGXGX$]. Given a SD event, what is the probability of having an extra central gap? It is found to be, for $\Delta\eta > 3$ gaps, about 25% (19%) at $\sqrt{s} = 1800$ (630) GeV. This is a factor 2 - 3 higher than the probablility of having such a gap in non-diffractive events [XGX]. Once one has a large gap in an event, the cost of having a second gap is less. This is understandable in terms of gap survival probability; if one gap is not killed by a secondary interaction another gap is more likely to survive.

2. Run 2 Goals

Hard non-diffractive processes, like W and $t\bar{t}$ production, can be calculated quite accurately with Monte Carlo simulations like PYTHIA, ISAJET, HERWIG etc. We can claim to understand these processes rather well (even though not all the input in such programs is perturbative QCD). If you run PYTHIA for W production it will not generate events with large rapidity gaps at the measured level. POMPYT is an example that includes diffraction but in an empirical, data-driven way. When these processes are better understood it will hopefully be possible to have an event generator for diffractive processes that is based on an underlying theory, and which can be tested by beating it against data. We need more data to point the way and to discover the features of this physics that must be incorporated into a Monte Carlo.

The Tevatron is a great place for diffractive physics, having a rapidity span of $\Delta y = ln\frac{s}{m_p^2} = 15.3$ (19.2 at LHC). So we can have longer gaps than anywhere; there is room for 2 or even 3 gaps > 3. We need data on many processes (SD, $D\mathbb{P}E$, $[PGXGX]$, etc) and with many probes $(JJ, JJJJ, s, c, b, W, Z, \gamma, ...)$ to beat against models. CDF has a complex trigger system with 64 different (Level 1) trigger channels. In *all* of these events one can look off-line for gaps on one side or both sides of the central system, and/or look for a forward $\bar{p}$. Of course gaps can be poisoned if another interaction took place in the same beam-bunch crossing. The average number $< n >$ of inelastic events per crossing is given by:

$$< n >= L\sigma_{inel}\Delta t$$

which is $< n > = 1.2$ events/crossing if $L = 5.10^{31}$ cm^{-2}s^{-1}, $\sigma_{inel} = 60.10^{-27}$ cm^2 and $\Delta t = 4.10^{-7}$ s. Still, about 30% of the events occur singly, but that will drop to about 9% at $L = 10^{32}$ cm^{-2}s^{-1}. Gap physics should be done before the luminosity gets too high, while pot physics can probably be done at higher luminosity (this is to be demonstrated).

Part of the answer to the question "What are our Run 2 goals?" is simply "more of the same". In particular we have only scratched the surface of hard $D\mathbb{P}E$ physics (dijets and heavy flavors). However there are many completely new things we can do, and I shall discuss charm and several $D\mathbb{P}E$ studies, some of which are related to the idea that the Higgs boson might be detectable (at the LHC if not at the Tevatron) in $D\mathbb{P}E$. But first I give an overview of the Run 2 detectors.

3. Detectors in Run 2

In the 5-year shut down from 1996 - 2001 most of CDF was rebuilt for the ability to run faster and with improved performance for tracking, calorimetry and triggering. The silicon precision tracking covers a larger solid angle and has more layers, substantially improving beauty and charm physics. Outside that the drift chamber tracking has many more wires and shorter drift times. A barrel of Time of Flight (TOF) counters has been added for π, K, p identification; this will be very valuable for low-mass exclusive $DIPE$ studies, discussed later. The forward ("Plug") calorimetry been replaced with scintillator tile read-out (it was proportional chambers), better shower maximum strip detectors, etc. The detectors for forward and diffractive physics are also improved. On both (E and W) sides the reach in η of the Plug calorimeters is extended with Miniplug calorimeters (built at Rockefeller University) which are lead plate stacks immersed in liquid scintillator and read out with wavelength-shifting fibers and multichannel PMTs; they cover $3.5 \leq |\eta| \leq 5.5$. They can be used for very forward jet physics (e.g. JGJ events with a gap up to 7 units) and for forward rapidity gap physics. In front of (part of) these are Cerenkov Luminosity Counters (CLC) which detect charged particles. Beyond the Miniplugs, for $5.5 \leq \eta \leq 7.5$ are a set of scintillation counters hugging the beam pipe to detect showers generated by very small angle particles hitting the beam pipe; these serve as rapidity gap counters, and are used in gap triggers. On the outgoing $\bar{p}$ side we have the same roman pot detectors as in Run 1, but with new electronics.

4. Charm

At the Tevatron we have not yet measured diffractive (SD and $DIPE$) production of charm. Charm will be produced almost exclusively by $gg \rightarrow c\bar{c}$, the $c\bar{c}$ content of the pomeron being small. This is therefore a good tool for the gluonic content of the pomeron. In Run 2 CDF is already accumulating very high statistics on inclusive (ND) charm production[15] thanks to the silicon vertex detector and a trigger based on two tracks with $p_T > 2$ GeV/c and a displaced (by $> 100\mu$m) vertex. In a first look at 10 pb^{-1} of data we find 56,000 $D^o \rightarrow K\pi$, 6,000 $D^o \rightarrow K^+K^-$ and 2,000 $D^o \rightarrow \pi^+\pi^-$ all with signal:background better than 1:1. The mass plot of $\phi\pi$ with $\phi \rightarrow K^+K^-$ shows two nice peaks corresponding to the D^+ and D_s^+ with 1350 and 2360 events respectively, already enough to make a state-of-the-art measurement of their mass difference. In 1 fb^{-1} which we could expect in the next 2 years, we will have 100 $\times$ that data, more than 5.10^6 D's, of which probably more than 10^4 will be diffractive and perhaps thousands will be $DIPE$. Quite possibly the ratio $D_s^+ : D^+$ will be different in ND, SD and $DIPE$ events. In $DIPE$ events with the whole central system contained and well measured, when there is charm there will be anticharm. So $D\mathbb{P}E$ production of charm-anticharm could give some valuable information on the nature of the process.

In addition to open charm, production of hidden charm $c\bar{c}$, with states such as J/ψ and χ_c^o, is very interesting, both inclusive and exclusive. Non-diffractive J/ψ production was theoretically underestimated by a factor of several at the Tevatron, and the "pre"dictions were adjusted upwards by including effects such as the $c\bar{c}$ being produced in a color octet state and becoming a colorless onium state by soft gluon radiation. How will it be in $D\mathbb{P}E$? Exclusive J/ψ production is forbidden in $D\mathbb{P}E$ because it has $J^{PC} = 1^{--}$ and if it were observed would be evidence for pomeron-odderon $\mathbb{P}\ \mathbb{O}$ exchange. However it will be difficult to be sure there was not a soft photon around! Exclusive χ_c^o production is particularly interesting as it has quantum numbers $I^G J^{PC} = O^+O^{++}$ like the Higgs, and this process relates to diffractive Higgs production.

5. Diffractive Higgs?

Most Standard Model Higgs bosons (if they exist!) at the Tevatron are produced by gg-fusion through a top loop. The cross section for a 120 GeV H via $gg \rightarrow H$ is 700 fb at $\sqrt{s} = 2$ TeV, 70% of which decay to $b\bar{b}$. Observation by this channel is hopeless because of the huge background from $gg \rightarrow b\bar{b}$ by the strong interaction and the very poor $b\bar{b}$ mass resolution, of order 12 GeV (the $H(120)$ has a width of only 3.5 MeV, so one would love to have such a resolution!). There have been proposals[16] that H could be produced exclusively $p\bar{p} \rightarrow pH\bar{p}$ by $D\mathbb{P}E$. The color removed from the protons by the hard g is cancelled by another gluon (SCI). It was predicted that as much as 1% of the $H(120)$ might be produced this way, which would be about 25 events in 5 fb^{-1}. If this were true, and one can measure the outgoing p and $\bar{p}$ very well in precision roman pot detectors[17], the missing mass resolution σ_{MM} can be $\approx$ 250 MeV, a factor 50 better than the dijet mass resolution σ_{JJ}. Furthermore the continuum background is not the huge $gg \rightarrow b\bar{b}$ process but $D\mathbb{P}E \rightarrow b\bar{b}$ which is probably a factor $\approx$ 1000 smaller. *If right* this might make a Higgs discovery (and excellent mass measurement) possible up to $\approx$130 GeV in 10 fb^{-1} (which, however, is now considered a "stretch" target before the LHC start).

However, others[18] predict much smaller cross sections (by factors more than 100!) in which case it is impossible at the Tevatron (although one might find a few such events at the LHC). There are three main factors that reduce the exclusive cross section. (1) One must include a survival probability T^2 against gluon radiation (Sudakov) which may be ≈ 0.1. (2) There will be soft rescattering of spectators with probability $S^2 \approx 0.1$. (3) The outgoing p

and/or $\bar{p}$ can diffractively dissociate. A recent paper by Khoze, Martin and Ryskin (KMR)[18] analyses this. They get an exclusive cross section 0.2 fb at the Tevatron and 3 fb at LHC. LHC may get 10 fb^{-1} in a year, and with good p tagging (which is cheap ... roman pots with tiny but very high precision tracking detectors) something may be seen. Even if the Higgs is discovered another way, this would be very valuable as M_H could be measured with good resolution. It would be great if diffraction becomes valuable for Higgs physics; after all the Higgs has the quantum numbers of the vacuum!

5.1. RELATED EXCLUSIVE MEASUREMENTS

Given the importance of firm predictions for central Higgs production, measurements of related processes are valuable. Higgs production proceeds through a top-quark loop; a very similar process replacing the top loop with a b-quark loop produces the χ_b^o, with a c-quark loop the χ_c^o, and with a u-quark loop can produce two photons $\gamma\gamma$. In the χ_b^o and χ_c^o cases the color flow is rather different because they are hadrons. In the $\gamma\gamma$ case it is the same, but $\gamma\gamma$ will be produced dominantly by $q\bar{q}$ annihilation except (presumably) in the *fully* exclusive case $p\bar{p} \rightarrow p\gamma\gamma\bar{p}$. For the χ states, as well as measuring the exclusive process, it would be useful to measure events where one or both beam particles dissociates diffractively (still with big rapidity gaps) and also allowing some additional central particles. This is because central Higgs production by $D\mathbb{P}E$ but not exclusive would still be very interesting, even if the missing mass method cannot then be used.

Some other related processes that need to be measured, and are of course of great interest quite independent of the Higgs issue, are $D\mathbb{P}E$ dijet production and $D\mathbb{P}E$ $b\bar{b}$ dijets. In these cases one can not say whether the process is "exclusive" but one can look for an excess of events where the jets take most of the central energy: $X = \frac{M_{JJ}}{M_{cen}} \approx 1$.

5.2. DOUBLE POMERON DIJETS

In CDF we observed[13] dijet production by $D\mathbb{P}E$, with the $\bar{p}$ in the pots, $0.035 \leq \xi \leq 0.095$, two central jets with $E_T \geq 7$ GeV, and a rapidity gap on the p side with $2.4 \leq \eta \leq 5.9$. From kinematics we can infer that $\approx 0.01 \leq \xi_p \leq 0.03$ using:

$$\xi_p = 1 - x_p = \frac{1}{\sqrt{s}} \sum_{all} E_T e^{-\eta}$$

summing over all particles. The shape of the jet spectrum is not visibly different from that in SD events, but the statistics are very modest. There are ≈ 120 signal events corresponding to a cross section $\sigma = 43.6 \pm 4.4 \pm$

21.6 nb. This data was taken in a very short run and with a SD trigger. At a luminosity of $L = 5.10^{31}$ cm^{-2}s^{-1} with a better trigger we *could* get 25,000 such events per day. In practice we are limited in the number of diffractive triggers we can record per second (about 2.5) and will make a more restrictive (higher threshold) trigger. In one year of running we should be able to extend the data above $M_{JJ} \approx 100$ GeV.

Of special interest is the subset of $b\bar{b}$ dijets. In $DIPE$ exclusive production of 2 (and only 2) jets nearly all events are gg, with $< 1\%$ being $b\bar{b}$[19] and even fewer $u\bar{u}, d\bar{d}, s\bar{s}$. Color factors, quarks being spin $\frac{1}{2}$ and gluon polarization selection rules lead to $\frac{b\bar{b}}{gg} < 10^{-2}$, while the $J_z = 0$ rule by which the Born amplitude vanishes for $M_q = 0$, gives an additional suppression $\approx 5.10^{-2}$. So the $DIPE \to b\bar{b}$ mass spectrum will be smaller than the di-jet mass spectrum by a factor ≈ 1000, in which case the Higgs signal should be bigger than the continuum background even for a mass resolution of several GeV (and thus not needing the forward pots to give the missing mass). However, according to KMR, the fully exclusive signal is too small to be seen, $\sigma(p\bar{p} \to pH\bar{p}) \approx 0.06$ fb, at the Tevatron with (say) 10 fb^{-1}. However if we allow low mass $p, \bar{p}$ dissociation and some limited central radiation, still with two large gaps, the cross section will be increased. This is something of a long shot, but has the great advantage of not needing any apparatus or triggers not already present in CDF (and D0). If one is optimistic the $DIPE$ $b\bar{b}$ dijet mass spectrum will have dropped out of sight before 100 GeV, with a handful of anomalous events around 110 - 130 GeV! The existing (one arm) roman pots can be used to check the mass scale; a good measurement of the $\xi_{\bar{p}}$, when an associated $\bar{p}$ is detected, can be compared with what is expected from the central dijets. Unfortunately this search, as it requires gaps, can only use single interactions in a bunch crossing, which reduces the effective luminosity.

5.3. DOUBLE POMERON χ AND $\gamma\gamma$

We consider χ_c^o and χ_b^o which have $I^G J^{PC} = 0^+0^{++}$ like the vacuum (and the Higgs!). The $\chi_c^o(1P)$ has $M = 3415$ MeV and $\Gamma = 16$ MeV. Most decay channels are not known, and those in the PDG have branching fractions $\approx$ 1-2%. Take e.g. $\chi_c^o \to K^+K^-$ (0.6%). In CDF the K are identified by Time-of-Flight (TOF) and by $\frac{dE}{dx}$ and the mass resolution is excellent. Another useful channel is $J/\psi\gamma$. The cross section at the Tevatron has been calculated to be about 735 nb[20] or 600 nb[19]. This is big! It means about 6.10^8 events in 1 fb^{-1}. However when we take into account the requirement of a single interaction, the branching fractions of the χ and, for the J/ψ, the branching to e^+e^- or $\mu^+\mu^-$ (12%), and the acceptance we are left with $\approx 2.10^4$ events. Even if this is optimistic by a factor 100 it still would mean a very impressive narrow peak! The problem is making a trigger that selects these events efficiently and yet runs at a low enough rate to be acceptable to the rest of CDF. The trigger that meets these criteria most effectively is to require forward gaps on both sides with "nothing" in $|\eta| \geq 2$ (tracks cannot be measured in this region) i.e. in the Plugs, Miniplugs, CLC and BSC, together with 2 - 4 charged particles in the TOF barrel (this covers azimuth ϕ with 216 counters, used in pairs in the trigger for technical reasons). We plan to test a subset of this trigger this year, refining it after studying the events.

The $\chi_b^o(9860)$ is very much harder. The exclusive cross section is 880 pb [20] or 120 pb[19]. No decay modes are given in the PDG except $\Upsilon\gamma$, said to be <6%, but this should be much smaller than $\chi_c^o \rightarrow J/\psi\gamma$ because there are so many more channels available, and the quark charge is less. When you put in all the factors you expect 1-2 events per fb^{-1}. At least it is easier to trigger on $\Upsilon \rightarrow \mu^+\mu^-$ than on J/ψ.

Most $\gamma\gamma$ pairs in $D\mathbb{P}E$ will be the result of $q\bar{q}$ annihilation from q in $\mathbb{P}$, which will leave other $\mathbb{P}$ fragments. POMWIG predictions for this have been made by Cox, Forshaw and Heinemann[21]. In the *exclusive* case, $gg \rightarrow \gamma\gamma$ through a u-loop is the dominant diagram and shares the feature with exclusive H that the (central) final state is colorless. The cross section at $M_{\gamma\gamma} = M_H$ is invisibly small, but it can be studied at much lower $M_{\gamma\gamma}$. Select the 2-photon $D\mathbb{P}E$ events, plot the central associated charged multiplicity n_{ass} distribution, and see if there is any peak at $n_{ass} = 0$. Quite independent of the Higgs question, $\gamma\gamma$ is an interesting test of $D\mathbb{P}E$ models.

6. Double Parton Scattering

Double parton scattering DPS has been observed in pp at the ISR[22] and in $p\bar{p}$ at the $Sp\bar{p}S$[23] and the Tevatron[24]. Measuring four jets (or a photon and three jets) one can distinguish this $2 \times (2 \rightarrow 2)$ process from double bremstrahlung DBS $2 \rightarrow 4$ through angular and E_T balance variables. The cross section (compared to the 2-jet cross section) depends on the size of the proton, or, if we study this in $D\mathbb{P}E$, on the size of the pomeron. Also in principle one can measure skewed parton distributions $F(\beta_1, \beta_2)$. Thus we address the questions: Is the pomeron compact? Are there correlations between the β's of two partons in a pomeron?

7. Rapidity Gaps between Jets

Both CDF[5] and D0[4] have observed JGJ events, by counting particles in a central region and observing an excess with $n = 0$. At $\sqrt{s} = 1800$ (630) GeV we find $\approx 1\%$ (2.7%) of the events in this JGJ class. The 4-momentum transfer across the gap is typically 40 GeV, which is clearly huge compared with a conventional pomeron; it is called "Color Singlet Exchange". We can think of these events as due to qq or gg scattering at high E_T as usual but with additional soft gluon exchange(s) cancelling the color (SCI). This costs a factor ≈ 10, with a similar additional factor, the so-called gap survival probability, to be sure that soft spectator interactions do not spoil the gap. The hard and soft exchanges have very different Q^2 and therefore time scales, and at no one time is there a color-singlet entity being exchanged. This may also be a good picture for diffractive dijet production.

This will be further studied in Run 2, in particular using the far forward Miniplugs. These can measure jets out to $\eta \approx 5$ and gaps up to $\Delta\eta \approx 8$. Note that $8 = 3 + 2 + 3$, so we can have central production between two rapidity gaps of 3 units each; $D\mathbb{P}E$ with very high $|t|$ pomerons! How does the central system compare with that between two low-$|t|$ pomerons?

8. Low Mass Exclusive Double Pomeron Exchange

By "Low Mass Exclusive $D\mathbb{P}E$" I mean events with central systems of mass less than about 5 GeV, completely reconstructed, and with large rapidity gaps ($\Delta y >\approx 4$) on both sides. Such events have never been measured above $\sqrt{s} = 63$ GeV (apart from $\alpha\alpha$ collisions in the ISR with $\sqrt{s} = 126$ GeV). At the ISR[25] and lower (fixed target) energies[26] the central mass spectra show remarkable structures. As the (soft) pomeron is mostly glue this is a "happy hunting ground" for glueballs[27] and hybrids, and is a good general tool for meson spectroscopy because of the "quantum number filter" $I^G J^{PC} = 0^+ even^{++}$. Glueballs are important for diffraction: Do glueballs lie on the pomeron trajectory and its daughters? (Clearly a scalar glueball cannot lie on the pomeron trajectory.) The $\pi^+\pi^-$ spectrum shows a striking drop ("cliff") at 980 MeV due to the $f^o(980)$ ... a resonance appearing as a drop rather than a peak! The ρ^o is absent at the higher ISR energy, evidence that pomeron exchange is dominant and (I=1) Regge exchange is very small. This is not true at the SPS(fixed target), showing that high energy (large rapidity gaps) is very important. At the Tevatron $\Delta y = 5.5$ gaps still allow ± 2 units for the central state. The only significant non-pomeron contribution might be the odderon. Seeing central C-odd states (such as the ϕ or J/ψ) would be a very interesting signature for the $\mathbb{O}$ as long as accompanying soft γ can be ruled out (which is difficult).

There are other good reasons for wanting to study low mass $D\mathbb{P}E$ at the Tevatron. Many channels have not been studied at all, e.g. $K^+K^-\pi^+\pi^-$ and $\phi\phi$, for glueballs. The observation of exclusive $p\bar{p}$ pairs[25, 28] means that hyperon pairs like $\Lambda\bar{\Lambda}$[28] up to even $\Omega\bar{\Omega}$ must be produced, depending only on their masses and the coupling of the pomeron to strangeness. Measuring the polarizations of the hyperons provides information on the spin structure of diffraction. The exclusive production of $D\bar{D}$ and χ_c states will determine the coupling to charm.

CDF is a fantastic detector for this physics, better than the ISR or fixed target detectors. It has excellent momentum (thus mass) resolution, π, K, p identification in the relevant region with both $\frac{dE}{dx}$ and TOF, and nearly 4π coverage with calorimeters and counters. On the negative side: (1) We have pots only on one side and the interesting $\bar{p}$ have $\xi \leq 0.01$ where there is no coverage. The events are thus GXG and the (t, ϕ) of the protons are unknown (2) We require single events, which means a factor 3 lower effective luminosity at $L = 5.10^{31}$ cm^{-2} s^{-1}; still, this is not small cross section physics (3) We need a good trigger, with a beam crossing and nothing forward, and something soft central (4) Last but not least, this physics gets very low priority in CDF compared with "hard" physics. But one might still get $\approx 10^6$ events per year (not the 10^8 one could get in a dedicated experiment!).

Many of the above remarks apply to D0 also. They have pots on both sides but the acceptance at low ξ is only at moderately large $|t|$. One could make a low missing mass (= central mass) trigger. Their central particle identification is lacking, but some useful physics can still be done.

9. Conclusions

The Tevatron has a larger rapidity range ($\Delta y = 15.3$) than anywhere until the LHC, and this was well exploited in Run 1. There is much more to do in hard (jets, W/Z, b, c, ...) and soft (down to $p\pi^+\pi^-\bar{p}$) diffraction. CDF has a very good central detector (better than in Run 1) and better triggers. In the forward directions we have better coverage (Miniplugs, BSC) and the old pots with new electronics which are being commissioned.

We should get much more diffractive physics in the coming years.

10. Acknowledgements

I would like to thank all my colleagues in CDF, especially Dino Goulianos, Michele Gallinaro, Koji Terashi and the rest of the Rockefeller group, without whom this diffractive program would not be possible. I thank the Department of Energy for support.

References

1. Abe, F. et al. (CDF) (1994) Measurement of small angle antiproton-proton elastic scattering at $\sqrt{s} = 546$ and 1800 GeV, *Phys. Rev.* **D50** 5518-5534.
2. Abe, F. et al. (CDF) (1994) Measurement of the antiproton-proton total cross section at $\sqrt{s} = 546$ and 1800 GeV, *Phys. Rev.* **D50** 5550-5561.

3. Abe, F. et al. (CDF) (1994) Measurement of $\bar{p}p$ single diffractive dissociation at $\sqrt{s}$ = 546 and 1800 GeV, *Phys. Rev.* **D50** 5535-5549.
4. Abachi, S. et al. (D0) (1996) Jet production via strongly interacting color singlet exchange in $\bar{p}p$ collisions, *Phys.Rev.Lett.* **76** 734-739.
5. Abe, F. et al. (CDF) (1995) Observation of rapidity gaps in $\bar{p}p$ collisions at 1.8 TeV, *Phys. Rev. Lett.* **74** 855-859; (1998) Dijet production by color-singlet exchange at the Fermilab Tevatron, *Phys.Rev.Lett.* **80** 1156-1161; (1998) Events with a rapidity gap between jets in $\bar{p}p$ collisions at $\sqrt{s} = 630$ GeV, *Phys.Rev.Lett.* **81** 5278-5283.
6. Bjorken, J.D. (1993) Rapidity gaps and jets as a new physics signature in very high energy hadron-hadron collisions, *Phys. Rev.* **D47** 101-113.
7. See e.g. Edin, A., Ingelman G. and Rathsman, J. (1995) Rapidity gaps in DIS through soft color interactions, DESY-95-145, hep-ph/9508244; Buchmuller, W. (1997) Soft color interactions and diffractive DIS, DESY-97-122, hep-ph/9706498; Enberg, R., Ingelman G. and Timneanu, N. (2001) Soft color interactions and diffractive hard scattering at the Tevatron, *Phys.Rev.* **D64** 114015-114050.
8. Abe, F. et al. (CDF) (1997) Measurement of diffractive dijet production at the Tevatron, *Phys.Rev.Lett.* **79** 2636-2641; Affolder, T. et al. (CDF) (2000) Diffractive dijets with a leading antiproton in $\bar{p}p$ collisions at $\sqrt{s} = 1800$ GeV, *Phys.Rev.Lett.* **84** 5043-5048.
9. Affolder, T. et al. (CDF) (2000) Observation of diffractive beauty production at the Fermilab Tevatron, *Phys.Rev.Lett.* **84** 232-237.
10. Affolder, T. et al. (CDF) (2001) Observation of diffractive J/ψ production at the Fermilab Tevatron, *Phys.Rev.Lett.* **87** 241802-241817.
11. Abe, F. et al. (CDF) (1997) Observation of diffractive W-boson production at the Fermilab Tevatron, *Phys.Rev.Lett.* **78** 2698-2703.
12. Affolder, T. et al. (CDF) (2000) Diffractive dijets with a leading antiproton in $\bar{p}p$ collisions at $\sqrt{s} = 1800$ GeV, *Phys.Rev.Lett.* **84** 5043-5048; Acosta, D. et al. (CDF) (2002) Diffractive dijet production at $\sqrt{s}$ = 630 and 1800 GeV at the Fermilab Tevatron, *Phys.Rev.Lett.* **88** 151802-151817.
13. Affolder, T. et al. (CDF) (2000) Dijet production by double pomeron exchange at the Fermilab Tevatron, *Phys.Rev.Lett.* **85** 4215-4220.
14. K.Goulianos (CDF) (2002) Multigap diffraction in CDF, hep-ph/0205217.
15. CDF (2002) http://www-cdf.fnal.gov/physics/new/bottom/bottom.html
16. Bialas, A. and Landshoff, P.V. (1991) Higgs production in pp collisions by double pomeron exchange, *Phys.Lett.* **B 256** 540-546; Cudell, J-R. and Hernandez, O.F (1996) Particle production in a hadron collider rapidity gap: the Higgs case, *Nucl.Phys.* **B471** 471-502; Kharzeev, D. and Levin, E. (2000) Soft double diffractive Higgs production at hadron colliders, hep-ph/0005311.
17. Albrow, M.G. and Rostovtsev, A. (2000) Searching for Higgs at hadron colliders using the missing mass method, hep-ph/0009336.
18. Khoze, V.A., Martin, A.D. and Ryskin, M.G. (2002) Diffractive Higgs production: myths and reality, hep-ph/0207313 and refs therein; Enberg, R., Ingelman, G., Kissavos, A. and Timneanu, N. (2002) Diffractive Higgs boson production at the Tevatron and LHC, hep-ph/0203267; De Roeck, A. and Royon, C. (2002) Discussion session on diffractive Higgs production, *Acta Phys.Polon.* **B33** 3491-3498.
19. Khoze,V.A., Martin, A.D. and Ryskin,M.G. (2000) Double diffractive processes in high resolution missing mass experiments at the Tevatron, *Eur.Phys.J* **C19** 477-484; De Roeck, A. et al. (2002) Ways to detect a light Higgs boson at the LHC, *Eur.Phys.J* **C25** 391-403.
20. Yuan, F. (2001) Heavy quarkonium production in double pomeron exchange processes in perturbative QCD, *Phys.Lett.* **B510** 155-160.
21. Cox, B., Forshaw, J. and Heinemann, B. (2002) Double diffractive Higgs and diphoton production at the Tevatron and LHC, *Phys.Lett.* **B540** 263-268.
22. Akesson, T. et al. (AFS) (1987) Double parton scattering in pp collisions at $\sqrt{s}$ = 63 GeV, *Z.Phys.* **C34** 163-184.

23. J.Alitti et al. (UA2) (1991) A study of multi-jet events at the CERN $\bar{p}p$ collider and a search for double parton scattering, *Phys.Lett.* **B268** 145-154.
24. F.Abe et al. (CDF) (1997) Double parton scattering in $\bar{p}p$ collisions at $\sqrt{s} = 1.8$ TeV, *Phys.Rev.* **D56** 3811-3832.
25. Akesson, T. et al.(AFS) (1986), A search for glueballs and a study of double pomeron exchange at the CERN Intersecting Storage Rings, *Nucl.Phys.* **B264** 154-188.
26. Kirk, A. (1997) A search for non-$Q\bar{Q}$ mesons in the WA 102 experiment at the CERN Omega spectrometer, hep-ex/9709024 and refs therein; Sosa, M. et al. (1999) Spin parity analysis of the centrally produced $K_s^o K^\pm \pi^\mp$ system at 800 GeV/c, *Phys.Rev.Lett.* **83** 913-916.
27. Klempt, E. (2001) Meson spectroscopy: glueballs, hybrids and $Q\bar{Q}$ mesons, hep-ex/0101031
28. Barberis, D. et al. (WA102) (1999) A study of centrally produced baryon-antibaryon systems in pp interactions at 450 GeV/c, *Phys.Lett.* **B446** 342-348.

DIFFRACTION AT THE TEVATRON IN PERSPECTIVE

K. GOULIANOS
The Rockefeller University
1230 York Avenue, New York, NY 10021, USA

Abstract. We review the results of measurements on soft and hard diffractive processes performed by the CDF Collaboration at the Fermilab Tevatron $\bar{p}p$ collider in run I and place them in perspective by internal comparisons, as well as by comparisons with results obtained at the HERA *ep* collider and with theoretical expectations.

1. Introduction

Diffractive $\bar{p}p$ interactions are characterized by a leading (high longitudinal momentum) outgoing proton or antiproton and/or a large *rapidity gap*, defined as a region of pseudorapidity, $\eta \equiv -\ln \tan \frac{\theta}{2}$, devoid of particles. The large rapidity gap is presumed to be due to the exchange of a Pomeron, which carries the internal quantum numbers of the vacuum. Rapidity gaps formed by multiplicity fluctuations in non-diffractive (ND) events are exponentially suppressed with increasing $\Delta\eta$, so that gaps of $\Delta\eta > 3$ are mainly diffractive. At high energies, where the available rapidity space is large, diffractive events may have more than one diffractive gap. The ratio of two-gap to one-gap events is unaffected by the presence of color exchanges in the same event that tend to "fill in" diffractive gaps and suppress the gap formation cross section (gap survival probability [1]), since the presence of one gap is sufficient to guarantee that no color exchange occurred. Thus, two-gap to one-gap ratios can be used to test QCD based models of diffraction without having to simultaneously address the problem of gap survival.

Diffractive events that incorporate a hard scattering are referred to as *hard diffraction*. In this paper, we first present a brief review of hard diffraction results obtained by CDF at the Tevatron and place them in perspective relative to results obtained at HERA and relative to theoretical expecta-

R. Fiore et al. (eds.), Diffraction 2002, 13–21.

tions. Then, we review the recent double-gap results reported by CDF, and finally conclude with remarks pertaining to both soft and hard diffraction.

2. Hard diffraction

The CDF results on hard diffraction fall into two classes, characterized by the signature used to identify and extract the diffractive signal: a large rapidity gap or a leading antiproton.

2.1. RAPIDITY GAP RESULTS

Using the rapidity gap signature to identify diffractive events, CDF measured the single-diffractive (SD) fractions of W [2], dijet [3], b-quark [4] and J/ψ [5] production in $\bar{p}p$ collisions at $\sqrt{s} = 1800$ GeV and the fraction of dijet events with a rapidity gap between jets (double-diffraction, DD) at $\sqrt{s} = 1800$ [6] and 630 [7] GeV. The results for the measured fractions are shown in Table 2.1.

TABLE 1. Diffractive fractions

Hard process	$\sqrt{s}$ (GeV)	$R = \frac{\text{DIFF}}{\text{TOTAL}}$ (%)	Kinematic region
SD			
$W(\rightarrow e\nu)$+G	1800	1.15 ± 0.55	E_T^e, $\not{E}_T > 20$ GeV
Jet+Jet+G	1800	0.75 ± 0.1	$E_T^{jet} > 20$ GeV, $\eta^{jet} > 1.8$
$b(\rightarrow e + X)$+G	1800	0.62 ± 0.25	$\lvert\eta^e\rvert < 1.1$, $p_T^e > 9.5$ GeV
$J/\psi(\rightarrow \mu\mu)$+G	1800	1.45 ± 0.25	$\lvert\eta^\mu\rvert < 0.6$, $p_T^\mu > 2$ GeV
DD			
Jet-G-Jet	1800	1.13 ± 0.16	$E_T^{jet} > 20$ GeV, $\eta^{jet} > 1.8$
Jet-G-Jet	630	2.7 ± 0.9	$E_T^{jet} > 8$ GeV, $\eta^{jet} > 1.8$

Since the different SD processes studied have different sensitivities to the gluon/quark ratio of the interacting partons, the approximate equality of the SD fractions at $\sqrt{s} = 1800$ GeV indicates that the gluon fraction of the diffractive structure fraction of the proton (gluon fraction of the Pomeron) is not very different from the proton's inclusive gluon fraction. By comparing the fractions of W, JJ and b production with Monte Carlo predictions, the gluon fraction of the Pomeron was found to be $f_g = 0.54^{+0.16}_{-0.14}$ [4]. This result was confirmed by a comparison of the diffractive structure functions obtained from studies of J/ψ and JJ production, which yielded a gluon fraction of $f_g^D = 0.59 \pm 0.15$ [5].

2.2. LEADING ANTIPROTON RESULTS

Using a Roman Pot pectrometer to detect leading antiprotons and determine their momentum and polar angle (hence the t-value), CDF measured the ratio of SD to ND dijet production rates at $\sqrt{s}$=630 [8] and 1800 GeV [9] as a function of x-Bjorken of the struck parton in the $\bar{p}$. In leading order QCD, this ratio is equal to the ratio of the corresponding structure functions. For dijet production, the relevant structure function is the color-weighted combination of gluon and quark terms given by $F_{jj}(x) = x[g(x) + \frac{4}{9}\sum_i q_i(x)]$. The diffractive structure function, $\tilde{F}^D_{jj}(\beta)$, where $\beta = x/\xi$ is the momentum fraction of the Pomeron's struck parton, was obtained by multiplying the ratio of rates by the known F^{ND}_{jj} and changing variables from x to β using $x \to \beta\xi$ (the tilde over the F indicates integration over t and ξ).

The CDF $\tilde{F}^D_{jj}(\beta)$ is presented in Fig. 1a and compared with a calculation based on diffractive parton densities obtained by the H1 Collaboration at HERA from a QCD fit to diffractive DIS data. The CDF result is suppressed by a factor of $\sim$ 10 relative to the prediction from from HERA data, indicating a breakdown of factorization of approximately the same magnitude as that observed in the rapidity gap data.

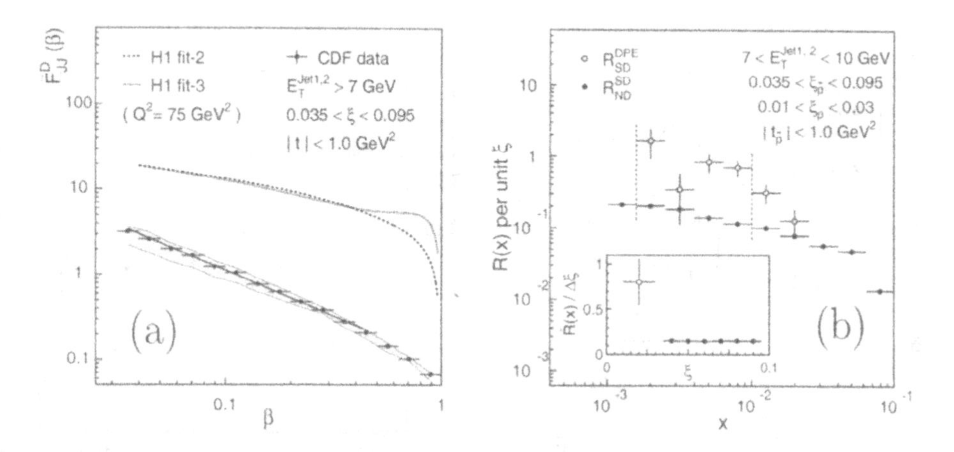

Figure 1. (a)The diffractive structure function measured by CDF (data points and fit) compared with expectations based on the H1 fit 2 (dashed) and fit 3 (dotted) on diffractive DIS data at HERA (a more recent H1 fit on a more extensive data set yields a prediction similar in magnitude to that of fit 2 but with a shape which is in agreement with that of the CDF measurement). (b) The ratio of DPE/SD rates compared with that of SD/ND rates as a function of x-Bjorken of the struck parton in the escaping nucleon. The inequality of the two ratios indicates a breakdown of factorization.

Factorization was also tested *within CDF data* by comparing the ratio of DPE/SD to that of SD/ND dijet production rates (Fig. 1b). The DPE events were extracted from the leading antiproton data by requiring a rapidity gap in the forward detectors on the proton side. At $\langle\xi\rangle = 0.02$ and $\langle x_{bj}\rangle = 0.005$, the ratio of SD/ND to DPE/SD rates normalized per unit ξ was found to be [10] 0.19 ± 0.07, violating factorization.

3. Double-gap soft diffraction

CDF has reported results on two double-gap soft diffraction processes, $\bar{p}p \to (\bar{p}' + gap1) + X + gap2 + Y$ and $\bar{p}p \to (\bar{p}' + gap1) + X + gap2$ [11]. Figure 2 shows the topology of these processes in pseudorapidity space.

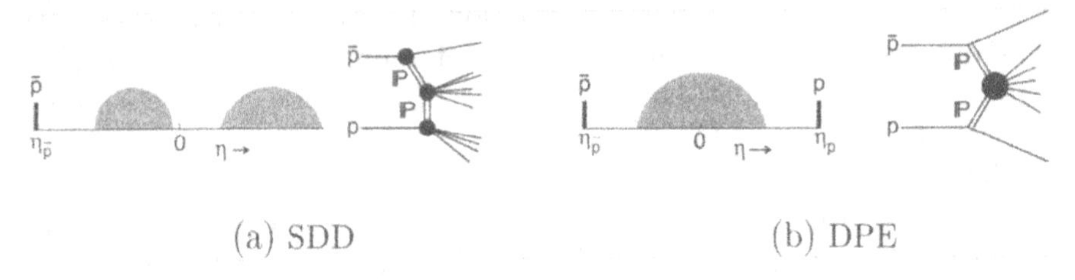

Figure 2. Schematic drawings of pseudorapidity topologies and Pomeron exchange diagrams for (a) single plus double diffraction and (b) double Pomeron exchange.

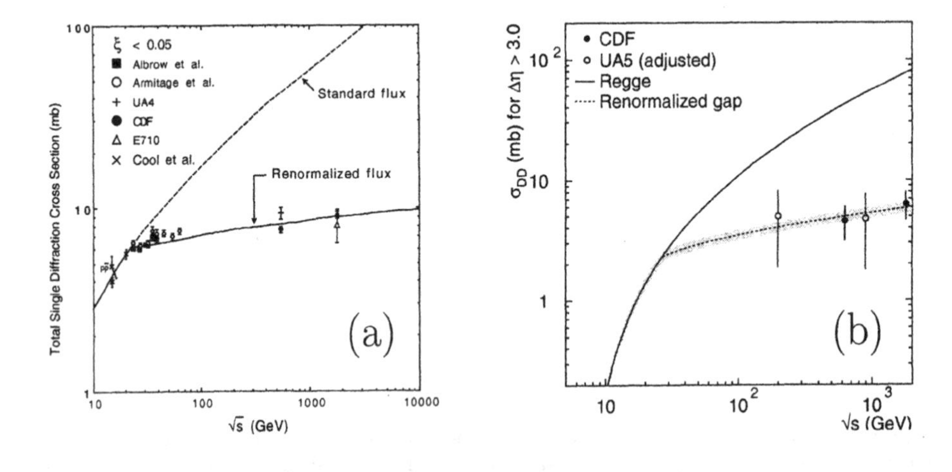

Figure 3. (a) The $\bar{p}p$ total SD cross section exhibits an s-dependence consistent with the renormalization procedure of Ref. [12], contrary to the $s^{2\epsilon}$ behavior expected from Regge theory (figure from Ref. [12]); (b) the $\bar{p}p$ total DD (central gap) cross section agrees with the prediction of the *renormalized rapidity gap* model [13], contrary to the $s^{2\epsilon}$ expectation from Regge theory (figure from Ref. [14]).

As mentioned in the introduction, the motivation for studying double-gap processes is their potential for providing further understanding of the

TABLE 2. Events used in the double-gap analyses

Process	ξ	Events at 1800 GeV	Events at 630 GeV
SDD	$0.06 < \xi < 0.09$	412K	162K
DPE	$0.035 < \xi < 0.095$	746K	136K

underlying mechanism responsible for the suppression of diffractive cross sections at high energies relative to Regge theory predictions. As shown in Fig. 3, such a suppression has been observed for both single diffraction (SD), $\bar{p}(p) + p \rightarrow [\bar{p}(p) + gap] + X$, and double diffraction (DD), $\bar{p}(p) + p \rightarrow X_1 + gap + X_2$.

Naively, the suppression relative to Regge based predictions is attributed to the spoiling of the diffractive rapidity gap by color exchanges in addition to Pomeron exchange. In an event with two rapidity gaps, additional color exchanges would generally spoil both gaps. Hence, ratios of two-gap to one-gap rates should be non-suppressed. Measurements of such ratios could therefore be used to test the QCD aspects of gap formation without the complications arising from the rapidity gap survival probability.

3.1. DATA AND RESULTS

The data used for this study are inclusive SD event samples at $\sqrt{s} = 1800$ and 630 GeV collected by triggering on a leading antiproton detected in a Roman Pot Spectrometer (RPS) [9, 8]. Table 3.1 lists the number of events used in each analysis within the indicated regions of antiproton fractional momentum loss $\xi_{\bar{p}}$ and 4-momentum transfer squared t, after applying the vertex cuts $|z_{vtx}| < 60$ cm and $N_{vtx} \leq 1$ and a 4-momentum squared cut of $|t| < 0.02$ GeV^2 (except for DPE at 1800 GeV for which $|t| < 1.0$ GeV^2):

In the SDD analysis, the mean value of $\xi = 0.07$ corresponds to a diffractive mass of ≈ 480 (170) GeV at $\sqrt{s} = 1800$ (630) GeV. The diffractive cluster X in such events covers almost the entire CDF calorimetry, which extends through the region $|\eta| < 4.2$. Therefore, the same method of analysis was employed as that used to extract the gap fraction in the case of DD [14]. In this method, one searches for *experimental gaps* overlapping $\eta = 0$, defined as regions of η with no tracks or calorimeter towers above thresholds chosen to minimize calorimeter noise contributions. The results, corrected for triggering efficiency of BBC_p (the beam counter array on the proton side) and converted to *nominal gaps* defined by $\Delta\eta = \ln \frac{s}{M_1^2 M_2^2}$, are shown in Fig. 4.

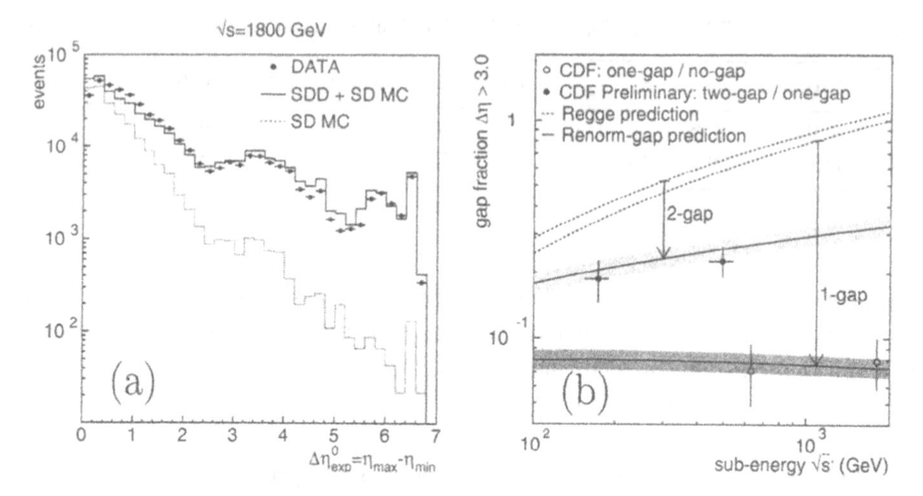

Figure 4. (a) The number of events as a function of $\Delta\eta^0_{exp} = \eta_{max} - \eta_{min}$ for data at $\sqrt{s} = 1800$ GeV (points), for SDD Monte Carlo generated events (solid line), and for only SD Monte Carlo events (dashed line); (b) ratios of SDD to SD rates (points) and DD to total (no-gap) rates (open circles) as a function of $\sqrt{s'}$ of the sub-process $I\!Pp$ and of $\bar{p}p$, respectively. The uncertainties are highly correlated among all data points.

The SDD Monte Carlo simulation is based on Regge theory Pomeron exchange with the normalization left free to be determined from the data. The differential $dN/d\Delta\eta^0$ shape agrees with the theory (Fig. 4a), but the two-gap to one-gap ratio is suppressed (Fig. 4b). However, the suppression is not as large as that in the one-gap to no-gap ratio. The bands through the data points represent predictions of the renormalized multigap parton model approach to diffraction [15], which is a generalization of the renormalization models used for single [12] and double [13] diffraction.

In the DPE analysis, the ξ_p is measured from calorimeter and beam counter information using the formula below and summing over all particles, defined experimentally as beam-beam counter (BBC) hits or calorimeter towers above η-dependent thresholds chosen to minimize noise contributions.

$$\xi_p^{\rm X} = \frac{M_{\rm X}^2}{\xi_{\bar{p}} \cdot s} = \frac{\sum_i E_{\rm T}^i \exp(+\eta^i)}{\sqrt{s}}$$

For BBC hits, the average value of η of the BBC segment of the hit was used and the E_T value was randomly chosen from the expected E_T distribution. The ξ^X obtained by this method was calibrated by comparing $\xi_{\bar{p}}^X$, obtained by using $\exp(-\eta^i)$ in the above equation, with the value of $\xi_{\bar{p}}^{RPS}$ measured by the Roman Pot Spectrometer.

Figure 5a shows the $\xi_{\bar{p}}^X$ distribution for $\sqrt{s}$ =1800 GeV. The bump at $\xi_{\bar{p}}^X \sim 10^{-3}$ is attributed to central calorimeter noise and is reproduced in

TABLE 3. Double-gap to single-gap event ratios

Source	R_{SD}^{DPE}(1800 GeV)	R_{SD}^{DPE}(630 GeV)
Data	0.197 ± 0.010	0.168 ± 0.018
P_{gap} renormalization	0.21 ± 0.02	0.17 ± 0.02
Regge $\oplus$ factorization	0.36 ± 0.04	0.25 ± 0.03
$I\!P$-flux renormalization	0.041 ± 0.004	0.041 ± 0.004

Monte Carlo simulations. The variation of tower E_T threshold across the various components of the CDF calorimetry does not affect appreciably the slope of the $\xi_{\bar{p}}^X$ distribution. The solid line represents the distribution measured in SD [16]. The shapes of the DPE and SD distributions are in good agreement all the way down to the lowest values kinematically allowed.

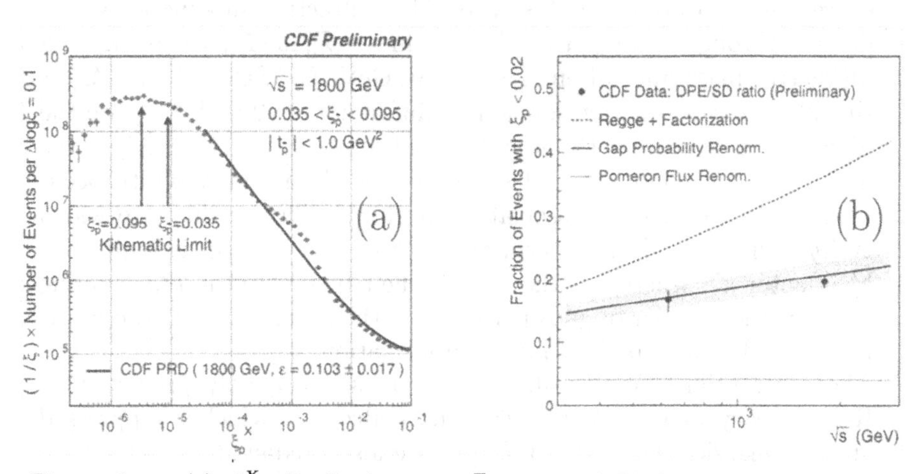

Figure 5. (a) $\xi_{\bar{p}}^X$ distribution at $\sqrt{s}$ =1800 GeV for events with a $\bar{p}$ of $0.035 < \xi_{\bar{p}}^{RPS} < 0.095$. The solid line is the distribution obtained in single diffraction dissociation. The bump at $\xi_{\bar{p}}^X \sim 10^{-3}$ is due to central calorimeter noise and is reproduced in Monte Carlo simulations. (b)Measured ratios of DPE to SD rates (points) compared with predictions based on Regge theory(dashed), Pomeron flux renormalization for both exchanged Pomerons (dotted) and gap probability renormalization (solid line).

The ratio of DPE to inclusive SD events was evaluated for $\xi_p^X < 0.02$. The results for $\sqrt{s}$ =1800 and 630 GeV are presented in Table 3.1 and shown in Fig. 5b. Also presented in the table are the expectations from gap probability renormalization [15], Regge theory and factorization, and Pomeron flux renormalization for both exchanged Pomerons [12]. The quoted uncertainties are largely systematic for both data and theory; the theoretical uncertainties of 10% are due to the uncertainty in the ratio of the triple-Pomeron to the Pomeron-nucleon couplings [17].

The data are in excellent agreement with the predictions of the gap renormalization approach.

4. Concluding remarks

The experimental data on soft and hard diffraction obtained at the Tevatron by the CDF and DØ [1] collaborations show the following remarkable characteristics:

- All diffractive differential cross sections can be factorized into two terms:

  (a) a term representing the rapidity gap dependence or *gap probability distribution*, and

  (b) a second term consisting of the $\bar{p}p$ total cross section at the sub-energy of the diffractive cluster(s). The diffractive sub-energy $\sqrt{s'}$ is defined through the equation $\ln s' = \ln s - \sum_i \ln M_i^2$, where M_i is a diffractive mass and all energies are in GeV. For SD, i=1 and M_1 is the $I\!P$-p collision energy; for double diffraction, i=2 and $M_{1,2}$ are the dissociation masses of the p and $\bar{p}$; and for DPE, i=1 and M_1 is the $I\!P$-$I\!P$ collision energy. In all cases, $\ln s'$ is the rapidity width occupied by the diffraction dissociation products. The rapidity gap width is $\Delta y = \ln s - \ln s'$. This term of the cross section, (b), is multiplied by a factor κ^n, where κ is the ratio of the triple-$I\!P$ to the $I\!Pp$ coupling and n is the number of gaps per event in the process under study; κ is interpreted as a color factor for gap formation [15].
- The differential shape of all cross sections is correctly predicted by Regge theory, apart from an overall normalization, which is suppressed at high energies. Differential shapes are also predicted correctly by a parton model approach to diffraction [15].
- The term representing the rapidity gap dependence does not depend explicitly on s. This feature represents a scaling law for diffraction and forms the basis of the renormalization model proposed to account for the suppression of the Regge cross sections at high energies [12, 13, 15].
- Two-gap to one-gap ratios are not suppressed, and thus can be used for testing QCD-based diffractive models without the complications arising from gap survival effects.

[1] The DØ results were not discussed in this paper, but are generally in good agreement with the CDF results

References

1. Bjorken, J.D. (1993) Rapidity gaps and jets as a new signature in very high-energy hadron-hadron collisions, *Physical Review D* **47**, 101-113.
2. F. Abe *et al.* (1997) Observation of diffractive W boson production at the Fermilab Tevatron, *Physical Review Letters* **78**, 2698-2703.
3. Abe, F. *et al.* (1997) Measurement of diffractive dijet production at the Fermilab Tevatron, *Physical Review Letters* **79**, 2636-2641.
4. Affolder, T. *et al.* (2000) Observation of diffractive b-quark production at the Fermilab Tevatron, *Physical Review Letters* **84**, 232-237.
5. Affolder, T. *et al.* (2001) Observation of diffractive J/ψ production at the Fermilab Tevatron, *Physical Review Letters* **87**, 241802 (6 pp).
6. Abe, F. *et al.* (1998) Dijet production by color-singlet exchange at the Fermilab Tevatron, *Physical Review Letters* **80**, 1156-1161.
7. Abe, F. *et al.* (1998) Events with a rapidity gap between jets in $\bar{p}p$ collisions at $\sqrt{s} = 630$ GeV, *Physical Review Letters* **81**, 5278-5283.
8. Acosta, A. *et al.* (2002) Diffractive dijet production at $\sqrt{=}$630 and 1800 GeV at the Fermilab Tevatron, *Physical Review Letters* **88**, 151802 (6 pp).
9. Affolder, T. *et al.* (2000) Diffractive dijets with a leading antiproton in $\bar{p}p$ collisions at $\sqrt{s} = 1800$ GeV, *Physical Review Letters* **84**, 5083-5048.
10. Affolder, T. *et al.* (2000) Dijet production by double Pomeron exchange at the Fermilab Tevatron, *Physical Review Letters* **85**, 4215-4222.
11. Goulianos, K. (2002) Multigap diffraction at CDF, *Acta Physica Polonica B* **33** 3467-3472; *e-Print Archive* hep-ph/0205217.
12. Goulianos, K. (1995) Renormalization of hadronic diffraction and the structure of the Pomeron, *Physics Letters B* **358**, 379-388.
13. Goulianos, K. Diffraction: results and conclusions, *e-Print Archive* hep-ph/9806384, 1-6.
14. Affolder, T. *et al.* (2001) Double diffraction dissociation at the Fermilab Tevatron collider, *Physical Review Letters* **87**, 141802 (6 pp).
15. Goulianos, K. (2001) The nuts and bolts of diffraction, *e-Print Archive* hep-ph/0110240, 1-3; ib. (2002) Diffraction ion QCD, *e-Print Archive* hep-ph/0203141, 1-11.
16. Abe. F. *et al.* (1994) Measurement of single diffraction dissociation at $\sqrt{s} = 546$ AND 1800 GeV, *Physical Review D* **50** 5535-5549.
17. Goulianos, K. and Montanha, J. (1999) Factorization and scaling in hadronic diffraction, *Physical Review D* **50**, 114017 (39 pp).

DIFFRACTION AT HERA: INCLUSIVE FINAL STATES AND EXCLUSIVE VECTOR MESON PRODUCTION

ALEXANDER A. SAVIN
University of Wisconsin-Madison,
1150 University Ave., Madison WI 53706-1390, USA

Abstract. Diffractive processes in photon-proton interactions at HERA offer the opportunity to improve the understanding of the transition between the soft, non-perturbative regime in hadronic interactions at $Q^2 = 0$ and the perturbative region at high Q^2. Recent experimental results from HERA on exclusive vector meson production and the properties of the hadronic final state in diffraction are reviewed. The results are discussed in the context of current theoretical models.

1. Introduction

One of the most important results from the *ep* collider HERA is the observation that about 10% of deep inelastic scattering (DIS) events exhibit a

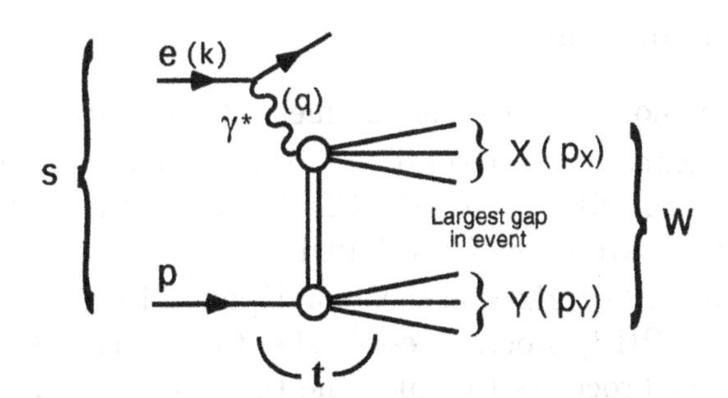

Figure 1. A generic diffractive process at HERA.

R. Fiore et al. (eds.), Diffraction 2002, 23–34.

large rapidity gap between the direction of the proton beam and that of the nearest significant energy deposit in the detector, thus showing a behavior typical for diffractive interactions[1, 2]. These events can be interpreted in terms of the exchange of a color-singlet object known as the Pomeron ($I\!P$), that can be described in QCD-inspired models as an object whose partonic composition is dominated by gluons. Alternatively, the diffractive process can be described by the dissociation of the virtual photon into a $q\bar{q}$ or $q\bar{q}g$ final state that interacts with the proton by exchange of a gluon ladder [3].

Figure 1 illustrates the generic diffractive process at HERA of the type $ep \rightarrow eXY$. The positron couples to a virtual photon $\gamma^*(q)$ which interacts with the proton (P). Together with the usual kinematic variables like Q^2, y, W and x, one has t, the 4-momentum transferred at the proton vertex and M_X and M_Y, the invariant masses of the photon and proton dissociative systems, respectively. $x_{I\!P} = \frac{q \cdot (P - P_Y)}{q \cdot P}$ is the fraction of proton momentum carried by the $I\!P$.

In a QCD interpretation in which a partonic structure is ascribed to the colorless (diffractive) exchange the lowest order contribution to the diffractive cross section is a quark scattering ($\gamma^* q \rightarrow q$). In this case $\beta = \frac{x}{x_{I\!P}}$ can be interpreted as the fraction of the $I\!P$ momentum carried by the struck quark. The next-to-leading (NLO) contributions are the boson-gluon fusion (BGF) ($\gamma^* g \rightarrow q\bar{q}$) and QCD-Compton ($\gamma^* q \rightarrow qg$) processes. In this case the invariant mass squared $\hat{s}$ of the partons emerging from the hard subprocess is non-zero. Therefore, the quantity $z_{I\!P} = \beta(1 + \hat{s}/Q^2)$ is introduced, which corresponds to the longitudinal momentum fraction of the $I\!P$ carried by the parton (quark or gluon) which enters the hard interaction.

The main HERA results can be divided into two parts: inclusive diffraction, such as $F_2^{D(3)}$, jets and hadronic final state studies, and exclusive diffraction, mainly the vector meson measurements.

2. Inclusive Diffraction

Recent high precision inclusive measurement of the diffractive DIS process $ep \rightarrow eXY$ presented in the form of a diffractive structure function $F_2^{D(3)}$ is discussed in [4]. A NLO DGLAP QCD fit was performed to the data to extract diffractive parton densities (DPDFs).

The proof of the factorization theorem [5] for diffractive DIS and direct photoproduction (PHP) processes establishes the universality of DPDFs [6, 7] for the class of processes to which the theorem applies. It means that QCD factorization in diffraction can be tested by taking the DPDFs and predicting diffractive final state observables such as dijet and charm cross sections at HERA or diffractive dijet production at the Tevatron.

A particular motivation for studying diffractive DIS charm production is that one expects charm production to form a substantial fraction of the diffractive DIS cross section. Moreover, since open charm production is dominated by the BGF process, it also provides a good test of the normalization of the diffractive gluon density.

Perturbative QCD (pQCD) -based models are expected to describe the data since the high jet transverse momentum or charm-quark mass provide the nessesary hard scale.

2.1. DIJET CROSS SECTION

The predictions are obtained using the RAPGAP MC program [8]. The renormalisation and factorization scales are set to $\mu^2 = Q^2 + p_t^2 + m_q^2$.

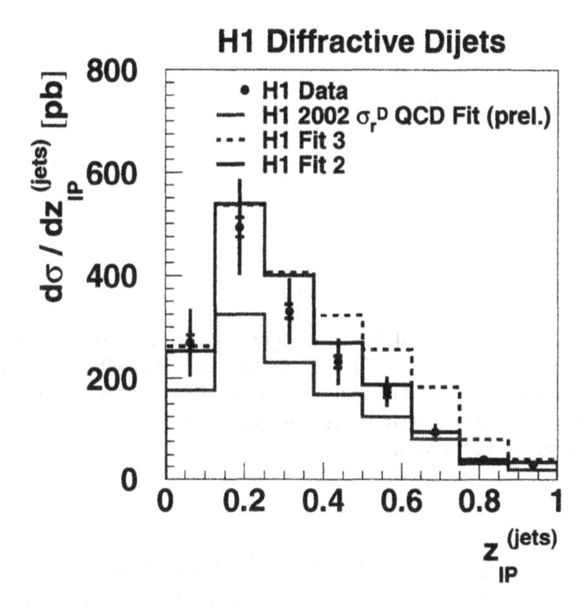

Figure 2. H1 measurement of the diffractive DIS dijet cross section differential in Z_{IP}^{jets}.

Figure 2 shows predictions [6] obtained using the new H1 LO DPDFs, predictions based on the previous H1 QCD fits [9] (H1 fit 2 and fit 3) and the H1 DIS dijet data [10] ($40 < Q^2 < 120$ GeV^2).

The prediction based on the recent QCD fit is generally below those for the previous fits due to the reduced diffractive gluon distribution. This leads to an improved description of the D^* cross section (not shown), but a worse description for the dijet cross section.

Using the old DPDFs to predict diffractive dijet production at the Tevatron leads to an overestimation of the observed rate by an order of magnitude. The new prediction is slightly closer to the Tevatron data but the overall level of disagreement is still large [6].

The transition from DIS to hadron-hadron scattering can be studied in PHP, where the processes in which an almost real photon participates directly in the hard scattering are expected to be similar to DIS. Processes where the photon is first resolved into partons which then initiate the hard scattering resemble hadron-hadron scattering.

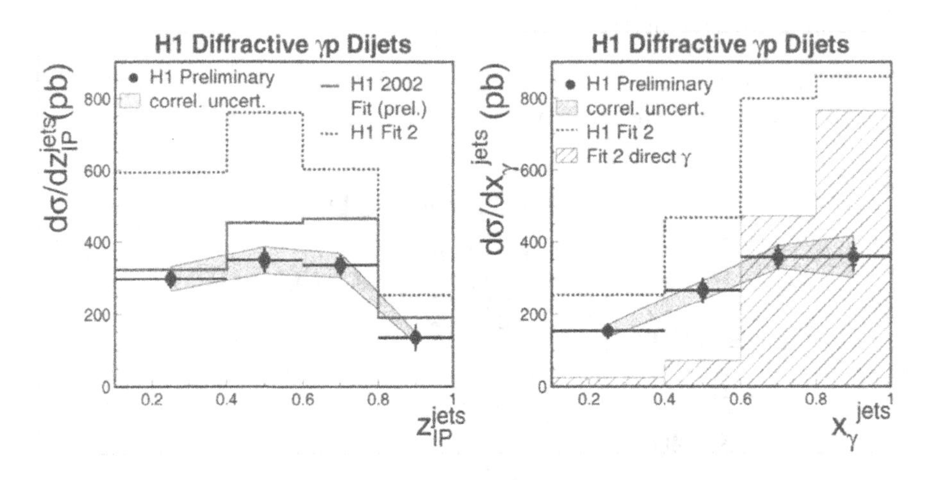

Figure 3. H1 measurements of diffractive dijet PHP [11]. Predictions of the RAPGAP MC model with different LO QCD fits are shown. The left figure shows the dijet cross-section differential in $Z_{I\!P}^{jets}$. The right figure shows the cross-section differential in x_γ^{jets}, where the direct photon contributions (boson-gluon fusion and QCD compton) are indicated by the hatched histogram.

Figure 3 (left) shows the diffractive dijet cross section differential in the estimator $Z_{I\!P}^{jets}$. The RAPGAP prediction based on the H1 QCD fit 2 overestimates the normalization of the data by a factor $\approx$2. The new H1 2002 QCD fit overestimates the cross section by a factor $\approx$1.3, but is compatible with the data given the uncertainty of the gluon density resulting from the fit. Figure 3 (right) shows the prediction of the H1 fit 2, with the contribution from direct photon processes displayed separately.

The same suppression of both resolved and direct photon processes is needed by an overall scale factor 1.8 $\pm$ 0.45(exp.) compare to the DIS dijet process. Although this factor is subject to considerable additional uncertainties, it is much smaller than the suppression factor of about 10 for single diffractive dijet production at the Tevatron.

2.2. DIS D^* (2010) PRODUCTION

Figures 4 and 5 show the differential $D^{*\pm}(2010)$ cross sections measured by ZEUS [12].

The predictions of three models are compared to the data: (1) the "saturation" model of Golec-Biernat and Wüsthoff [13], as implemented in the SATRAP MC generator; (2) the two-gluon-exchange model of Bartels et al. (BJLW) [14]; (3) the resolved Pomeron model, as implemented in the NLO fits to HERA data performed by Alvero et al.(ACTW) [7].

The SATRAP model describes the data, both in shape and normalization.

The $c\bar{c} + c\bar{c}g$ BJWL predictions, obtained for m_c=1.45 GeV and $\mu = \sqrt{p_{c,t}^2 + 4m_c^2}$ and value of cutoff on the gluon transverse momentum of $K_{T,g}^{out}$ = 1.5 GeV, describes the shapes of the distributions reasonably well except for the x_{IP} distribution (the reasons are discussed in [14]).

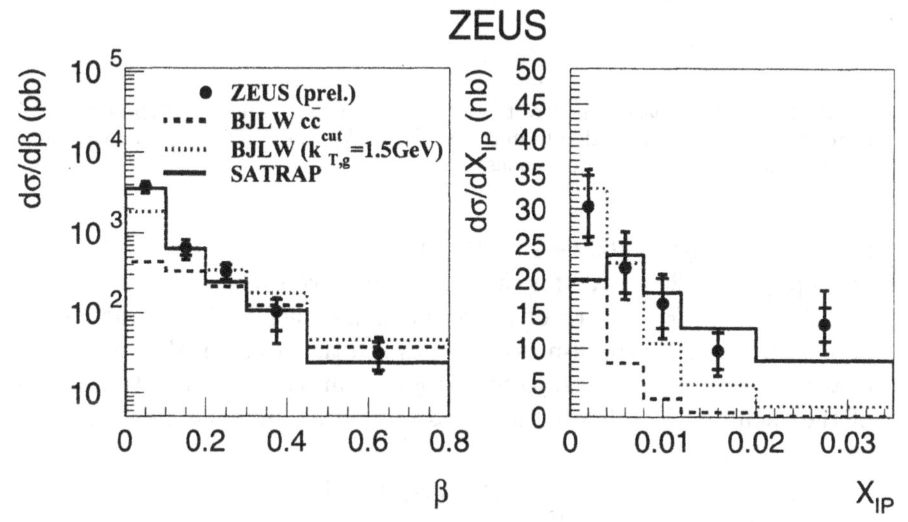

Figure 4. ZEUS measurement of the differential cross section for $D^{*\pm}(2010)$ production as a function of the kinematic variables β and X_{IP}. The histograms correspond to the different models described in the text.

The ACTW NLO QCD prediction (gluon-dominated fit B) is also in reasonable agreement with the measured differential distributions supporting the notion of hard scattering factorization for diffractive lepton-hadron processes.

3. Exclusive Vector Meson Production

The exclusive diffractive production of vector mesons (VM), $ep \to eVp$, where $V = (\rho^0;\ \omega;\ \phi;\ J/\psi;\ \psi';\ \Upsilon)$ is measured at HERA in the kinematic

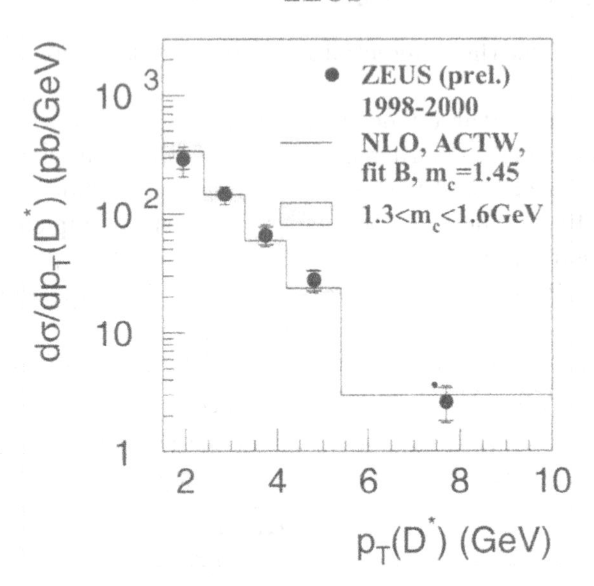

Figure 5. ZEUS measurement of the differential cross section for D* production. The histogram shows the prediction from the NLO ACTW model. The shaded area shows the effect of varying the charm-quark mass.

region of $0 < Q^2 < 100$ GeV^2, $20 < W_{\gamma P} < 290$ GeV, $0 < |t| < 1.5$ GeV^2 and up to $|t| = 20$ GeV^2 for proton-dissociative events.

Within the framework of Regge phenomenology [15], diffractive interactions at large centre-of-mass energies are the result of the t-channel exchange of the Pomeron trajectory, $\alpha_{I\!P}(t)$, carrying the quantum numbers of the vacuum. The differential cross section at high energies is expressed as

$$\frac{d\sigma}{dt} \propto F(t) \cdot W^{4\cdot[\alpha_{I\!P}(t)-1]}, \tag{1}$$

where $F(t)$ is a function of t only. If $d\sigma/dt$ decreases exponentially and the trajectory is linear in t, $\alpha_{I\!P}(t) = \alpha_{I\!P}(0) + \alpha'_{I\!P}t$, the cross section can be expressed as

$$\frac{d\sigma}{dt} \propto W^{4[\alpha_{I\!P}(0)-1]} \cdot e^{b(W)t}, \tag{2}$$

where the slope parameter is

$$b(W) = b_0 + 4\alpha'_{I\!P} \ln(W/W_0) \tag{3}$$

and W_0 is an arbitrary energy scale parameter. A fit to hadronic data [16] yields the soft-Pomeron parameter $\alpha'_{I\!P} = 0.25$ GeV^{-2}.

Exclusive PHP of a heavy VMs (J/ψ; ψ'; Υ) and VM production at high Q^2 or t are expected to be described by pQCD-models, since the mass of the heavy quarks, Q^2 or t provide a hard scale. In such models, the cross section is proportional to the square of the gluon density.

The models predict a rapid rise of the cross section with W [17, 18], where W is the photon-proton centre-of-mass energy, which is caused by the fast increase of the gluon density in the proton at small values of Bjorken x. The effective $\alpha'_{I\!P}$ in the perturbative regime [19, 20] is predicted to be much smaller than 0.25 GeV^{-2}, and the slope parameter b must have little variation with W [18, 21], i.e. much smaller "shrinkage", than observed for the soft processes.

3.1. W DEPENDENCE OF THE CROSS SECTION

The γp cross sections for exclusive VM PHP from HERA are displayed as a function of W in Fig. 6 together with data from fixed-target experiments.

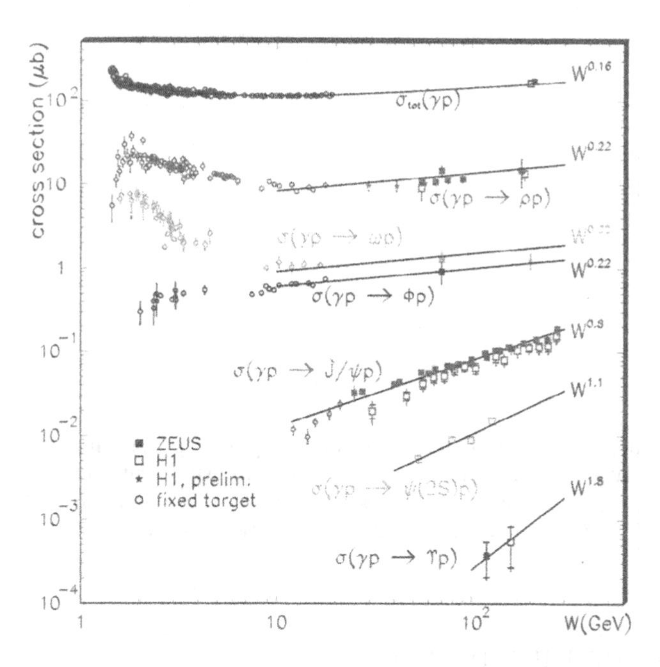

Figure 6. Total photoproduction and VM photoproduction cross sections, for fixed target and HERA experiments. The fit to the total cross section corresponds to the "soft" Pomeron parametrisation, with decreasing contribution of Reggeon exchange at low energy; other lines represent the fit of the form $\sigma \propto W^{\delta}$, and the results of the fit are shown in the figure.

The results of fits of the form $\sigma \propto W^{\delta}$ are also shown. A striking effect is the rapid energy dependence of heavy VM PHP, much stronger than for the light once. This is an effect of the large charm mass, which implies that the process takes place over short distances and probes directly the hard gluon content of the proton: $\sigma \propto |xG(x)|^2$. The recent data on J/ψ PHP are precise enough to distinguish between different PDF's, but theoretical uncertainties make this extraction impossible at present. Figure 7 shows the energy dependence of ρ electroproduction for different values of Q^2. The power of the W dependence, δ, increases with Q^2 with the absolute value of δ at high Q^2 corresponding to the one predicted by pQCD-models. It also agrees well with the value measured in J/ψ PHP.

The similarity of the Q^2 and M_{VM}^2 W-dependences suggests the use of a universal scale $(Q^2 + M_{VM}^2)$ for description of VM production. Such a scale was already discussed in many papers [22] and is logical since both Q^2 and M_{VM}^2 define the size of the $q\bar{q}$ fluctuation. The latter can be directly observed by comparing the b-slopes of the t-dependences.

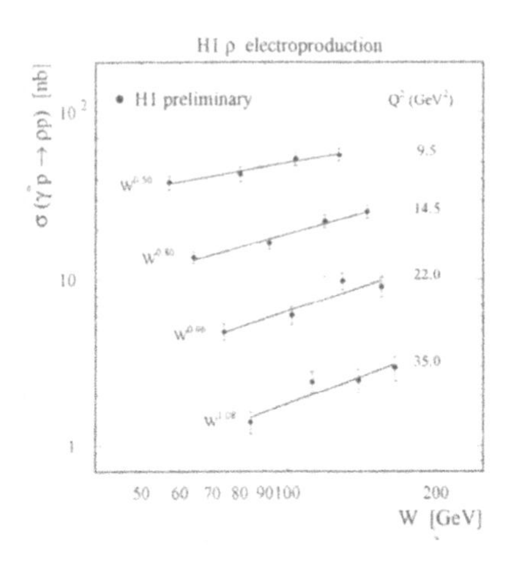

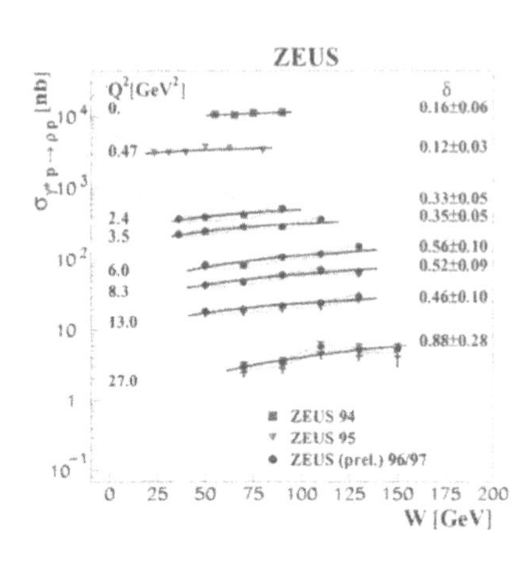

Figure 7. W dependence of the cross section $\sigma(\gamma^* p \to \rho^0 p)$ for various Q^2 values as denoted in the figure from H1 (left) and ZEUS (right) measurements. The lines represent the results of fitting $\sigma \propto W^{\delta}$ and the results of the fit are shown in the figure. The shaded area in ZEUS case indicates additional uncertainties due to proton dissociation background.

3.2. B-SLOPE OF THE T DEPENDENCE

To measure the b-slope, the differential cross-section $d\sigma_{\gamma p \to Vp}/dt$ is calculated in bins of W and fitted to the form $d\sigma/dt \propto e^{bt}$. The value of b is proportional to the radius squared of the scattered objects, and thus can be used to measure the effective size of the VM-proton interaction.

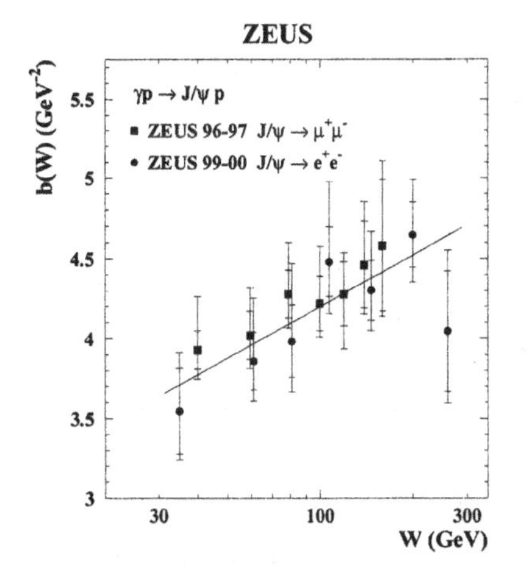

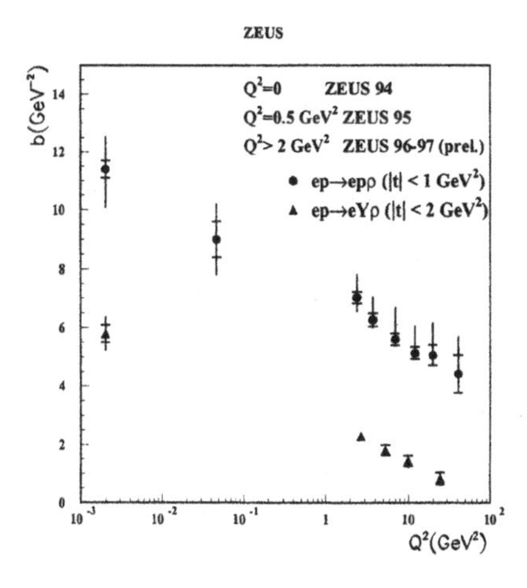

Figure 8. The left figure shows the values of the slope , b, of the t distributions, plotted as a function of W. The line is the result of a fit of the form $b(W) = b_0 + 4\alpha'_{I\!P} \ln(W/W_0)$, where b_0 is taken at 90 GeV, and W_0 =90 GeV. The right figure shows the slope b of exclusive and proton-dissociative electroproduction of ρ^0 mesons, plotted as a function of Q^2.

The b-slope as a function of W for J/ψ PHP is shown in Fig. 8 . The line is a fit to the form (3), where

$$b_0 = 4.15 \pm 0.05(stat.)^{+0.30}_{-0.18}(syst.)\ \text{GeV}^{-2}$$

$$\alpha'_{I\!P} = 0.116 \pm 0.026(stat.)^{+0.010}_{-0.025}(syst.)\ \text{GeV}^{-2}$$

The b-slope increases with W and is approximately equal to that expected from the size of the proton [23], which suggests that the size of the J/ψ is small compared to that of the proton. The value of $\alpha'_{I\!P}$ is different from zero, but still much smaller than that for the "soft" Pomeron.

The b-slopes as a function of Q^2 for elastic and proton-dissociative ρ production are also sown in Fig. 8. Both slopes decrease with increasing Q^2, although the elastic one approaches the value of about 4, at high Q^2, similar to J/ψ PHP, thus confirming the similarity of the high Q^2 and high M^2_{VM} behavior.

The proton-dissociative b-slope shows that the scattering objects are much smaller then the proton. The ratio of the proton-dissociative events to the elastic ones is also independent of Q^2. This is important because the proton-dissociative events are background to the elastic channel (at HERA it is 15-30 % of the total VM production cross section). The recent ZEUS measurement confirmed that this ratio can be precisely measured at low Q^2 with high statistics and extrapolated to the whole Q^2 range.

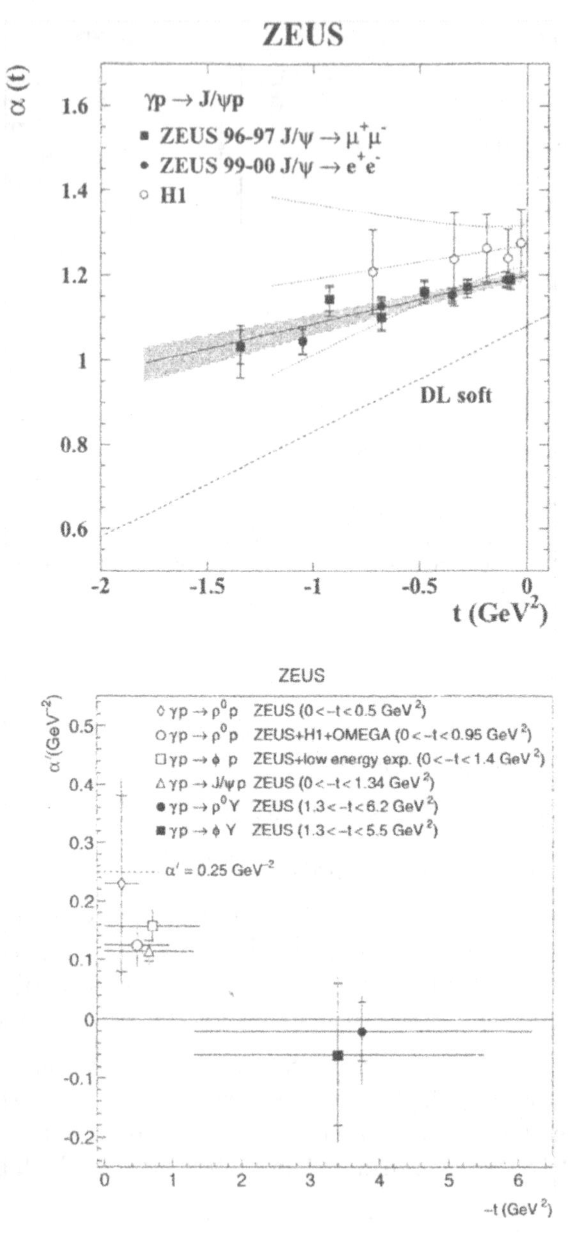

Figure 9. The top figure shows the Pomeron trajectory as a function of t as obtained for J/ψ PHP in the two leptonic decay channels, $J/\psi \to \mu^+\mu^-$ and $J/\psi \to e^+e^-$. The solid and dotted lines are the result of linear fits to the ZEUS and H1 data, respectively. The one standard deviation contour is indicated for the ZEUS (shaded area) and H1 (dotted lines) measurements. The dashed line shows the DL soft-Pomeron trajectory. The bottom figure presents the comparison of results on $\alpha'_{I\!P}$ for VM production. The results for the proton-dissociative production are shown with solid symbols and those for elastic photoproduction with the open symbols. The horizontal bars correspond to the $-t$ range in which $\alpha'_{I\!P}$ is measured.

3.3. POMERON TRAJECTORY

The Pomeron trajectory is determined directly by measuring the variation of the W dependence of the elastic cross section at fixed t, as parameterized in Eq. (1). This method is insensitive to the proton-dissociative background, since the latter was measured to be independent of W. The resulting values of $\alpha_{I\!P}(t)$ are shown in Fig. 9, as a function of t for J/ψ PHP. They were fitted to the linear form $\alpha_{I\!P}(t) = \alpha_{I\!P}(0) + \alpha'_{I\!P}t$. The combined ZEUS measurement gives $\alpha_{I\!P}(0) = 1.200 \pm 0.009(stat.)^{+0.004}_{-0.010}(syst.)$ and $\alpha'_{I\!P} = 0.115 \pm 0.018(stat.)^{+0.008}_{-0.015}(syst.)$ GeV^{-2}. The value of $\alpha'_{I\!P}$ is not consistent with zero which indicates a small shrinkage. In contrast to the PHP of light VM this trajectory is not consistent with the "soft" Pomeron one, but is much closer to the DIS VM production result.

The momentum transfer t itself is another hard scale used in some pQCD-models. The recent measurements of the proton-dissociative PHP of VM at high t are compared with low t data in Fig. 9. in form of the slope $\alpha'_{I\!P}$ of the Pomeron trajectory. $\alpha'_{I\!P}$ decreases with t. At high t the value of $\alpha'_{I\!P}$ is consistent with zero, as expected from the models predictions. This indicates, together with other observations, that the scale t plays a different role then Q^2 and M^2_{VM} in diffraction.

4. Conclusions

The recent data obtained by HERA shows that the experimental uncertainties for the most of the observables are much smaller than the theoretical ones. For example the DPDFs extracted from the HERA data fail to reproduce the Tevatron measurements and the extraction of the gluon PDFs from the PHP J/ψ data is not yet possible. Thus the HERA results are playing an important role in understanding the structure of diffractive exchange.

References

1. ZEUS Coll., Derrick, M.*et al.* (1993), Observation of Events with a Large Rapidity Gap in Deep Inelastic Scattering at HERA, *Phys. Lett.*, **Vol. no. B315**, pp. 481–493;
2. H1 Coll., Ahmed, T. *et al.* (1994), Deep Inelastic Scattering Events with a Large Rapidity Gap at HERA, *Nucl. Phys.*, **Vol. no. B429**, pp. 477–502;
3. See e.g. Ingelman, G., DeRoeck, A. and Klanner, R. (eds.), *Proc. of the Workshop on Future Physics at HERA* **Vol.2**, DESY, Hamburg, 1996;
4. Capua, M. *Talk presented at this Conference* ;
5. Collins, J.C. (1998), Proof of Factorization for Diffractive Hard Scattering, *Phys. Rev*, **Vol. no. D57**, pp. 3051–3056;
6. H1 Coll. (2002), Measurement and NLO DGLAP QCD Analysis of Inclusive Diffractive Deep Inelastic Scattering, *Abstract 980 31 ICHEP*, Amsterdam, July 2002;

7. Alvero, L. *et al.* (1999), Diffractive Production of Jets and Weak Bosons, and Tests of Hard Scattering Factorization, *Phys. Rev.*, **Vol. no. D59**, pp. 074022–074058;
8. Jung, H. (1995), Hard Diffractive Scattering in High-Energy ep Collisions and the Monte Carlo Generator RAPGAP, *Comp. Phys. Commun.*, **Vol. no. 86**, pp. 147–161;
9. H1 Coll., Adloff, C. *et al.* (1997), Inclusive Measurement of Diffractive Deep Inelastic ep Scattering, *Z. Phys.*, **Vol. no. C76**, pp. 613–629;
10. H1 Coll. Adloff, C. *et al.* (2001), Diffractive Jet Production in Deep Inelastic ep Collisions at HERA, *Eur. Phys. J.*, **Vol. no. C20**, pp. 29–49;
11. H1 Coll. (2002), Diffractive Photoproduction of Jets at HERA, *Abstract 987 31 ICHEP*, Amsterdam, July 2002;
12. ZEUS Coll. (2002), Measurement of D^* Photoproduction in Diffractive Deep Inelastic Scattering at HERA, *Abstract 780 31 ICHEP*, Amsterdam, July 2002;
13. Golec-Biernat, K. and Wüsthoff, M. (1999), Saturation Effects in Deep Inelastic Scattering at Low Q^2 and Its Implications on Diffraction, *Phys. Rev.*, **Vol. no. D59**, pp. 014017–014041;
14. Bartels, J., Jung, H. and Wüsthoff, M. (1999), Quark-anti-quark Gluon Jets in DIS Diffractive Dissociation, *Eur. Phys. J.*, **Vol. no. C11**, pp. 111–125;
15. Collins, P.D.B. *An Introduction to Regge Theory and High Energy Physics.* Cambridge University Press, 1977;
16. Jaroszkiewicz, G.A. and Landshoff, P.V. (1974), Model for Diffraction Excitation, *Phys. Rev.*, **Vol. no. D10**, pp. 170–174;
17. ZEUS Coll, Chekanov, S. *et al.* (2002), Exclusive Photoproduction of J/ψ mesons at HERA, *Europ. Phys. Journal*, **Vol. no. C24**, pp. 345–360;
18. H1 Coll, Adloff, C. (2000), Elastic Photoproduction of J/ψ and Υ mesons at HERA, *Phys. Lett.*, **Vol. no. B483**, pp. 23–35;
19. Frankfurt, L., McDermott, M. and Strikman, M. (2001), A Fresh Look at Diffractive J/ψ Photoproduction at HERA, with Predictions for THERA, *Journal of High Energy Phys.*, **Vol. no. 103**, pp. 45–74;
20. Nikolaev, N.N., Zakharov, B.G. and Zoller, V.R. (1996), Exploratory Study of Shrinkage of the Diffraction Cone for the Generalized BFKL Pomeron, *Phys. Lett.*, **Vol. no. B366**, pp. 337–344;
21. Levy, A. (1998), Evidence for No Shrinkage in Elastic Photoproduction of J/ψ, *Phys. Lett.*, **Vol. no. B424**, pp. 191–194;
22. ZEUS Coll. (2002), Energy Dependence of Exclusive Vector-Meson Production in *ep* Interactions at HERA. *Abstract 820 31 ICHEP*, Amsterdam, July 2002;
23. Akimov, Y. *et al.* (1976), Analysis of Diffractive $pD \to XD$ and $pp \to XP$ Interactions and Test of the Finite Mass Sum Rule, *Phys. Rev.*, **Vol. no. D14**, pp. 3148–3161.

EXCLUSIVE PROCESSES AT INTERMEDIATE ENERGIES: HERMES RESULTS AND PROSPECTS

A. BORISSOV
Department of Physics and Astronomy, University of Glasgow, Glasgow, G128QQ, United Kingdom

Abstract. The measurement of the production of exclusive vector mesons, pions and photons in deep-inelastic lepton scattering gives a means to access the recently introduced generalized parton distributions (GPDs). Recent data on these exclusive reactions, collected by the HERMES collaboration at DESY, are compared to calculations which are based on a description of the reactions in terms of GPDs. In addition, prospects of a new HERMES recoil detector which will clearly identify exclusive events in the next running period are presented.

1. Introduction

Recent interest in hard exclusive processes has emerged since the introduction of generalized parton distributions (GPDs) [1] which provide a unified formalism for the description of inclusive deep-inelastic scattering, exclusive meson production, electromagnetic form factor measurements, and deeply-virtual Compton scattering (DVCS).

Schematic diagrams of exclusive processes are presented in Fig.1. There, each structure function depends on the three variables x, ξ and t, see [2] for details. First, x is the mean value of the longitudinal momentum fraction, which is unmeasured. ξ is the exchanged longitudinal momentum fraction, often called the skewedness parameter, which is related to x_{Bj} in the Bjorken limit by $\xi \rightarrow \frac{x_{Bj}}{2-x_{Bj}}$. And t is four-momentum transfer $t = \Delta^2 = (p - p')^2$ to the nucleon. Four different GPDs are used per flavor. The properties of those are such that $H^q(x, \xi, t)$, $\tilde{H}^q(x, \xi, t)$ conserve nucleon helicity, while $E^q(x, \xi, t)$, $\tilde{E}^q(x, \xi, t)$ flip it. In addition, $H^q(x, \xi, t)$ and $E^q(x, \xi, t)$ correspond to unpolarized virtual gamma states,

R. Fiore et al. (eds.), Diffraction 2002, 35–45.

while $\tilde{H}^q(x,\xi,t), \tilde{E}^q(x,\xi,t)$ correspond to polarized virtual gamma states. In the limit $\xi = 0$ and $t = 0$, $H^q(x,0,0) = q(x_B), \tilde{H}^q(x,0,0) = \Delta q(x)$ where $q(x)$ and $\Delta q(x)$ are the ordinary quark density and quark helicity distributions. No corresponding relationship exists for E^q and $\tilde{E}^q$. Different exclusive reactions are sensitive to different combinations of GPDs, as shown in Tab. 1.

Interest in GPDs was stimulated by the paper of Ji [1] that the second moment of the unpolarized GPDs in the limit of $t = 0$ is sensitive to the total quark angular momentum J_q where $J_q = 0.5 \int_{-1}^{1} dx x (H^q(x,\xi) + E^q(x,\xi))$. Such data will complement results of the measurement of the quark spin in nucleon and in several experiments to measure the gluon spin. A theoretical understanding of the possibility of the extraction of quark orbital angular momentum based on the formalism of GPDs is under progress now, see for example [2], but it is well understood that the quark orbital angular momentum is at present inaccessible in the QCD framework of ordinary parton distributions.

Experimental information on GPDs can be obtained by measuring cross sections or asymmetries of the exclusive reactions. A measurement of the lepton charge asymmetry in DVCS using unpolarized e^{+-} beams and targets allows access to the real part of the DVCS amplitudes $H_1, \tilde{H}_1$, [3, 4]:

$$A_{ch} \approx \cos(\phi_\gamma) \cdot Re\Big\{F_1 H_1 + \frac{x_B}{2 - x_B}(F_1 + F_2)\tilde{H}_1 - \frac{t}{4M^2} F_2 E_1\Big\} \quad (1)$$

From the measurement of the lepton helicity asymmetry of a longitudinally polarized beam and an unpolarized target one can get get access to the imaginary part of $H_1, \tilde{H}_1$:

$$A_{LU} \approx \sin(\phi_\gamma) \cdot Im\Big\{F_1 H_1 + \frac{x_B}{2 - x_B}(F_1 + F_2)\tilde{H}_1 - \frac{t}{4M^2} F_2 E_1\Big\} \quad (2)$$

So far, very little experimental data exists on exclusive processes that would allow one to gain general information on GPDs. The reasons are

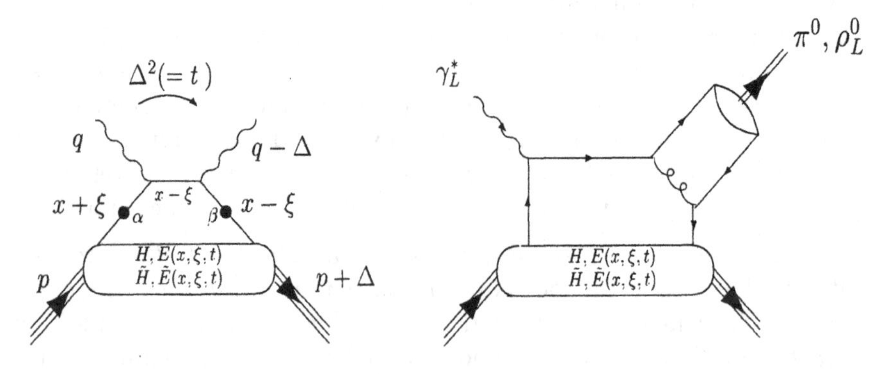

Figure 1. 'Handbag' diagrams for DVCS (left) and exclusive meson production (right).

TABLE 1. Hard exclusive processes accessible at HERMES in conjunction with the GPDs involved.

DVCS	$\gamma^* p \rightarrow \gamma p$	$H, \tilde{H}, E, \tilde{E}$
exclusive pseudoscalar meson production	$\gamma^* p \rightarrow \pi^+ n$ $\gamma^* p \rightarrow \pi^0 p$	$\tilde{H}, \tilde{E}$
exclusive vector meson production	$\gamma^* p \rightarrow \rho^o p$ $\gamma^* p \rightarrow \omega p$ $\gamma^* p \rightarrow \phi p$	H, E

the small cross sections involved and the high energy resolution required. Furthermore, in meson production one needs to extract the longitudinal cross section from the total cross section.[1]

In this paper first experimental data is presented on the exclusive processes mentioned above. These data have been collected by the HERMES experiment at DESY in Hamburg. A detailed description of the HERMES experiment can be found in [6]. It is sufficient to note here that the experiment consists of a forward angle spectrometer with an angular acceptance ranging from 40 to 140 mrad in the vertical direction, and from -170 to + 170 mrad in the horizontal direction. A dipole magnet, various sets of tracking chambers, and various particle identification detectors provide information on the direction, momentum, charge and type of the particles detected. Behind these detectors, a large segmented electromagnetic calorimeter is located, which is used for photon identification and calorimetry, and electron-hadron discrimination.

The exclusive data have been interpreted within a GPD framework. The aim of these first results is to demonstrate the feasibility of future experiments at higher luminosity, where it is hoped to gain detailed information on GPDs. This paper is organized as follows. In the subsequent sections, experimental results on vector-meson production, deeply-virtual Compton scattering, and pseudoscalar-meson production are presented. Following this, plans to install a recoil detector at HERMES are given. The paper concludes with a short summary.

2. ρ^o, ω and ϕ Longitudinal Cross Sections

In the unpolarized longitudinal cross section only quadratic combinations of GPDs appear. An evaluation of the electroproduction amplitudes in the leading order for γ, π^o, and ρ^o_L have been performed [7] for a proton target.

[1]The factorization proof [5] only applies to exclusive reactions induced by longitudinal photons.

There it was found that different final states correspond to the different combinations of GPDs:

$$H^P_{DVCS}(x,\xi,t) = 4/9H^{u/p} + 1/9H^{d/p} + 1/9H^{s/p},$$
$$H^P_{\rho^0}(x,\xi,t) = 1/\sqrt{2}\{2/3H^{u/p} + 1/3H^{d/p}\},$$
$$H^P_{\pi^0}(x,\xi,t) = 1/\sqrt{2}\{2/3\tilde{H}^{u/p} + 1/3\tilde{H}^{d/p}\}.$$

Under the assumption that H^q is proportional to the unpolarized quark distribution, one can use the following parameterization to define the GDPs via the usual structural functions:

$$H^{u/p}(x,\xi,t) = 1/2u(x)F^{u/p}(t),$$
$$H^{d/p}(x,\xi,t) = d(x)F^{d/p}(t),$$
$$H^{s/p}(x,\xi,t) = 0.$$

Similar expressions hold for $\tilde{H}^{q/p}$. For the calculation of the absolute cross sections improvements have been done in a subsequent paper of the authors of [7] where higher twist effects have been included in a phenomenological fashion, and where the Q^2-dependence of the strong coupling constant α_s has been accounted for.

The HERMES data on light vector meson production has been compared with the calculations of [7]. In order to extract information on the longitudinal ρ^0 production cross section, data on r^{04}_{00}, the longitudinal fraction of ρ^0 cross section, are used. See an analysis of vector meson at HERMES production for more details [8]. Assuming s-channel helicity conservation (SCHC) the r^{04}_{00} matrix element is related to the ratio R of the longitudinal to transverse production cross sections $R = \frac{\sigma_L}{\sigma_T} = \frac{r^{04}_{00}}{\epsilon(1-r^{04}_{00})}$, where ϵ represents the virtual-photon polarization parameter. Using a parameterization of R [8], the longitudinal cross section σ_L for ρ^0, ω, and ϕ production has been determined $\sigma_L = \frac{R}{1+\epsilon R}\sigma_{total}$, where σ_{total} represents the total measured cross section. The resulting values for ρ^0 [8], ω and ϕ production are shown in Fig. 2 and Fig. 3, respectively, and are compared to the calculations of Vanderhaeghen et al [7]. The calculations for ρ^0 and ω production are in agreement with the data if both the 'quark-exchange' and the 'gluon-exchange' contributions are included. Further evidence in support of the dominance of quark-exchange in ρ^0 production at $W \approx 5$ GeV came from a measurement of the target spin-asymmetry at HERMES [9]. The data yield a positive asymmetry, which can be attributed to unnatural parity contributions such as quark-exchange.

Note that for ϕ-meson production only the 'gluon-exchange' mechanism has been used for the calculation of the longitudinal cross section since the proton contains an almost negligible amount of s-quark. The agreement of the theoretical calculations [7] in such approach with data is presented in the right panel of Fig.3.

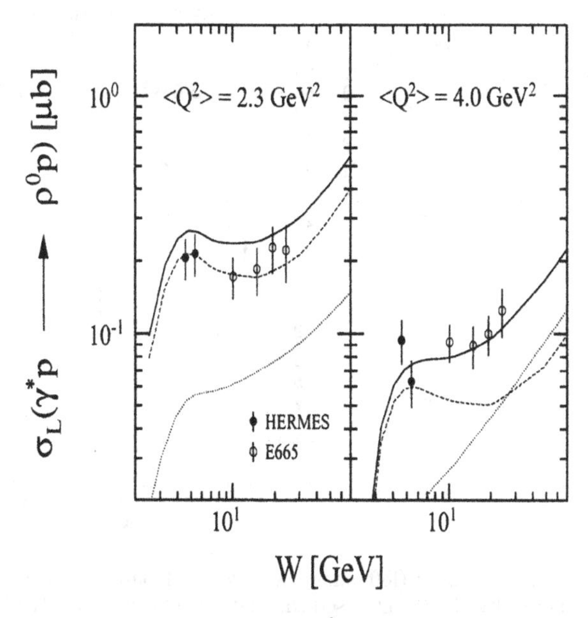

Figure 2. The derived longitudinal ρ^0 production cross section is compared to the results of a GPD-based calculation. The dotted curves represent the gluon-exchange contribution, the dashed curves the quark-exchange contribution, and the solid curves their sum.

3. Deeply Virtual Compton Scattering (DVCS)

In order to study GPDs in exclusive processes it is more advantages to have a photon instead of a meson in the final state, because there is no remaining uncertainty due to the wave function of the meson in the final state. Moreover, while vector-meson production and pseudoscalar-meson production are each sensitive to two different sets of GPDs, DVCS is sensitive to four (see Tab.1). However, it is extremely difficult to identify photons associated with DVCS, because the competing Bethe-Heitler process clearly dominates at HERMES kinematics as seen in Fig.4. By exploiting the interference between the DVCS and BH processes, one may gain access to the imaginary part of the DVCS amplitude [10] via a measurement of the lepton helicity asymmetry using a longitudinally polarized beam and an unpolarized target. It is related to the GPDs via Eq.2. The leading order term which depends on the helicity of the incident lepton is determined by a linear combination of DVCS helicity amplitudes only. This beam-spin asymmetry A_{LU} in hard electroproduction of photons has been measured for the first time by the HERMES experiment at DESY using the HERA 27.6 GeV longitudinally polarized positron beam and an unpolarized hydrogen gas target. The asymmetry with respect to the helicity state of

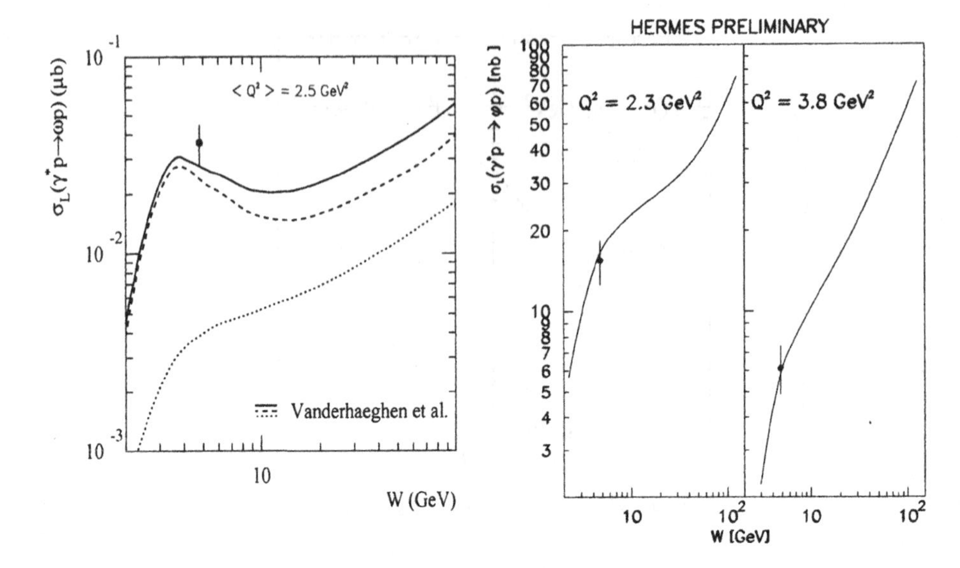

Figure 3. The derived longitudinal ω (left panel) and ϕ (right panel) production cross section is compared to the results of a GPD-based calculation. The curves in the left panel correspond to the same processes as in Fig.2. In the right panel only the gluon-exchange contribution has been used in the GPD based calculations.

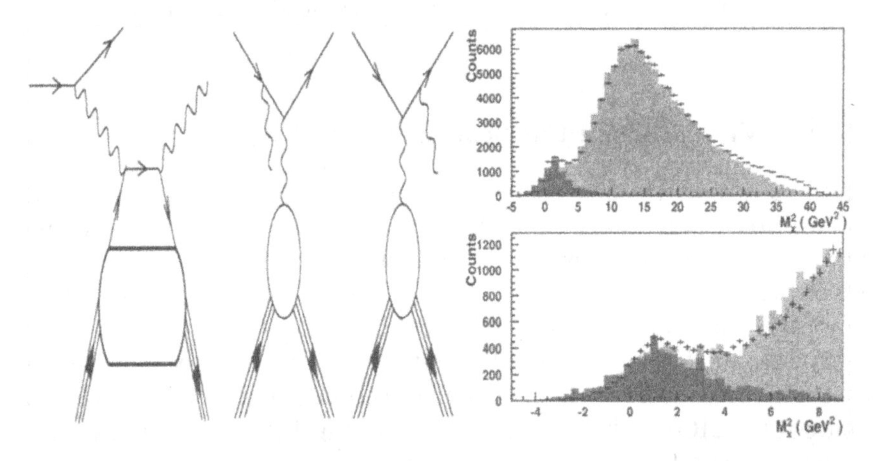

Figure 4. In the left panel Feynmann diagrams for DVCS and photon radiation from the incident and scattered lepton in the Bethe-Heitler processes are shown correspondently. In right panel the measured distributions of photons as observed in hard electroproduction versus the missing mass squared M_x^2 are displayed. The light-gray histogram represents the results of a Monte-Carlo simulation in which fragmentation processes and the Bethe-Heitler process are included, while the dark-shaded histogram represents only the Bethe-Heitler contribution. The Monte-Carlo simulation includes the effect of the detector resolution which explains the negative values of M_x^2.

the incoming positron beam is plotted as a function of ϕ in Fig. 5 (left panel) for events with a missing mass close to the proton mass. The missing mass (M_x) dependence of the effect has been studied by evaluating the beam-spin analyzing power: $A_{LU}^{\sin\phi} = \frac{2}{N}\sum_{i=1}^{N}\frac{\sin\phi_i}{(P_l)_i}$, where $(P_l)_i$ represents the beam polarization measured during the beam burst of the ith event. The results are shown in the right panel of Fig. 5. As the missing mass resolution of the HERMES spectrometer for DVCS-like events is about 0.8 GeV, the M_x-bins left and right of $M_x = m_{proton}$ also show non-zero values of $A_{LU}^{\sin\phi}$. In the missing-mass range below 1.7 GeV the analyzing power in the $\sin\phi$ moment was measured to be -0.23 $\pm$ 0.04(stat) $\pm$ 0.03(syst). The beam-spin averaged data can be used to determine an upper limit of a possible false asymmetry due to instrumental effects which - averaged for M_x between -1.5 and $+1.7$ - amounts to -0.03 ± 0.04. The average values of the kinematic variables corresponding to this measurement are $\langle x \rangle = 0.11$, $\langle Q^2 \rangle = 2.6$ GeV^2, and $\langle -t \rangle = 0.27$ GeV^2. The observed analyzing power is somewhat smaller than the value of -0.37 quoted in Ref. [11] for kinematics close to those of the present experiment. Note, that the results presented above include associated DVCS with final states such as $e'\gamma\Delta^+$. To separate these events a recoil detector would be extremely useful. For more precise measurement of the charge asymmetry presented in Eq.5, (right panel) more data acquired with the electron beam are needed.

4. Exclusive pseudoscalar-meson production

It has been shown [12] that GPD functions can describe longitudinal electroproduction of exclusive π^+ from a transversely polarized nucleon. If, in contrast, the target is polarized longitudinally with respect to the lepton beam direction, a small transverse target polarization component ($P_\perp$) appears in the virtual photon frame in addition to the main longitudinal target polarization component (P_L). In this case the polarized cross section can be written as $\sigma_{pol} \approx \sin\phi \cdot \{S_\perp \sigma_L + S_L \sigma_{LT}\}$, where ϕ is the angle between the hadronic and the leptonic scattering planes. The average value of $P_\perp/|P|$ is approximately 0.17 for HERMES kinematics. Theoretical calculations [12] for transversely polarized targets are in agreement with each other and predict a positive asymmetry. While factorization has been proven for longitudinally polarized photons, at present no proof exists for transversely polarized photons. Hence the interference term σ_{LT} and the total cross section σ_{pol} can not be calculated on the basis of the factorization theorem. In the case of a transversely polarized target the second term of the total cross section is suppressed. Thus future HERMES data of exclusive π^+ production, collected with a transverse target, can be directly compared with GPDs based calculations. The yield of exclusive π^+

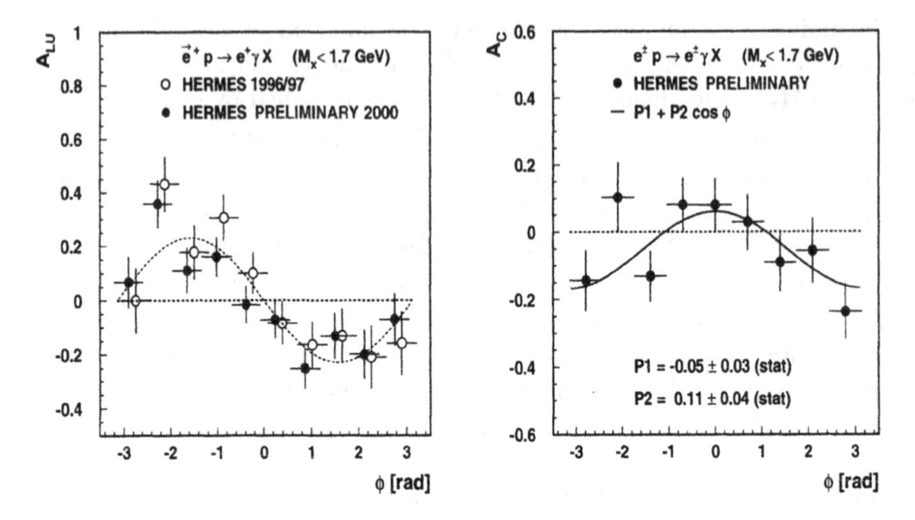

Figure 5. In the left panel the beam-spin asymmetry A_{LU} for hard electroproduction of photons as a function of the azimuthal angle ϕ is shown. The data correspond to the missing mass region between -1.5 and $+1.7$ GeV, . The dashed curve represents a $\sin\phi$ dependence with an amplitude of -0.23. In the right panel the beam-charge analyzing power $A_c^{sin\phi}$ of photons produced on a hydrogen target is displayed as a function of the azimuthal angle ϕ.

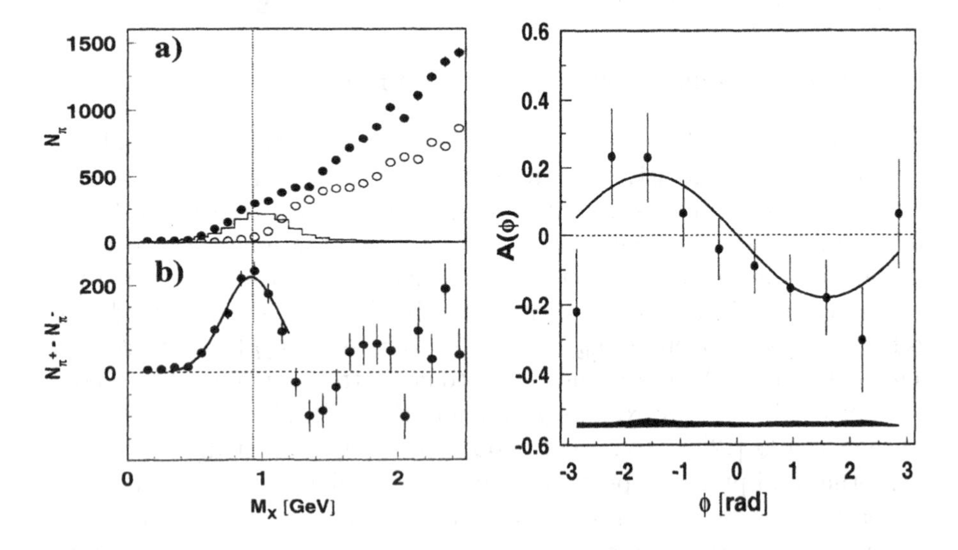

Figure 6. In the left panel the missing mass spectrum for exclusive production of π^+ mesons on hydrogen is shown. On the right panel the target spin azimuthal asymmetry for exclusive electroproduction of π^+ mesons. The curve represents a $\sin\phi$ fit to the data.

mesons produced on ^1H target was determined by using the π^- yield as a measure of the non-exclusive background. The π^- spectrum was normalized be a factor of 1.4 using the missing mass spectrum above 2 GeV. A clear peak at a missing mass close to the nucleon mass arises after subtraction of the normalized π^- yield from the π^+ yield, as shown in Fig.6 (left panel). The average kinematics of the exclusive sample are given by $< Q^2 >= 2.2$ GeV^2, $< x_{Bj} >= 0.15$, and $< -t >= 0.46$ GeV^2. Note that acceptance corrections cancel in the ratio and background corrections have been applied based on the asymmetry $A_{bg} = 2.5 \pm 2.5\%$ measured from the adjacent region at larger missing mass. Uncertainty of the background correction procedure has been accounted for in the total systematic error. The data show a clear $\sin\phi$ dependence of the polarized cross section which has been fitted to $A(\phi) = A_{UL}^{\sin\phi} \cdot \sin\phi$, where $A_{UL}^{\sin\phi} = -0.18 \pm 0.05 \pm 0.01$, as shown in Fig. 6 (right panel). The sign of the asymmetry is opposite to that of the predictions for a transversely polarized target, indicating that the asymmetry gets a significant contribution from the second term of the total cross section. The band represents the systematic uncertainty from the background yield and from target polarization which dominated the overall uncertainty.

5. Prospects for HERMES

Run II of the HERMES experiment will start in the beginning of 2002 with three main physics objectives for the next four to five years: a measurement of transversity distributions, an improved measurement of helicity distributions, and a measurement of exclusive reactions to access GPDs using a new recoil detector [13].

At present the missing mass resolution of the HERMES spectrometer is not sufficient to exclude experimentally the low-lying excited nucleon states (Δ isobars). A new project to add a recoil detector to the experiment has been started in order to allow one to clearly identify exclusive events. The recoil detector will be able to identify the recoil proton through a combination of tracking, charge determination, and angular separation from possible π^+ background. It is also planned to measure the momentum of the recoil proton in the range from 100 to 1200 MeV/c. In addition the recoil detector is designed to detect and identify π^+, π^-, and π^0 to help reduce the background from associated Δ-production. The large acceptance recoil detector consists of three active detector parts: a silicon detector around the target cell inside the vacuum chamber, a scintillating fibre tracker in a longitudinal magnetic field and a layer of scintillator strips (or pads) that uses the return yoke as preshower material. A Monte-Carlo study showed that for DVCS the resolution of the t variable will be on the level of 0.15

GeV^2, and that strong background suppression is expected for associated DVCS with $e^{'}\gamma\Delta^{+}$ in the final state. Thus the $(e^{'}\pi^{+}\Delta^{o})$ background for exclusive π^{+}, and double-dissociative background for exclusive vector meson production are suppressed.

High statistics data sets on exclusive photon, pion, and meson production are expected from future running with a high density unpolarized target with an annual integrated luminosity of 2 fb^{-1}. In this scenario, for example for DVCS, both the lepton charge asymmetry and the lepton helicity asymmetry will be measured in a several bins of x_{Bj} and t. This will allow us to get access to the real and imaginary parts of DVCS amplitudes with good accuracy [4]. It is however important to note that at present no proven recipe exists for a complete determination of all four GPDs for each quark and gluon flavor.

6. Summary

We have presented three results of exclusive reactions performed at the HERMES experiment which are related to GPDs. First, the data of the measured longitudinal cross sections of light vector mesons are in agreement with GPD calculations based on the sum of 'quark exchange' and 'gluon exchange' mechanisms for ρ^0 and ω and only 'gluon exchange' for ϕ meson production. Second, a measurement of a significant single-spin asymmetry relative the longitudinal polarization of the beam has been performed which is sensitive to DVCS through the interference between DVCS and BH processes. Third, a clear π^{+} exclusive peak has been observed. A significant negative single-spin asymmetry of exclusive π^{+} has been measured for the first time with a longitudinal polarized target.

We also have presented prospects for the next running period. There, a transversely polarized proton target will complement the results of SSA of exclusive pion production. And, significant improvements in resolution and systematic uncertainties are expected with the operation of the recoil detector in measuring hard exclusive reactions.

References

1. X. Ji (1997) Deeply Virtual Compton Scattering, *Phys. Rev. D* , **55**, 7114-7125; Radyushkin, A.V. (1996) Asymmetric Gluon Distributions and Hard Diffractive Electroproduction, *Phys. Lett.B* , **385**, 333-342.
2. Goeke, K., Poluakov M.V., Vanderhaeghen M. (2001) Hard Exclusive Reactions and the Structure of Hadrons, *Progr. Part. Nucl. Phys.*, **47**, 401-515; hep-ph/0106012.
3. Belitsky A.V. *et al.* (2001) Leading Twist Asymmetries in Deeply Virtual Compton Scattering, *Nucl.Phys.* , **B593**, 289-310.
4. Korotkov V.A. and Nowak W.-D. (2001) Future Measurement of the Deeply Virtual Compton Scattering at HERMES, *Eur.Phys.J.*, **C23**, 455-461; hep-ph/0108077.

5. Collins J., Frankfurt L. and Strikman M. (1997) Factorization for Hard Exclusive Electroproduction of Mesons in QCD, *Phys.Rev.D,* **56**, 2982-3006.
6. HERMES Collab. (Ackerstaff K. *et al.*) (1998) The HERMES Spectrometer, *Nucl.Instr.Meth.A* , **417**, 230-265; hep-ex/9806008.
7. Vanderhaeghen M., Guichon P.A.M., Guidal M. (1998) Hard Electroproduction of Photons and Mesons on the Nucleon, *Phys.Rev.Lett.* , **80**, 5064-5067; M.Vanderhaeghen M., Guichon P.A.M., Guidal M. (1999) Deeply Virtual Electroproduction of Photons and Mesons on the Nucleon: Leading Order Amplitudes and Power Corrections, *Phys.Rev.D,* **60**, 094017; Vanderhaeghen M., Guidal M., *priv.comm.* 2000.
8. HERMES Collab. (Ackerstaff K. *et al.*) (2000) Exclusive Leptoproduction of rho0 Mesons from Hydrogen at Intermediate Virtual Photon Energies, *Eur.Phys.J. C,* **17**, 389-398
9. HERMES Collab. (Airapetian A. *et al.*) (2001) Double Spin Asymmetry in the Cross-Section for Exclusive Rho0 Production in Lepton-Proton Scattering, *Phys.Lett.B,* **513**, 301-310; hep-ex/0102037.
10. Diehl M. *et al.* (1997) Testing the Handbag Contribution to Exclusive Virtual Compton Scattering, *Phys. Lett.B* , **411**, 193-202.
11. Kivel N., Polyakov M. and Vanderhaeghen M. (2001) DVCS on the Nucleon: Study of the Twist-Three Effects, *Phys.Rev.D* , **63**, 114014; hep-ph/0012136.
12. Frankfurt L.L. *et al.* (2000) Hard Exclusive Electroproduction of Decuplet Baryons in the Large N(C) Limit, *Phys.Rev.Lett.* , **84**, 2589-2592; Frankfurt L.L. *et al.* (1999) Hard Exclusive Pseudoscalar Meson Electroproduction and Spin Structure of a Nucleon, *Phys.Rev.D* , **60**, 014010; Belitsky A.V. and Mueller D. (2001) Hard Exclusive Meson Production at Next-to-Leading Order, *Phys.Lett.B,* **513**, 349-360; hep-ph/0105046.
13. Kaiser R., Contact Person (HERMES Collaboration) (2002) A Large Acceptance Recoil Detector for HERMES, Addendum to the Proposal DESY PRC 97-07, DESY PRC 01-01.

OVERVIEW OF THE COMPETE PROGRAM

COMPETE * *collaboration*

V.V. EZHELA, YU.V. KUYANOV, K.S. LUGOVSKY, S.B. LUGOVSKY, V.S. LUGOVSKY, E.A. RAZUVAEV, M.YU. SAPUNOV, N.P. TKACHENKO AND O.V. ZENIN

COMPAS group, IHEP, Protvino, Russia

K. KANG

Physics Department, Brown University, Providence, RI, U.S.A.

M.R. WHALLEY

HEPDATA group, Durham University, Durham, United Kingdom

A. LENGYEL

Institute of Electron Physics, Universitetska 21, UA-88000 Uzhgorod, Ukraine

S.K. KANG

Department of Physics, Seoul National University, Seoul 131-717, Korea.

J.R. CUDELL, E. MARTYNOV[1] AND O. SELYUGIN[2]

Institut de Physique, Bât. B5, Université de Liège, Sart Tilman, B4000 Liège, Belgium

[1] *on leave from the Bogolyubov Institute for Theoretical Physics, 03143 Kiev, Ukraine*

[2] *on leave from Bogoliubov Theoretical Laboratory, JINR, 141980 Dubna, Moscow Region, Russia*

AND

P. GAURON AND B. NICOLESCU

LPNHE[‡]*-LPTPE, Université Pierre et Marie Curie, Tour 12 E3, 4 Place Jussieu, 75252 Paris Cedex 05, France*

*COmputerised Models Parameter Evaluation for Theory and Experiment.
‡Unité de Recherche des Universités Paris 6 et Paris 7, Associée au CNRS.

R. Fiore et al. (eds.), Diffraction 2002, 47–61.

Introduction

Nowadays, scientific databases have become the bread-and-butter of particle physicists. They are used not only for citation and publication [1, 2, 3, 4], but also for access to data compilations [5, 6, 7] and for the determination of the best parameters of currently accepted models [5, 6, 8]. These databases provide inestimable tools as they organize our knowledge in a coherent and trustworthy picture. They have lead not only to published works such as the Review of Particle Physics [9], but also to web interfaces, and reference data compilations available in a computerized format readily usable by physicists. It should be pointed out at this point that one is far from using the full power of the web, as cross-linking between various databases and interactive interfaces are only sketchy. Part of the problem comes from the absence of a common repository or environment.

One must also stress that this crucial activity is not a given, and that these databases must be maintained and checked repeatedly to insure the accuracy of their content. In fact, we run the risk to loose some or all of the information contained in them, as the maintainers are getting older. There is a need for teaching the systematization and evaluation of data in the standard physics curriculum, need which so far has totally been overlooked. We stress the importance of summer schools and workshops dedicated to data systematization.

The COMPETE collaboration aims at motivating data maintenance through the interfacing of theory and experiment at the database level, thus providing a complete picture of the phenomenology describing a given subclass of phenomena. The database concept then needs to be supplemented by a "model-base" [10]. Such an object enables one not only to decide what the best description may be, but also to discern what potential problems exist in the data. The systematization of such a cross-fertilization between models and data, which is at the core of physics, results in what we shall call an "object of knowledge", containing both factual and theoretical information, and presumably becoming the point at which all existing information resources on a given problem could converge.

There are many advantages to such a global approach. First of all, the maintenance of a data set is not a static task: it needs to be motivated by physics. Discrepancies between models and data call for checks, and often those checks lead to a new data set, where published errata in data are fixed and preliminary data are removed. A clear example of such improvements can be found in the total cross section data set. Furthermore, at times such studies show that there may be problems in the experimental analysis itself. For instance, in the analysis of the ρ parameter, the systematic error resulting from the use of a specific model is usually neglected. A general re-analysis of these data is therefore needed, or a different treatment of systematic errors may be brought in.

The second advantage is obviously that one can have a common testing ground for theories and models, so that all the details of the comparison are under control. This means that it becomes possible, for a given set of assumptions, to define the best models reproducing a given set of data. In this respect, as many models have to be tested, and as the usual "best fit" criterion, i.e. lowest χ^2/dof, is not fully satisfactory, we have developed a set of procedures that enable artificial intelligence decisions, simulating to some extent a physicist's intuition and taste.

Thirdly, it is obvious that an extensive theoretical database can be used to plan new experiments, and to predict various quantities. The automated treatment of a large number of models and theories enables us to quote a theoretical error, which gives the interval in which existing models can reproduce experimental results, and to determine the sensitivity needed to discriminate between various models.

Finally, as new data come in, one can very quickly decide on their theoretical impact, and hence immediately evaluate the need for new physics ideas.

As we want to treat a large amount of data and many models, computer technology constitutes an important part of our activity. We have concentrated on the elaboration of artificial intelligence decision-making algorithms, as well as on the delivery of computer tools for the end-user: these include web summaries of results, web calculators of various quantities for the best models, and of course computer-readable data-sets and Fortran codes. Finally, the consideration of several different physics problems brings in the need to interface various objects of knowledge. The interconnection and compatibility of these is an important constraint. Further linkage with existing databases, such as PDG [5], COMPAS [6], and HEPDATA [7] is being developed or planned.

METHODOLOGY

Our work is based on the following information model: theoretical descriptions and data are arranged in bases, which are then interfaced. This cross assessment leads to a ranking of models, and to an evaluation of data: some models globally reproduce data and some don't, some data seem incompatible with all models or with similar data. At this point, a selection of acceptable models and of acceptable data is made. The models are then ranked according not only to their χ^2/dof, but also to their number of parameters, stability, extendability to other data, etc. The best models are kept, and organized into an object of knowledge. One can then make predictions based on these few best models, and evaluate theoretical errors from all acceptable models. The data set can also be re-evaluated, and after a new data set is produced, one can re-iterate the above procedure. The next step is then to find models that can accommodate more data and once such new models are proposed, one can iterate again.

It is worth pointing out here the problems directly linked with data and parameters. First of all, contradictory data lead to sizable uncertainties. One way to handle these is to use a Birge factor, renormalizing the χ^2/dof to 1. Another way is to re-normalize a given data set, and assign to this operation a penalty factor. Finally, it is also possible to shift data sets within their systematic errors in order to obtain the best data set overall. Each method has its problems and advantages, and no overall best method has been found so far. Secondly, one must stress that, besides the usual statistical and systematic errors, one should independently quote a theoretical error which may not be combined with the experimental systematics. Finally, it is important for a given set of parameters to indicate their area of applicability, *e.g.* often high-Q^2 or high-s models are used outside their area of validity.

The organization of work within the collaboration is similar to that of the Particle Data Group: each member of the collaboration has access to the current object of knowledge which gets released to the community once a year, in the form of computer-usable files, and web-accessible notes. Part of the collaboration is devoted to the finding of new data and models in the literature, as well as to their encoding. Other people check the accuracy of the encoding. The study and elaboration of the object of knowledge is done under the guidance of a few developers, whose work then gets partially or fully verified by the rest of the team in charge of that study. The web interface and tools are then developed or updated by another part of the collaboration.

The organization of the object of knowledge itself goes as follows: a compilation of data is interfaced with a compilation of models, typically kept as a set of Mathematica routines (which can then be used to produce Fortran or C). The conclusions of the cross-assessment are then fed into a program devoted to predictions, and freely executable via the web. Tables of predictions then become available. Another module gathers the information obtained from the cross-assessment and makes it available to the collaboration for cross-checks. Finally, another module uses the citation databases to track new work that refers to our existing databases.

Results

The results we have obtained so far fall within two main categories: the first concerns the tools that we have developed, which could be used by others in a wide variety of tasks, the second concerns the physics conclusions which we have reached.

TOOLS

Elements of the artificial intelligence
The usual indicator χ^2/dof is certainly an important measure of the quality of a fit. However, it does not give us all the relevant information to choose the best models. We have developed [11, 12], in the context of fits to soft data, a series of other indicators that enable us to study numerically some of the aspects of fits which so far had only received a qualitative treatment. Models usually rely on some approximation which breaks down in some region. For instance, in DIS, the starting value of Q_0 is an indication of the area of applicability of a given parameterisation. Within a common area of applicability, fits with a better χ^2/dof are to be preferred. If the parameters of a fit have physical meaning, then their values must be stable when one restricts the fit to a sub-set of the full data set, or if one limits the area of applicability, *e.g.* but modifying the starting Q_0 of a DIS fit. Similarly, fits that use a handful of parameters are usually preferred to those that use many. All these features can be studied numerically, and details can be found in refs. [11, 12]. The use of these indicators then enables one to decide which model may be preferred to describe some set of data.

Web
We have also developed an automatic generation of results which are then gathered in postscript files available on the web [13], as well as a calculational interface that predicts values of observables for the first few best parameterisations [14]. Furthermore, computer-readable files [15], as well as Fortran code for the best models [16], are also given. As we shall see, this is only a first step as a full interface between different objects of knowledge still needs to be built.

PHYSICS

SOFT FORWARD DATA: FORWARD2.1

We started our activities a few years ago [17], concentrating on analytic fits to total cross sections and to the ρ parameter. Such studies first revealed a few problems with the data set, and then proved the equivalence of simple, double and triple pole parameterisations in the region $\sqrt{s} \geq 9$ GeV. This resulted in the first version of the object of knowledge concerned with soft forward physics, FORWARD1.0. Its second version [11], dating from last year, came when it was realized that some fits could be extended down to $\sqrt{s} \geq 4$ GeV. The latter models thus became favored, and constitute the second version of the object of knowledge, FORWARD2.0. It now contains 3092 points (742 above 4 GeV), and 37 adjusted and ranked models. We have recently used it to produce predictions at present and future colliders [18], and included cosmic ray data to obtain FORWARD2.1, which is detailed in J.R. Cudell's contribution to these proceedings.

This object of knowledge has demonstrated that ρ parameter data were poorly reproduced. Some experiments at low energy seem to have systematic shifts with respect to other experiments, and the χ^2/dof of the ρ data is very bad for some data sets (pp, $p\pi^+$, pK^-). Although some of the problems can be understood as coming from the use of derivative dispersion relations (see O.V. Selyugin's contribution to these proceedings), discrepancies between data nevertheless make a good fit impossible.

The only clean way out is to perform the experimental analysis again, or part of it, either through a check (and correction) of the theoretical input used, or through a re-analysis of the data in the Coulomb-nuclear interference region. One thus needs a common parameterisation of electromagnetic form factors, a common procedure to analyse data in the Coulomb-nuclear interference region, a common set of strong interaction elastic scattering parameterisations, and a common study of Regge trajectories. The next few objects of knowledge are devoted to the systematization of such information.

REGGE TRAJECTORIES: RT1.0β

First of all the long way of modelling the forms of Regge trajectories for positive and negative values of t should be systemized. Even the good old idea of linearity at positive t should be tested and maintained.

We have extracted from the RPP-2002 database a new set of hadronic states (213 mesons and 123 baryons), including their masses, widths and quantum numbers. Corresponding isotopic multiplets (one isomultiplet – one point) are all presented in a log-linear Chew-Frautschi plot with different markers for different flavors to show the similarities and differences of the (M^2, J) populations for different hadron classes, see Fig.1.

There is only one linear meson trajectory (a_2, a_4, a_6) that give the acceptable fit quality with weights constructed from the errors in masses. This trajectory is placed on the Fig. 1 together with longest baryon trajectory (5 Δ members) with good fit quality. Preliminary fits of the RPP-2000 data to linear trajectories (in the approximation where weights in the fits are constructed from $\Delta(M^2) = M\Gamma$, instead of $\Delta(M^2) = 2M\Delta M$) show a clear systematic flavor dependence of the slope for mesons, as shown in Table I. Such a dependence of the slopes on flavor does not seem to be present in the baryon case.

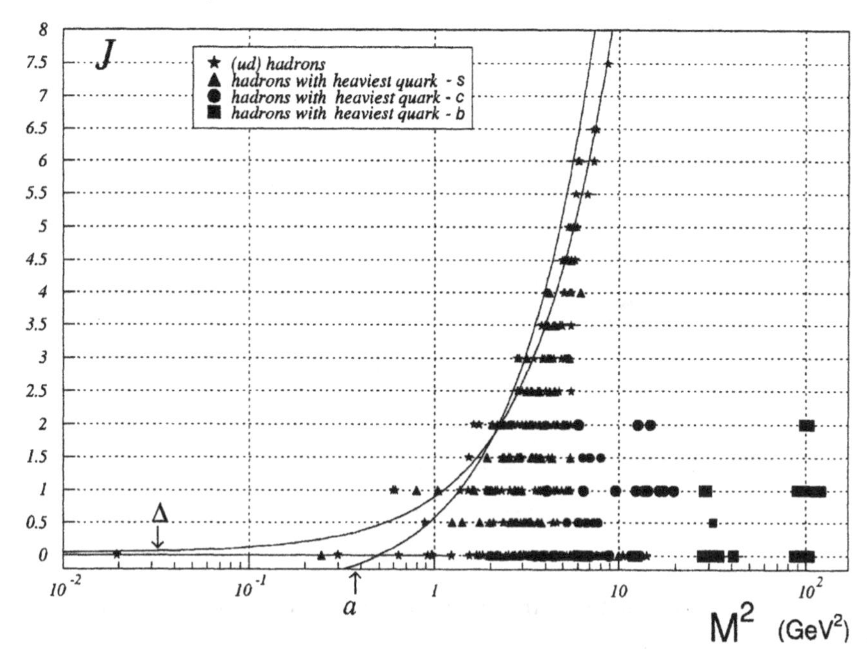

Figure 1. Chew-Frautschi plot for all hadrons from RPP-2002

TABLE 1. Slopes of meson Regge trajectories as a function of their flavor content (obtained from the 2000 data)

	q	s	c	b
$\bar{q}$	0.84 ± 0.09	0.86 ± 0.02	0.49 ± 0.08	0.22 ± 0.01
$\bar{s}$		0.82 ± 0.01	0.55 ± 0.01	0.22 ± 0.02
$\bar{c}$			0.40 ± 0.01	
$\bar{b}$				0.11 ± 0.01

We plan to reiterate fits in this approximation on the 2002 RPP data to see if the regularity is stable and we will then proceed to collect and compare different functional forms of the trajectories on a regular basis using the spectroscopic data together with the elastic scattering data, data on two body reactions, decay properties data to see if the decision rule for the dichotomy "quark model hadrons – exotic hadrons" can be constructed.

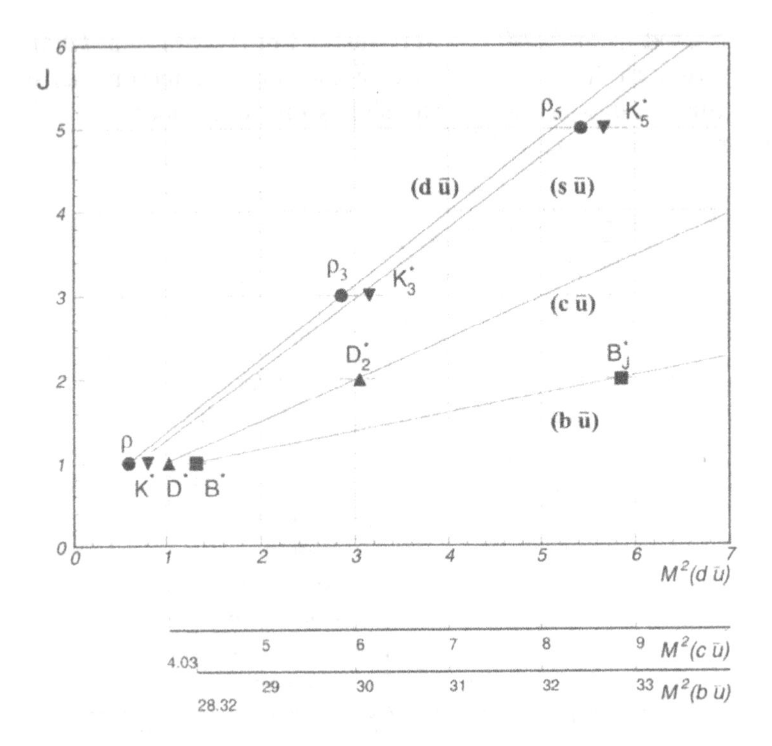

Figure 2. Linear Regge trajectories slopes for different meson flavors

ELECTROMAGNETIC FORM FACTORS OF HADRONS: EFFH1.0β

This object of knowledge is also under construction. So far only nucleon emff were considered. Its data set consists of 785 values of $d\sigma_{ep}/d\Omega$, 29 values of G_e/G_m and 31 values of σ_{tot} for $\bar{p}p \to e^+e^-$, for a total of 845 points. We ignore derived data on emff and produce fits only to the directly measured observables and then compare fits. The base of models consists of 4 adjusted and maintained parameterizations.

Recently the extended Gari-Kruempelmann [19] parameterization for the nucleon emff were fitted [20] to the most complete data set of the derived data with inclusion of the new data on G_E/G_M [21]. To include this extension of the Gari-Kruempelmann parameterization to the model base we started to check if it could be reasonably fitted to our database.

It turns out that in the VMD part of parameterization it is enough to include only one vector meson ($\rho(770)$) to obtain a reasonable fit to the $d\sigma/dt$ and G_E/G_M data.

However it leads to the determination of the electric and magnetic radii of the nucleons that are incompatible with that determined from the Lamb shift in hydrogen atom measurements.

TABLE 2. Fit to the $d\sigma/dt$ data: $\chi^2/d.o.f. = 0.91$

$\langle r^2 \rangle$ fm^2	Value	σ^2 fm^2	Correlations			
$\langle (r_E^p)^2 \rangle$	0.6906	2.7E-03	1.00	−0.01	0.22	−0.26
$\langle (r_M^p)^2 \rangle$	0.6926	3.1E-03	−0.01	1.00	−0.44	0.95
$\langle (r_E^n)^2 \rangle$	−0.4266	3.2E-03	0.22	−0.44	1.00	−0.65
$\langle (r_M^n)^2 \rangle$	0.9003	5.3E-03	−0.26	0.95	−0.65	1.00

TABLE 3. Fit to the $d\sigma/dt$ and G_E/G_M data: $\chi^2/d.o.f. = 1.03$

$\langle r^2 \rangle$ fm^2	Value	σ^2 fm^2	Correlations			
$\langle (r_E^p)^2 \rangle$	0.6650	1.7E-03	1.00	0.32	−0.24	0.69
$\langle (r_M^p)^2 \rangle$	0.7153	4.7E-03	0.32	1.00	0.73	0.28
$\langle (r_E^n)^2 \rangle$	−0.3411	15.6E-03	−0.24	0.73	1.00	−0.46
$\langle (r_M^n)^2 \rangle$	0.8965	4.7E-03	0.69	0.28	−0.46	1.00

We see from Tables , that estimates of the physical parameters changed markedly with the addition of new observables measured in the same range of kinematic variables.

It should be noted that the mean square proton radii $\langle (r_E^p)^2 \rangle = 0.61$ fm^2, $\langle (r_M^p)^2 \rangle = 1.82$ fm^2 calculated from the fit obtained in [20] for the same model but with a VMD part containing ρ, ω, and ϕ contributions are even worse in comparison with estimates from the Lamb shift data. This is a signal for possible problems with the database and/or with parameterizations (see also [22]). Further cross-assessment iterations with models ranking are needed.

FORWARD ELASTIC SCATTERING OF HADRONS: FESH1.0β

This database contains the measured differential distribution $d\sigma/dt$ for $|t| < 0.6$ GeV^2 for $\pi^\pm p$ (438 points at 73 energies), $K^\pm p$ (204 points at 34 energies) and $\bar{p}p$, pp (564 points at 94 energies). The associated object of knowledge is interfaced with the FORWARD, EFFH and RT objects of knowledge as

$$\frac{d\sigma}{dt} = \pi \left(|f_c|^2 + 2Re(f_c^* f_h) + |f_h^2|^2 \right) \tag{1}$$

with f_c the Coulomb amplitude, which depends on the form factors of EFFH, f_h the hard-interaction amplitude, which depends on σ_{tot} and ρ

(from FORWARD), and on Regge trajectories (from RT).

So far, we have done a preliminary study trying to find regularity in energy of the several claimed evidences for oscillations on the diffraction cone. The method to reveal the oscillations is illustrated in the Figure 3

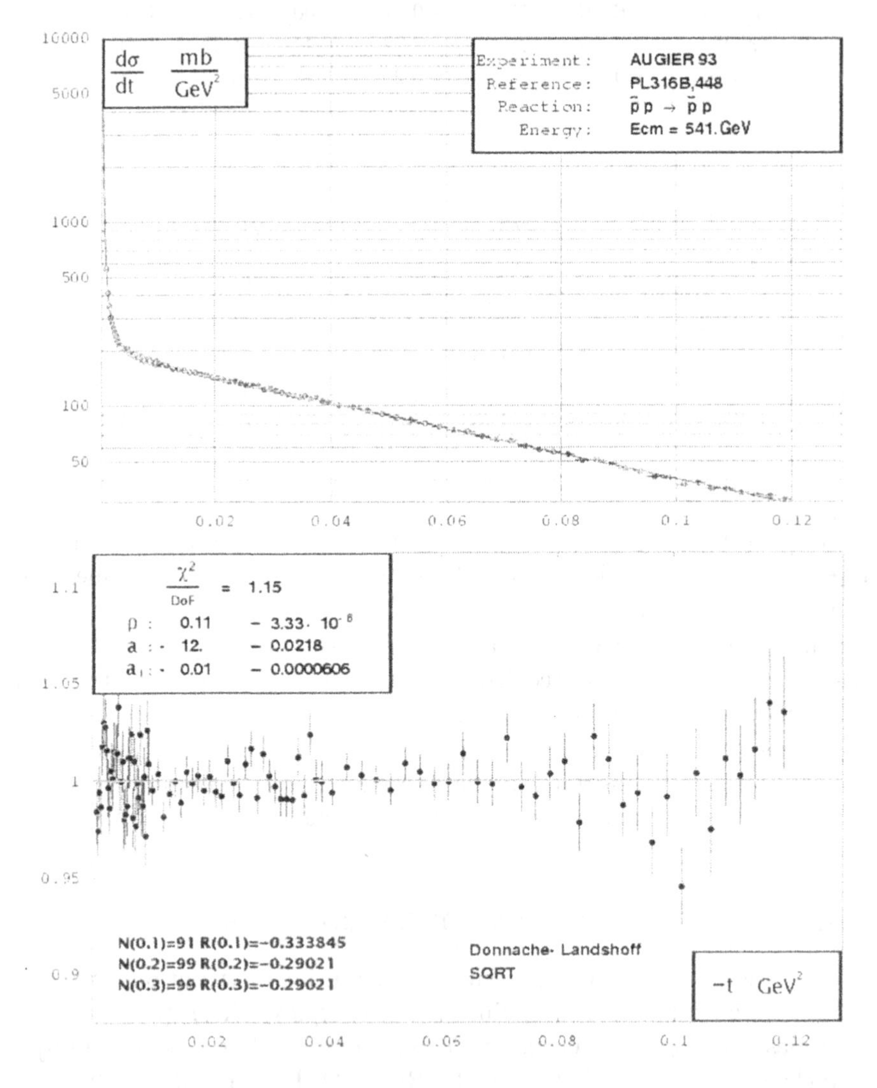

Figure 3. Fit to the $d\sigma/dt$ and results of autocorrelation function calculation

Using the naïve models for the diffractive cone description $(A(s)e^{B(s)\alpha(t)})$ and the standard Coulomb amplitude with popular dipole(pole) charge form-factors for nucleons(mesons), we calculate the normalized autocorrelation $R(s)$ of the difference $T(s,t) = (\frac{d\sigma^{data}/dt}{d\sigma^{theory}/dt} - 1)$ for 195 experimental distributions $d\sigma/dt$ at different values of $P_{lab} \geq 10$ GeV/c and having more than 7 data points in the region $|t| < 0.6$ GeV^2.

$$R(s) = \sum_i \frac{T(s,t_i)T(s,t_{i+1})}{\sigma_i^2\sigma_{i+1}^2}\left(\sum_i \frac{1}{\sigma_i^2\sigma_{i+1}^2}\right)^{-1},$$

where $\sigma_i = \frac{\sigma_i^{data}}{d\sigma^{theory}(s,t_i)/dt}$.

Large (> 1) values of the autocorrelator are signals for oscillations or fit biases, large negative values (< -1) are signals that we have problems with data (the errors are over-estimated).

Figure 4 shows that almost all values of the autocorrelations for the three intervals $|t| < 0.1$ GeV^2, $|t| < 0.2$ GeV^2, and $|t| < 0.3$ GeV^2 are close to the normal distribution. There are some outstanding points but the data on the corresponding scatter plots do not show any stable regularity in energy dependence of the scatter plots structures. These outstanding autocorrelation values may be due hidden t-dependent systematic effects, or to biases in the fits on the diffraction cone.

Furthermore, the possible oscillation pattern seems to be model-dependent, as seen on the same figure. For example, the Figure 3 clearly show absence of the oscillations claimed in some phenomenological papers. Our preliminary conclusion is based on the analyses of two different forms of the t-dependence of the pomeron trajectory: linear and square root dependence with branch point at $t = 4m_\pi^2$. The optical points also were calculated with use different models of the energy dependence of the total cross sections.

To make unambiguous conclusions we need more iterations of cross-assessments to find a parameterization that give good description of the diffractive cones and their evolution with energy. Having such a parameterisation it will be possible to clarify the situation with claimed oscillations.

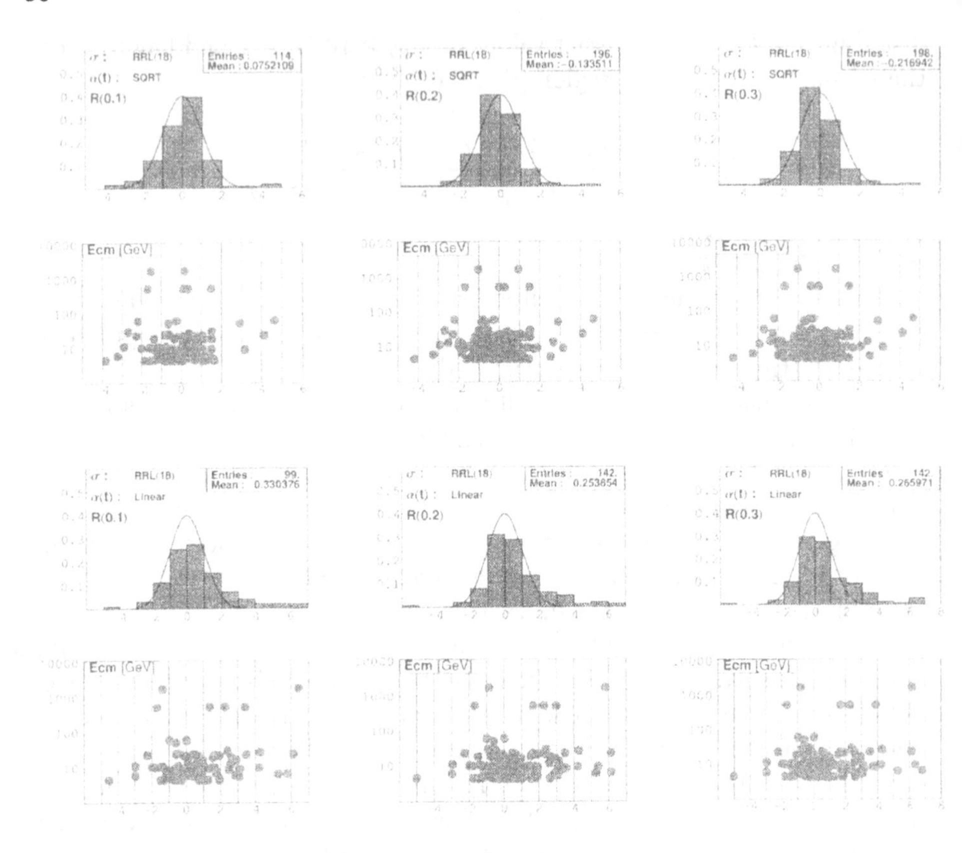

Figure 4. "Statistical pattern" of the autocorrelations as the indicator for the fine structure on the diffractive cone

CROSS SECTION IN $E^+E^- \to$ HADRONS, R, QCD TESTS: CSEE1.0

As a parallel activity, we have gathered [23] the data for the annihilation cross sections $\sigma_{e^+e^-\to hadrons}$ and for their ratio R to $\sigma_{ee\to\mu\mu}$ for 0.36 GeV$<\sqrt{s}<$ 188.7 GeV. The database consists of 1066 points rescaled to the hadronic R. The QCD fit to the hadronic part clearly shows that a 3-loop calculation is preferred with respect to the naïve Born formula, and leads to the following value of α_S:

$$\alpha_S(M_Z^2) = 0.128 \pm 0.032 \tag{2}$$

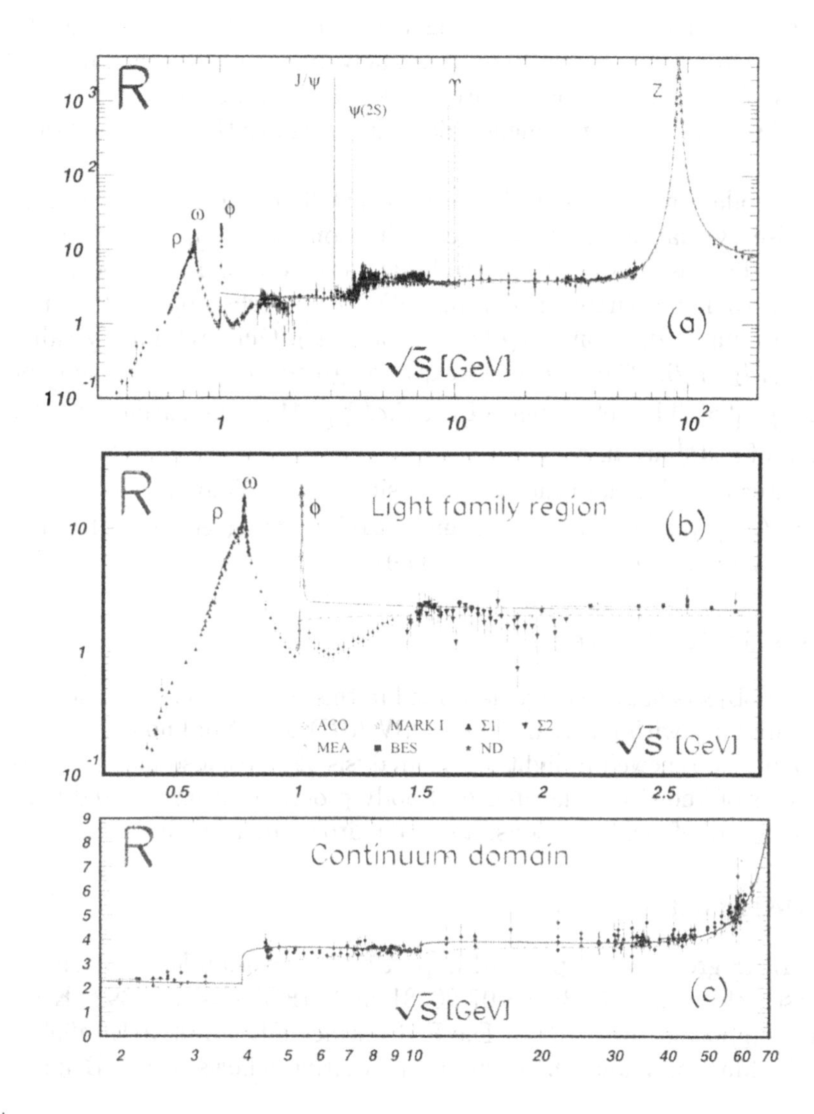

Figure 5. (a) World data on the ratio $R = \frac{\sigma(e^+e^- \to q\bar{q} \to \text{hadrons})}{\sigma(e^+e^- \to \mu^+\mu^-)}$. (b) Low $\sqrt{s}$ region crucial for the evaluation of $\Delta\alpha_{QED}^{had}(M_Z)$, $a_\mu{}_{LO}^{had}$, *etc.* (c) Applicability domain of 3-loop pQCD. Solid curves are 3-loop pQCD predictions (plus Breit-Wigner for narrow resonances on (a) and (b)). Broken curves show the "naïve" parton model prediction. Masses of c and b quarks are taken into account. The full set of radiative corrections is applied to all data. Further details and references to the original experimental data can be found in [23]. See also [24].

All available data on the total cross section and the R ratio of $e^+e^- \to$ hadrons are compiled from the PPDS(DataGuide, ReacData) (IHEP, Protvino, Russia) and HEPDATA(Reaction) (Durham, UK) databases and transformed to a compilation of data on the ratio $R = \frac{\sigma(e^+e^- \to q\bar{q} \to \text{hadrons})}{\sigma(e^+e^- \to \mu^+\mu^-)}$, with the full set of radiative corrections. This compilation is the most complete set of evaluated hadronic R ratio data publicly available to date. The current status of the data is shown in Fig. 5. The compilation is continuously maintained so that new experimental data are added as they become available.

The compilation is intended for tests of pQCD calculations as well as for a precise evaluation of hadronic contributions to $\Delta\alpha_{QED}(M_Z)$, $a_\mu = (g_\mu - 2)$, *etc.* The results we obtained so far are as follows: current theoretical predictions from the parton model and pQCD are well supported by the world "continuum" data on $\sigma_{tot}(e^+e^- \to \text{hadrons})$. Our preliminary value of $\Delta\alpha_{QED}^{had}(M_Z)$ is $0.02736 \pm 0.00040(\text{exp})$, in agreement with the results of other groups [25]. The refinement of the $\Delta\alpha_{QED}^{had}(M_Z)$ calculation and the evaluation of $a_\mu{}_{LO}^{had}$ are in progress.

Computer-readable data files are accessible on the Web at
`http://pdg.lbl.gov/2002/contents_plots.html` (see also [24]) and
`http://wwwppds.ihep.su:8001/eehadron.html`

Prospects for the future

The various objects of knowledge described in this report should be released to the community within a year. The FORWARD object of knowledge will also probably be renewed in light of the analysis of ρ. We also plan soon to build objects of knowledge devoted to 2-body processes at large s and t, to photoproduction of vector mesons, and to hadronic multiplicities.

Acknowledgments

The COMPAS group was supported in part by the Russian Foundation for Basic Research grants RFBR-98-07-90381 and RFBR-01-07-90392. K.K. is in part supported by the U.S. D.o.E. Contract DE-FG-02-91ER40688-Task A. E. Martynov and O. Selyugin are visiting fellows of the Belgian FNRS.

References

1. SLAC library http://www-spires.slac.stanford.edu/spires/.
2. Automated e-print archives http://www.arxiv.org/.
3. Physical review Online Archive, http://prola.aps.org/.
4. Russian Foundation for Basic Research, http://www.rfbr.ru/.

5. The Review of Particle Physics, http://pdg.lbl.gov/.
6. BAFIZ, a distributed network of knowledge bases in particle and nuclear physics, http://dbserv.ihep.su/.
7. Durham RAL Databases, http://durpdg.dur.ac.uk/HEPDATA/.
8. Committee on Data for Science and Technology, http://www.codata.org/.
9. Hagiwara, K., *et al.*[Particle Data Group] (2002) "Review of Particle Physics," *Phys. Rev.* **D66**, pp. 010001-1–974.
10. Ezhela, V. V. [COMPAS Group] (1984) "Particle Physics Data Systematization at IHEP" *Comput. Phys. Commun.* **33**, pp. 225-229.
11. Cudell, J. R., *et al.* [COMPETE collaboration] (2002) "Hadronic scattering amplitudes: Medium-energy constraints on asymptotic behaviour," *Phys. Rev.*, **D65**, pp. 074024-1–14, [arXiv:hep-ph/0107219].
12. Cudell, J. R., *et al.* [COMPETE collaboration] (2001) "New measures of the quality and of the reliability of fits applied to forward hadronic data at t = 0," [arXiv:hep-ph/0111025].
13. http://nuclth02.phys.ulg.ac.be/compete/publications/benchmarks_details/
14. http://nuclth02.phys.ulg.ac.be/compete/predictor.html/
15. http://pdg.lbl.gov/2002/contents_plots.html#tablepage
16. http://nuclth02.phys.ulg.ac.be/compete/predictor.html/RRPL2u21/-With_Rel/5_GeV/Sources_lab.html and http://nuclth02.phys.ulg.ac.be/-compete/predictor.html/RRPL2u21/With_Rel/5_GeV/Prediction_Ecm.html.
17. Cudell,J. R., Ezhela, V., Kang, K., Lugovsky, S. and Tkachenko, N. (2000) "High-energy forward scattering and the pomeron: Simple pole versus unitarized models," *Phys. Rev.* **D61**, pp. 034019-1–14, [Erratum-ibid. D **D63** (2001) 059901-1], [arXiv:hep-ph/9908218].
18. Cudell, J. R., *et al.* [COMPETE Collaboration] (2002) "Benchmarks for the forward observables at RHIC, the Tevatron-run II and the LHC," *Phys. Rev. Lett.* **89**, pp. 201801-1–4, [arXiv:hep-ph/0206172].
19. Gari, M. F. and Kruempelmann, W. (1992) "The Electric neutron form-factor and the strange quark content of the nucleon," *Phys. Lett.* **B274**, pp. 159; **B282**, pp. 483(E).
20. Lomon, E. L. (2001) "Extended Gari-Kruempelmann model fits to nucleon electromagnetic form-factors," *Phys. Rev.* **C64**, pp. 035204-1–9.
21. Jones, M. K. *et al.* (2000) "G_E^p/G_M^p ratio by polarization transfer in e(pol.) p $\rightarrow$ e p(pol.)," *Phys. Rev. Lett.* **84**, pp. 1398-1402.
22. Brash, E. J. and Kozlov, A. and Li, S. and Huber, G. M. (2002) "New empirical fits to the proton electromagnetic form-factors," *Phys. Rev.* **C65**, pp. 051001(R)-1–5.
23. Zenin, O.V. *et al.* [COMPAS (IHEP) and HEPDATA (Durham) Groups] (2001) "A Compilation of Total Cross Section Data on $e^+e^- \rightarrow$ hadrons and pQCD Tests," [arXiv:hep-ph/0110176].
24. Hagiwara, K. *et al.* [Particle Data Group] (2002) *Phys. Rev. D*, **D66**, pp. 010001-259–260.
25. Erler, J. and Langacker, P. (2002) "Electroweak Model and Constraints on New Physics," in Review of Particle Physics, *Phys. Rev.*, **D66**, p. 010001-98–112.

FORWARD OBSERVABLES AT RHIC, THE TEVATRON RUN II AND THE LHC

*COMPETE * collaboration update*

J. R. CUDELL AND E. MARTYNOV
Institut de Physique, Bât. B5, Université de Liège,
Sart Tilman, B4000 Liège, Belgium §

K. KANG
Physics Department, Brown University,
Providence, RI, U.S.A.

V. V. EZHELA, YU. V. KUYANOV, S. B. LUGOVSKY, E. A. RAZUVAEV AND N. P. TKACHENKO
COMPAS group, IHEP,
Protvino, Russia

AND

P. GAURON AND B. NICOLESCU
LPNHE¶-Theory Group, Université Pierre et Marie Curie,
Tour 12 E3, 4 Place Jussieu, 75252 Paris Cedex 05, France

Abstract. We present predictions on the total cross sections and on the ratio of the real part to the imaginary part of the elastic amplitude (ρ parameter) for present and future pp and $\bar{p}p$ colliders, and on total cross sections for $\gamma p \to$ hadrons at cosmic-ray energies and for $\gamma\gamma \to$ hadrons up to $\sqrt{s} = 1$ TeV. These predictions are based on a study of many possible analytic parametrisations and invoke the current hadronic dataset at $t = 0$. The uncertainties on total cross sections, including the systematic theoretical errors, reach 1% at RHIC, 3% at the Tevatron, and 10% at the LHC, whereas those on the ρ parameter are respectively 10%, 17%, and 26%.

This report is based on ref. [1], which constitutes the conclusion of an exhaustive study [2] of analytic parametrisations of soft forward data at

*COmputerised Models, Parameter Evaluation for Theory and Experiment.

§E. M. is on leave from Bogolyubov Institute for Theoretical Physics, 03143 Kiev, Ukraine

¶Unité de Recherche des Universités Paris 6 et Paris 7, Associée au CNRS.

R. Fiore et al. (eds.), Diffraction 2002, 63–72.

$t = 0$. As explained in V. V. Ezhela's contribution to these proceedings, this study has three main purposes. First of all, it helps maintain the dataset of cross sections and ρ parameters available to the community. Secondly, it enables us to decide which models are the best, and in which region of s. Finally, and this will be the main object of this report, it enables us to make predictions based on a multitude of models, and on all available data.

The dataset of this study includes all measured total cross sections and ratios of the real part to the imaginary part of the elastic amplitude (ρ parameter) for the scattering of pp, $\bar{p}p$, $\pi^{\pm}p$, $K^{\pm}p$, and total cross sections for γp, $\gamma\gamma$ and $\Sigma^{-}p$. Compared with the 2002 Review of Particle Properties dataset [3], it includes the latest ZEUS points[4] on total cross sections, as well as cosmic ray measurements [5]. The number of points of each sub-sample of the dataset is given in Table 1.

The base of models is made of 256 different analytic parametrisations. We can summarize their general form by quoting the form of total cross sections, from which the ρ parameter is obtained via derivative dispersion relations. The ingredients are the contribution M^{ab} of the highest meson trajectories (ρ, ω, a and f) and the rising term H^{ab} for the pomeron.

$$\sigma_{tot}^{ab} = (M^{ab} + H^{ab})/s \tag{1}$$

The first term is parametrised via Regge theory, and we allow the lower trajectories to be partially non-degenerate, *i.e.* we allow one intercept for the $C = +1$ trajectories, and another one for the $C = -1$ [6]. A further lifting of the degeneracy is certainly possible, but does not seem to modify significantly the results [7]. Hence we use

$$M^{ab} = Y_{+}^{ab}\left(\frac{s}{s_0}\right)^{\alpha_+} \pm Y_{-}^{ab}\left(\frac{s}{s_0}\right)^{\alpha_-} \tag{2}$$

with $s_0 = 1$ GeV^2. The contribution of these trajectories is represented by RR. As for the pomeron term, we choose a combination of the following possibilities:

$$\mathrm{H}^{ab} = X^{ab}\left(\frac{s}{s_0}\right)^{\alpha_{\wp}} + sP^{ab} \tag{3}$$

$$\mathrm{H}^{ab} = s\left[B^{ab}\ln\left(\frac{s}{s_0}\right) + P^{ab}\right] \tag{4}$$

$$\mathrm{H}^{ab} = s\left[B^{ab}\ln^2\left(\frac{s}{s_1}\right) + P^{ab}\right] \tag{5}$$

with $s_0 = 1$ GeV^2 and s_1 to be determined by the fit. The contribution of these terms is marked PE, PL and PL2 respectively. Note that the pole structure of the pomeron cannot be directly obtained from these forms, as

TABLE 1. Summary of the quality of the fits for different scenarios considered in this report, for $\sqrt{s} \geq 5$ GeV: DB02Z – the 2002 Review of Particle Properties database with new ZEUS data, DB02Z-CDF – with the CDF point removed; DB02Z-E710/E811 – with E710/E811 points removed. The first line gives the overall χ^2/dof for the global fits, the other lines give the χ^2/nop for data sub-samples, the last line gives in each case the parameter controlling the asymptotic form of cross sections.

Sample	Number of points	DB02Z	DB02Z −CDF	DB02Z −E710/E811
total		0.965	0.964	0.951
	total cross sections			
pp	111	0.84	0.90	0.90
$\bar{p}p$	57–59	1.15	1.12	1.05
π^+p	50	0.71	0.71	0.71
π^-p	95	0.96	0.96	0.96
K^+p	40	0.71	0.71	0.71
K^-p	63	0.62	0.62	0.61
Σ^-p	9	0.38	0.38	0.38
γp	37	0.58	0.58	0.58
$\gamma\gamma$	38	0.64	0.64	0.63
	elastic forward Re/Im			
pp	64	1.83	1.83	1.80
$\bar{p}p$	11	0.52	0.52	0.53
π^+p	8	1.50	1.52	1.46
π^-p	30	1.10	1.09	1.14
K^+p	10	1.07	1.10	0.98
K^-p	8	0.99	1.00	0.96
	values of the parameter B			
		0.307(10)	0.301(10)	0.327(10)

multiple poles at $J = 1$ produce constant terms which mimic simple poles at $t = 0$. Furthermore, we have considered several possible constraints on the parameters of Eqs. (2-5):

- degeneracy of the reggeon trajectories $\alpha_+ = \alpha_-$, noted $(\mathrm{RR})_d$;
- universality of rising terms (B^{ab} independent of the hadrons), noted $\mathrm{L2}_u$, L_u and E_u [8];
- factorization for the residues in the case of the $\gamma\gamma$ and γp cross sec-

tions. If not otherwise indicated by the subscript nf, we impose $H_{\gamma\gamma} = \delta H_{\gamma p} = \delta^2 H_{pp}$;
- quark counting rules [9] to predict the Σp cross section from pp, Kp and πp, indicated by the subscript qc;
- Johnson-Treiman-Freund [10] relation for the cross section differences, noted R_c.

All possible variations of Eqs. (2-5), using the above constraints, amount to 256 variants.

These variants are then fitted to the database, allowing for the minimum c.m. energy $\sqrt{s_{min}}$ of the fit to vary between 3 and 10 GeV. For $\sqrt{s} \geq 9$ GeV, 33 variants have an overall $\chi^2/d.o.f. \leq 1.0$ if one fits only to total cross sections, whereas 21 obeyed this criterion when one includes the ρ parameters in the data to be fitted to. One can try to lower the minimum energy of the fit, and one finds that for 11 models one can extend the minimum energy of the cross section fit to 4 GeV, and that of the combined fit of σ_{tot} and ρ to 5 GeV. Several parametrisations based on triple poles (RRPL2), double poles (RRPL) or simple poles (RRPE) are kept. The only notable candidate which seems to be ruled out is the popular simple-pole model (RRE) [11]. Its predictions for pp and $\bar{p}p$ nevertheless fall within our errors[1].

After this selection is made, the remaining models are ranked. We measure some characteristics of the fits, namely: the number of parameters, the confidence level in the considered region, the size of the region where the model achieves a $\chi^2/dof \leq 1$ and the value of that χ^2/dof, the stability of the parameters when the minimum c.m. energy is changed, their stability with respect to the inclusion of the ρ data, the uniformity of the χ^2/dof for different processes and quantities, and finally the quality of the correlation matrix. All these features are important, and we have managed to measure them, introducing new statistical indicators [2, 13]. The ideal fit would be the one with the least number of parameters, the biggest region of applicability, the best χ^2, etc. Unfortunately, a single fit does not concentrate all these virtues. As the new indicators do not have (yet) a probabilistic interpretation and as all the parametrisations which fit are *a priori* acceptable, we choose the "best" model through a ranking procedure: for each feature, the models are ordered according to how well they perform. One then sums the position of each model for each indicator, and the model with least points is preferred. The advantage of this method, besides the fact that it automatically looks at many qualities of each fit, is that the best model is decided on the basis of automatic criteria, which do not depend on our own prejudice.

[1]This conclusion could be affected by a re-calculation of the ρ parameter from *integral* dispersion relations – see O. V. Selyugin's contribution to these proceedings.

Following that procedure, the triple-pole parametrisation $\mathrm{RRP}_{nf}\mathrm{L2}_u$ [8, 12] gives the most satisfactory description of the data. This parameterization has a universal (u) $B\ln^2(s/s_0)$ term, a non-factorizing (nf) constant term and non-degenerate lower trajectories.

We are now in a position to evaluate several quantities of interest for future measurements. First of all, our best parametrisation can of course be used to predict σ_{tot} and ρ, with their statistical errors. We choose for this the parameters determined for a minimum c.m. energy $\sqrt{s_{min}} = 5$ GeV. For pp and $\bar{p}p$, the central value of this fit gives

$$\sigma_{tot}^{\bar{p}p,pp} = 43.\ s^{-0.46} \pm 33.\ s^{-0.545} + 35.5 + 0.307 \ln^2\left(\frac{s}{29.}\right) \tag{6}$$

with all coefficients in mb and s in GeV^2. We assign errors by using the full error matrix E_{ij} from the fit, and define

$$\Delta Q = \sum_{ij} E_{ij} \frac{\partial Q}{\partial x_i \partial x_j} \tag{7}$$

with $Q = \sigma_{tot}$ or ρ and x_i the parameters of the model. Our predictions are given in Table 2 and the corresponding 1 σ region is shown as a dark band in Figs. 1 and 2.

One can now concentrate on the evaluation of systematic errors. The first source of these is the presence of contradictory data points in the database. One can see from Table 1 that $\sigma_{\bar{p}p}$, ρ_{pp}, $\rho_{\pi p}$ and ρ_{K^+p} are not well fitted to. For the ρ parameters, this can partially be attributed to contradictions in the data, and partially to our use of derivative dispersion relations. For $\sigma_{\bar{p}p}$, this comes entirely from problems with the data. Of these, the most notable one is the disagreement at the Tevatron between the measurements of CDF and those of E710/E811. We show in Table 1 the effect of the removal of the CDF or of the E710/E811 measurements. First of all, one can see that the coefficient of the $\log^2 s$ changes by more than 1σ, and hence the predictions for the LHC are muddled by this discrepancy. We also see that our preferred parametrisation favours the CDF measurement, as its global χ^2/dof goes down when the E710/E811 point is removed. It must be noted that a similar situation exists for all $\log^2$ and simple pole parametrisations[2]. On the other hand, the dipole RRPL ($\log s$)

[2]The original Donnachie-Landshoff parametrisation did predict a low cross section at the Tevatron, but this was due to the use of a non-conventional (and non-probabilistic) χ^2. Using the statistical definition of χ^2 leads to a rejection of the E710/E711 point.

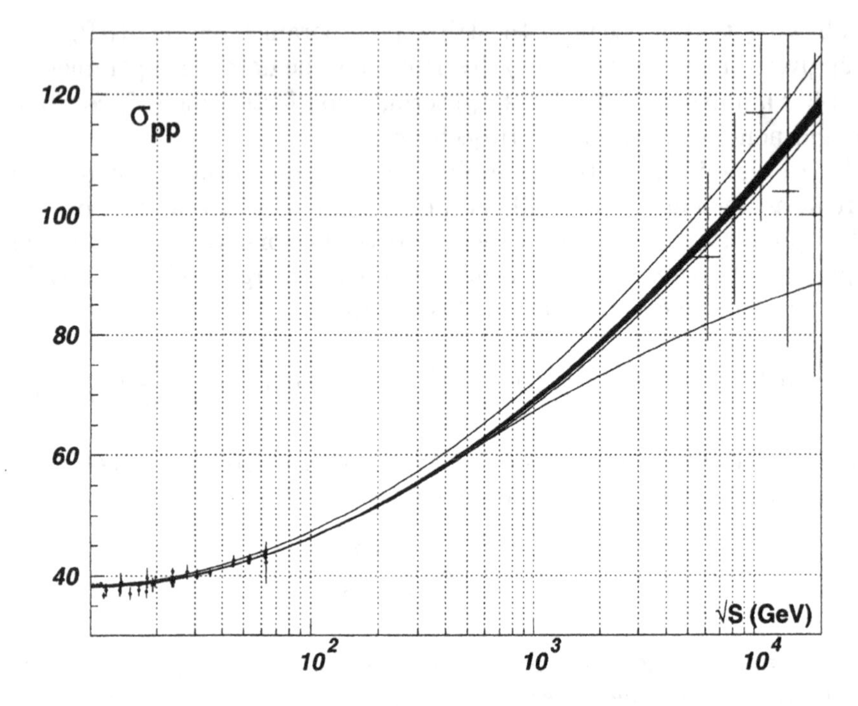

Figure 1. Predictions for total cross sections.

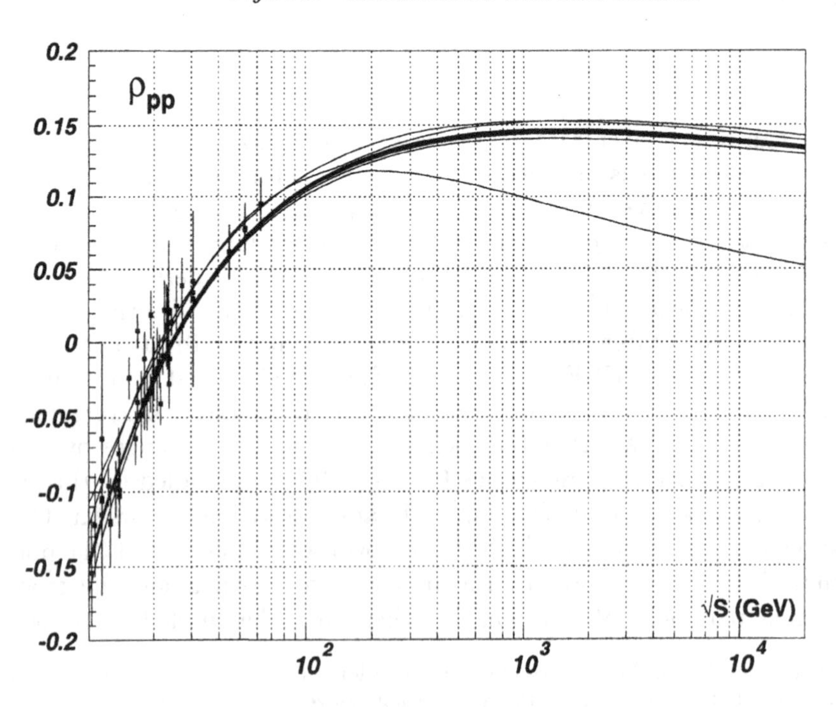

Figure 2. Predictions for the ρ parameter.

TABLE 2. Predictions for σ_{tot} and ρ, for $\bar{p}p$ (at $\sqrt{s}$ = 1960 GeV) and for pp (all other energies). The central values and statistical errors correspond to the preferred model RRP$_{nf}$L2$_u$, fitted for $\sqrt{s_{min}}$ = 5 GeV. The first systematic errors come from the consideration of two choices between CDF and E-710/E-811 $\bar{p}p$ data in the simultaneous global fits. The second systematic error corresponds to the consideration of the 21 parametrisations compatible with existing data.

$\sqrt{s}$ (GeV)	σ (mb)			ρ		
100	46.37 ± 0.06	$^{+0.11}_{-0.03}$	$^{+0.31}_{-0.06}$	0.1058 ± 0.0012	$^{+0.0028}_{-0.0009}$	$^{+0.0024}_{-0.0019}$
200	51.76 ± 0.12	$^{+0.27}_{-0.08}$	$^{+0.43}_{-0.15}$	0.1275 ± 0.0015	$^{+0.0035}_{-0.0011}$	$^{+0.0000}_{-0.0023}$
300	55.50 ± 0.17	$^{+0.39}_{-0.12}$	$^{+0.39}_{-0.20}$	0.1352 ± 0.0016	$^{+0.0038}_{-0.0012}$	$^{+0.0000}_{-0.0059}$
400	58.41 ± 0.21	$^{+0.49}_{-0.16}$	$^{+0.28}_{-0.23}$	0.1391 ± 0.0017	$^{+0.0039}_{-0.0013}$	$^{+0.0002}_{-0.0087}$
500	60.82 ± 0.25	$^{+0.58}_{-0.19}$	$^{+0.15}_{-0.25}$	0.1413 ± 0.0017	$^{+0.0040}_{-0.0013}$	$^{+0.0006}_{-0.0109}$
600	62.87 ± 0.28	$^{+0.66}_{-0.21}$	$^{+0.03}_{-0.26}$	0.1427 ± 0.0018	$^{+0.0040}_{-0.0013}$	$^{+0.0008}_{-0.0125}$
1960	78.26 ± 0.55	$^{+1.30}_{-0.42}$	$^{+0.08}_{-1.95}$	0.1450 ± 0.0018	$^{+0.0038}_{-0.0013}$	$^{+0.0022}_{-0.0226}$
10000	105.1 ± 1.1	$^{+2.6}_{-0.82}$	$^{+0.60}_{-8.30}$	0.1382 ± 0.0016	$^{+0.0032}_{-0.0011}$	$^{+0.0028}_{-0.0324}$
12000	108.5 ± 1.2	$^{+2.7}_{-0.87}$	$^{+0.70}_{-9.20}$	0.1371 ± 0.0015	$^{+0.0031}_{-0.0011}$	$^{+0.0030}_{-0.0332}$
14000	111.5 ± 1.2	$^{+2.9}_{-0.92}$	$^{+0.80}_{-10.2}$	0.1361 ± 0.0015	$^{+0.0030}_{-0.0011}$	$^{+0.0030}_{-0.0337}$

parametrisations do prefer the lower value. The only thing we can do at present is indicate what shift the adoption of one point or the others would cause on our central value. This is given in Table 2 as a systematic experimental error, corresponding to the shift in the upper and lower 1σ allowed values[3], and shown as the two curves closest to the central band in Figs. 1 and 2.

We can now, and maybe for the first time, give a reasonable estimate of the theoretical error. The idea is to choose a less constraining minimum energy for the fits (we take 9 GeV), and to consider the results of the 21 models that succeed in reproducing both σ_{tot} and ρ. This gives us 21 predictions with error bars. We can then define the theoretical systematic error by taking the distance between the highest (resp. lowest) values in

[3]This definition differs from that of [1] by about 1 statistical error

TABLE 3. Predictions for σ_{tot} for $\gamma p \to hadrons$ and for $\gamma\gamma \to hadrons$ for cosmic ray energies. We quote the central values, the statistical errors and the experimental systematic errors, defined as in Table 2.

p^{γ}_{lab} (GeV)	$\sigma_{\gamma p}$ (mb)	$\sqrt{s}$ (GeV)	$\sigma_{\gamma\gamma}$ (μ b)
$0.5 \cdot 10^6$	$0.24 \pm 0.01 \begin{smallmatrix} +0.00 \\ -0.11 \end{smallmatrix}$	200	$0.55 \pm 0.03 \begin{smallmatrix} +0.00 \\ -0.29 \end{smallmatrix}$
$1.0 \cdot 10^6$	$0.26 \pm 0.01 \begin{smallmatrix} +0.00 \\ -0.11 \end{smallmatrix}$	300	$0.61 \pm 0.04 \begin{smallmatrix} +0.00 \\ -0.29 \end{smallmatrix}$
$1.0 \cdot 10^7$	$0.33 \pm 0.02 \begin{smallmatrix} +0.00 \\ -0.11 \end{smallmatrix}$	400	$0.66 \pm 0.04 \begin{smallmatrix} +0.00 \\ -0.29 \end{smallmatrix}$
$1.0 \cdot 10^8$	$0.42 \pm 0.02 \begin{smallmatrix} +0.00 \\ -0.11 \end{smallmatrix}$	500	$0.70 \pm 0.05 \begin{smallmatrix} +0.00 \\ -0.29 \end{smallmatrix}$
$1.0 \cdot 10^9$	$0.52 \pm 0.03 \begin{smallmatrix} +0.00 \\ -0.11 \end{smallmatrix}$	1000	$0.84 \pm 0.07 \begin{smallmatrix} +0.00 \\ -0.29 \end{smallmatrix}$

the 1σ intervals with the 1 σ central region. We give the resulting numbers as a third error in Table 2, and as the outer curves of Figs. 1 and 2.

Note that the systematic errors cannot be added in quadrature, and that the theoretical systematic error is an absolute shift from model to model, and does not have any probabilistic interpretation.

One can see that the errors on total cross sections are of the order of 1% at RHIC, 3% at the Tevatron and as large as 10% at the LHC. At RHIC, the systematic errors dues to the Tevatron discrepancy and those due to theory are of comparable order. The value of the cross section is constrained by the $\bar{p}p$ data, and by the fact that we allow only one $C = -1$ contribution, which is well constrained by the overall fit. Very precise RHIC measurements, at the level of one in a thousand could shed light on the Tevatron discrepancy, and discriminate between models. Of course, the extrapolation to LHC energies presents the largest uncertainty and is dominated by systematic theoretical errors, with the double pole models (RRPL) giving a cross section significantly lower than the triple poles or the simple poles. A determination of the cross section at the 5% level could rule out one half of the models or the other.

The errors on the ρ parameter are much larger, reaching 10% at RHIC, 17% at the Tevatron and 26% at the LHC. This is due to the fact that experimental errors are bigger, hence less constraining, but this also stems from the incompatibility of some low-energy determinations of ρ [2], and from our use of derivative dispersion relations. Although integral dispersion relations have the potential to reduce the χ^2/dof, they have the inconvenient of introducing extra parameters (because they necessitate subtractions). Hence it is unlikely that a different theoretical treatment can reduce the errors. On the other hand, a re-analysis of some of the data could be envisaged. It should involve a combination of the information on total cross section with that on elastic hadronic cross sections, on electromagnetic form factors and on Regge trajectories (see V. V. Ezhela's contribution to these proceedings).

Finally, we can use the same approach to predict cross sections for cosmic photon studies. We show the results in Table 3, where we have given only the experimental systematic error[4].

To conclude, we believe that we have pushed the database technology to the point where it can make predictions, and decide on which models or theories are the best. This is an example proof of the feasibility of the COMPETE program, and of its utility.

We have given here the best possible estimates for present and future pp and $\bar{p}p$ facilities. The central values of our fits and the corresponding statistical error give our "best guess" estimate. The systematic experimental errors tell us how much this guess could be affected by incompatible data. The theoretical systematic errors will tell us directly whether an experiment can be fitted by one of the standard analytic parametrisations, or whether it calls for new ideas.

Acknowledgments

The COMPAS group was supported in part by the Russian Foundation for Basic Research grants RFBR-98-07-90381 and RFBR-01-07-90392. K.K. is in part supported by the U.S. D.o.E. Contract DE-FG-02-91ER40688-Task A. E. Martynov is a fellow of the Belgian FNRS. We thank Professor Jean-Eudes Augustin for the hospitality at LPNHE-Université Paris 6, where part of this work was done.

References

1. Cudell, J. R. *et al.* [COMPETE Collaboration]: 2002, 'Benchmarks for the forward observables at RHIC, the Tevatron-run II and the LHC', *Phys. Rev. Lett.* **89 (20)**, pp. 201801-1 – 201801-4 [arXiv:hep-ph/0206172].
2. Cudell, J. R. *et al.* [COMPETE Collaboration]: 2002, 'Hadronic scattering amplitudes: Medium-energy constraints on asymptotic behaviour', *Phys. Rev. D* **65**, pp. 074024-1 - 074024-14 [arXiv:hep-ph/0107219].
3. Hagiwara, K. et al.: 2002, 'Review Of Particle Physics', *Phys. Rev. D* **66**, p. 010001, available on the PDG WWW pages (URL: http://pdg.lbl.gov/).
4. Chekanov, S. *et al.* [ZEUS Collaboration]: 2002, 'Measurement of the photon-proton total cross section at a center-of-mass energy of 209 GeV at HERA', *Nucl. Phys. B* **627**, pp. 3-28 [arXiv:hep-ex/0202034].

[4]note that the corresponding table in ref. [1] has a typo in the error bars which are systematically a factor 10 too small.

5. Honda, M. *et al.*: 1993, 'Inelastic cross-section for $p - air$ collisions from air shower experiment and total cross-section for pp collisions at SSC energy', *Phys. Rev. Lett.* **70**, pp. 525-528; Baltrusaitis, R. M. *et al.*: 1984, 'Total Proton Proton Cross-Section At $\sqrt{s} = 30$-Tev', *Phys. Rev. Lett.* **52**, pp. 1380-1383.
6. Cudell, J. R., Kang, K. and Kim, S. K. : 1997, 'Bounds on the soft pomeron intercept', *Phys. Lett. B* **395**, pp. 311-331 [arXiv:hep-ph/9601336].
7. Desgrolard, P., Giffon, M., Martynov, E. and Predazzi, E.: 2001, 'Exchange-degenerate Regge trajectories: A fresh look from resonance and forward scattering regions', *Eur. Phys. J. C* **18**, pp. 555-561 [arXiv:hep-ph/0006244].
8. Soloviev, L. D.: 1973, *Pisma v ZHETF* **18**, pp. 455-7 (in Russian); 1973, 'Increase of total cross sections and slope of diffraction peak', *JETP Lett.* **18**; 1974, *Pisma v ZHETF* **19**, pp. 185-188 (in Russian); 1974, 'Growth of total cross sections and elastic scattering', *JETP Lett.* **19**, pp. 116-118.
9. Levin, E. M. and Frankfurt, L. L.: 1965, *Pisma v ZHETF* **3**, pp. 652 (in Russian); 1965, 'The quark hypothesis And relations between cross-sections at high energies', *JETP Lett.* **2**, pp. 65-67.
10. Johnson, K. and Treiman, S. B.: 1965, 'Implications of SU(6) Symmetry for Total Cross Sections', *Phys. Rev. Lett.* **14**, pp. 189-191; Freund, P. G. O.: 1965, 'Relation Between πp, pp, and $\bar{p}p$ Scattering at High Energies', *Phys. Rev. Lett.* **15** , pp. 929-930.
11. Donnachie, A. and Landshoff, P. V.: 1992, 'Total cross-sections', *Phys. Lett.* **B296**, pp. 227-232 [hep-ph/9209205]
12. Gauron, P. and Nicolescu, B.: 2000, 'A possible two-component structure of the non-perturbative pomeron', *Phys. Lett. B* **486**, pp. 71-76 [arXiv:hep-ph/0004066].
13. Cudell, J. R. *et al.* [COMPETE collaboration]: 2001, 'New measures of the quality and of the reliability of fits applied to forward hadronic data at $t = 0$', in the *Proceedings of the 6th Workshop on Non-Perturbative QCD*, Paris, France, 5-9 Jun 2001, (World Scientific: 2002), pp. 107-112 [arXiv:hep-ph/0111025].

CONSEQUENCES OF T-CHANNEL UNITARITY FOR THE INTERACTION OF REAL AND VIRTUAL PHOTONS AT HIGH ENERGIES

E. MARTYNOV[†,*] , J.R CUDELL[**] AND G. SOYEZ[***]
Université de Liège, Bât. B5-a, Sart Tilman, B4000 Liège, Belgium
[†] *on leave from Bogolyubov Institute for Theoretical Physics, Kiev, Ukraine.*
[*] E.Martynov@guest.ulg.ac.be
[**] J.R.Cudell@ulg.ac.be
[***] G.Soyez@ulg.ac.be

Abstract. We analyze the consequences of t-channel unitarity for photon cross sections and show what assumptions are necessary to allow for the existence of new singularities at $Q^2 = 0$ for the γp and $\gamma\gamma$ total cross sections. For virtual photons, such singularities can in general be present, but we show that, apart from the perturbative singularity associated with $\gamma^*\gamma^* \to q\bar{q}$, no new ingredient is needed to reproduce the data from LEP and HERA, in the Regge region.

1. Introduction

It is well known [1] that due to unitarity one can relate the amplitudes describing three hadronic elastic processes $aa \to aa, ab \to ab, bb \to bb$. Namely, if a simple Regge pole at $j = \alpha(t)$ contributes in the t-channel for each of the above-mentioned processes, the residues of the poles are factorized

$$\beta_{aa\to aa}(t)\beta_{bb\to bb}(t) = (\beta_{ab\to ab}(t))^2.$$

However it is difficult to check directly such a relation. Firstly, there are no experimental data for all three processes (for example, $\pi\pi$ is missing, if one considers $\pi\pi, \pi p$ and pp scattering). Secondly, the best fit to the hadronic cross section data is achieved in the models with multiple Regge

R. Fiore et al. (eds.), Diffraction 2002, 73–83.

poles rather than with simple ones [2]. Factorization properties of multiple poles are to be determined.

On the other hand, the DIS and total cross-section data [3] as well as the measurements of the $\gamma\gamma$ total cross section and of the off-shell photon structure function F_2^γ [4] are available now. If factorization is valid in the case of the photon amplitudes then it can be checked for another set of related processes: $pp, \gamma p, \gamma\gamma$ and, probably, for $pp, \gamma^* p, \gamma^*\gamma^*$.

In this talk, we show how to derive the generalized factorization for the partial amplitudes of the related processes at an arbitrary, but common for these amplitudes, t-channel Regge singularity. We also give arguments in favor of its validity in the photon case and apply the new factorization relations to describe $\gamma\gamma$ and $\gamma^*\gamma^*$ cross sections.

2. t-channel unitarity

It is an old result that one can relate the amplitudes describing three elastic processes $aa \to aa$, $ab \to ab$, $bb \to bb$. The trick is to continue these to the crossed channels $a\bar{a} \to a\bar{a}$, $a\bar{a} \to b\bar{b}$, $b\bar{b} \to b\bar{b}$, where they exhibit discontinuities because of the a and b thresholds. One then obtains a nonlinear system of equations, which can be solved. Working in the complex j plane above thresholds ($t > 4m_a^2,\ 4m_b^2$), and defining the matrix

$$T_0 = \begin{pmatrix} A_{aa\to aa}(j,t) & A_{ba\to ba}(j,t) \\ A_{ab\to ab}(j,t) & A_{bb\to bb}(j,t) \end{pmatrix} \tag{1}$$

one obtains

$$T_0 = \frac{D}{\mathbb{1} - RD} \tag{2}$$

with $R_{km} = 2i\sqrt{\frac{t-4m_k^2}{t}}\delta_{km}$ for the case of two thresholds and $D = T_0^\dagger$. The latter is made of the amplitudes on the other side of the cut. For any D, equation (2) is enough to derive factorization: the singularities of T_0 can only come from the zeroes of

$$\Delta = \det(1 - RD). \tag{3}$$

Taking the determinant of both sides of eq. (2), we obtain in the vicinity of $\Delta = 0$

$$A_{aa\to aa}(j,t)A_{bb\to bb}(j,t) - A_{ab\to ab}(j,t)A_{ba\to ba}(j,t) = \frac{C}{\Delta}, \tag{4}$$

where C is regular at the zeroes of Δ. As the l.h.s. is of order $1/\Delta^2$ we obtain the well-known factorization properties from eqs. (2) and (4):

- The elastic hadronic amplitudes have common singularities;
- At each singularity in the complex j plane, these amplitudes factorise.

For isolated simple poles one obtains the usual well-known factorization relations for the residues. However, it is appropriate to mention here that the relation

$$\lim_{j \to \alpha(t)} \left[A_{aa \to aa}(j,t) - \frac{A_{ab \to ab}(j,t) A_{ba \to ba}(j,t)}{A_{bb \to bb}(j,t)} \right] = \text{finite terms}, \quad (5)$$

where $\alpha(t)$ is the position of a zero of Δ, is valid not only for a simple pole but also for any common j-singularity in the amplitudes. Moreover, it has a more general form than just a relation between residues. To avoid a misunderstanding we would like to note that "any j-singularity" means a singularity of the full unitarized amplitude rather than those partial singularities which are produced *e.g.* by n-pomeron exchange.

These equations are used to extract relations between the residues of the singularities, which can be continued back to the direct channel.

2.1. EXTENSION IN THE HADRONIC CASE

We have extended [5] the above argument including all possible thresholds, both elastic and inelastic. The net effect is to keep the structure (2), but with a matrix D that includes multi-particle thresholds. Furthermore, we have shown that one does not need to continue the amplitudes from one side of the cuts to the other, but that the existence of complex conjugation for the amplitudes is enough to derive (2) and consequently the factorization relations (4,5).

Hence there is no doubt that the factorization of amplitudes in the complex j plane is correct, even when continued to the direct channel.

If $A_{pq}(j)$ has coinciding simple and double poles (at any t or *e.g.* colliding simple poles at $t = 0$),

$$A_{pq} = \frac{S_{pq}}{j - z} + \frac{D_{pq}}{(j - z)^2}, \quad (6)$$

one obtains the new relations

$$\begin{aligned} D_{11} D_{22} &= (D_{12})^2, \\ D_{11}^2 S_{22} &= D_{12}(2 S_{12} D_{11} - S_{11} D_{12}). \end{aligned} \quad (7)$$

In the case of triple poles

$$A_{pq} = \frac{S_{pq}}{j - z} + \frac{D_{pq}}{(j - z)^2} + \frac{F_{pq}}{(j - z)^3}, \quad (8)$$

the relations become

$$\begin{aligned} F_{11}F_{22} &= (F_{12})^2, \\ F_{11}^2 D_{22} &= F_{12}(2D_{12}F_{11} - D_{11}F_{12}), \\ F_{11}^3 S_{22} &= F_{11}F_{12}\,(2S_{12}F_{11} - S_{11}F_{12}) + D_{12}F_{11}\,(D_{12}F_{11} - 2D_{11}F_{12}) \\ &+ D_{11}^2 F_{12}^2. \end{aligned} \tag{9}$$

Although historically one has used t-channel unitarity to derive factorization relations in the case of simple poles, it is now clear [6] that a soft pomeron pole is not sufficient to reproduce the γ^*p data from HERA [3]. However, it is possible, using multiple poles, to account both for the soft cross sections and for the DIS data [7, 8]. We shall see later that relations (7, 9) enable us to account for the DIS photon-photon data from LEP.

2.2. THE PHOTON CASE

For photons, due to the fact that an undetermined number of soft photons can be emitted, two theoretical possibilities exist:

i) The photon cross sections are zero for any fixed number of incoming or outgoing photons [9]. In this case, it is impossible to define an S matrix, and one can only use unitarity relations for the hadronic part of the photon wave function. Because of this, photon states do not contribute to the threshold singularities, and the system of equations does not close. The net effect is that the singularity structure of the photon amplitudes is less constrained. γp and $\gamma\gamma$ amplitudes must have the same singularities as the hadronic amplitudes, but extra singularities are possible: in the γp case, these may be of perturbative origin, but must have non perturbative residues. In the $\gamma\gamma$ case, these singularities have their order doubled. It is also possible for $\gamma\gamma$ to have purely perturbative additional singularities.

ii) It may be possible to define collective states in QED for which an S matrix would exist [10]. In this case, we obtain the same situation for on-shell photons as for hadrons. However, in the case of DIS, virtual photons come only as external states. Because they are virtual, they do not contribute to the t-channel discontinuities, and hence the singularity structure for off-shell photons is as described in i).

Let us consider the second possibility in more details and define virtuality of photons as shown in Fig.1.

In the case of real photons ($Q^2 = P^2 = 0$) we have obtained for the matrix T ($a \equiv p, b \equiv \gamma$) (see [5]) the same expression as in the hadron case (eq.(2)). It means that there are no extra singularities in the photon amplitudes besides those contributing to pp amplitude.

In the DIS case ($Q^2 \neq 0, P^2 = 0$) we have

$$T(Q^2, 0)(\mathbb{I} - RD(0,0)) = D(Q^2, 0), \tag{10}$$

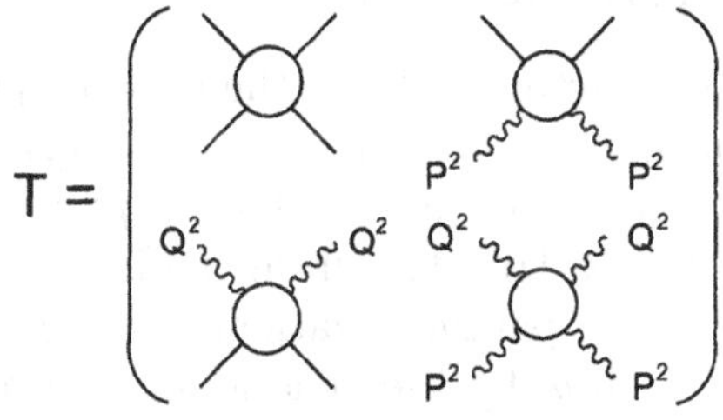

Figure 1. Graphic representation of the matrix T in the case of photons with virtualities Q^2 and P^2.

where R is a diagonal matrix with elements $R_{11} = 2i\sqrt{(t-4m_p)^2/t}$ and $R_{22} = 2i$, $D(Q^2, P^2)$ is expressed through $T^\dagger(Q^2, P^2)$. Hence we see that all the on-shell singularities must be present in the off-shell case, but we can have new ones coming from the singularities of $D(Q^2, 0)$. These singularities can be of perturbative origin (*e.g.* the singularities generated by the DGLAP evolution) but their coupling will depend on the threshold matrix R, and hence they must know about hadronic masses, or in other words they are not directly accessible by perturbation theory.

In the case of $\gamma^*\gamma^*$ scattering, we take $Q^2 \neq 0$ and $P^2 \neq 0$, and obtain [5]

$$T(Q^2, P^2) = D(Q^2, P^2) + \frac{D(Q^2, 0)RD(0, P^2)}{\mathbb{I} - RD(0,0)}. \tag{11}$$

This shows that the DIS singularities will again be present, either through $\Delta = \det(\mathbb{I} - RD(0,0))$, or through extra singularities present in DIS (in which case their order will be different in $\gamma\gamma$ scattering, at least for $Q^2 = P^2$).

It is also possible to have extra singularities purely from $D(Q^2, P^2)$. *A priori* these could be independent from the threshold matrix, and hence be of purely perturbative origin (*e.g.* $\gamma^*\gamma^* \to \bar{q}q$ or the BFKL pomeron coupled to photons through a perturbative impact factor).

We also want to point out that the intercepts of these new singularities can depend on Q^2, and as the off-shell states do not enter unitarity equations, these singularities can be fixed in t. However, their residues must vanish as $Q^2 \to 0$.

In the following, we shall explore the possibility that no new singularity is present for on-shell photon amplitudes, and show that it is in fact possible to reproduce present data using pomerons with double or triple poles at $j = 1$.

3. Application to HERA and LEP

For a given singularity structure, a fit to the $C = +1$ part of proton cross sections, and to $\gamma^{(*)}p$ data enables one, via relations (2), to predict the $\gamma^{(*)}\gamma^{(*)}$ cross sections. Hence we have fitted [5] pp and $\bar{p}p$ cross sections and ρ parameters, as well as DIS data from HERA [3].

The general form of the parametrizations which we used is given, for total cross sections of a on b, by the generic formula $\sigma_{ab}^{tot} = (R_{ab} + H_{ab})$. The first term, from the highest meson trajectories (ρ, ω, a and f), is parametrized via Regge theory as

$$R_{ab} = Y_{ab}^{+} (\tilde{s})^{\alpha_{+}-1} \pm Y_{ab}^{-} (\tilde{s})^{\alpha_{-}-1} \tag{12}$$

with $\tilde{s} = 2\nu/(1 \text{ GeV}^2)$. Here the residues Y_{+} factorize. The second term, from the pomeron, is parametrized either as a double pole [7, 11]

$$\begin{aligned} H_{ab} &= D_{ab}(Q^2)\Re e\left[\log\left(1 + \Lambda_{ab}(Q^2)\tilde{s}^{\,\delta}\right)\right] \\ &+ C_{ab}(Q^2) + (\tilde{s} \to -\tilde{s}) \end{aligned} \tag{13}$$

or as a triple pole [8]

$$H_{ab} = t_{ab}(Q^2)\left[\log^2\left(\frac{\tilde{s}}{d_{ab}(Q^2)}\right) + c_{ab}(Q^2)\right]. \tag{14}$$

It may be noted, in the double-pole case, that the parameter δ is close to the hard pomeron intercept of [6]. At high Q^2, because the form factor Λ falls off, the logarithm starts looking like a power of $\tilde{s}$, and somehow mimics a simple pole. It may thus be thought of as a unitarized version of the hard pomeron, which would in fact apply to hard and soft scatterings.
In the triple-pole case, this is accomplished by a different mechanism: the scale of the logarithm is a rapidly falling function of Q^2, and hence the $\log^2$ term becomes relatively more important at high Q^2.

3.1. RESULTS

The details of the form factors entering (13, 14) can be found in [5]. Such parametrizations give χ^2/dof values less than 1.05 in the region $\cos(\vartheta_t) \geq \frac{49}{2m_p^2}$, $\sqrt{2\nu} \geq 7$ GeV, $x \leq 0.3$, $Q^2 \leq 150$ GeV^2. What is really new is that these forms can be extended to photon-photon scattering, using relations (7, 9). The total $\gamma\gamma$ cross section is well reproduced (see Fig. 2) and the de-convolution using PHOJET is preferred.

The fit to F_2 has quite a good χ^2 as well. We have checked that one can easily extend it to $Q^2 \approx 400$ GeV^2 for the triple pole, and to $Q^2 \approx 800$ GeV^2 in the double-pole case. It is interesting that one cannot go as high

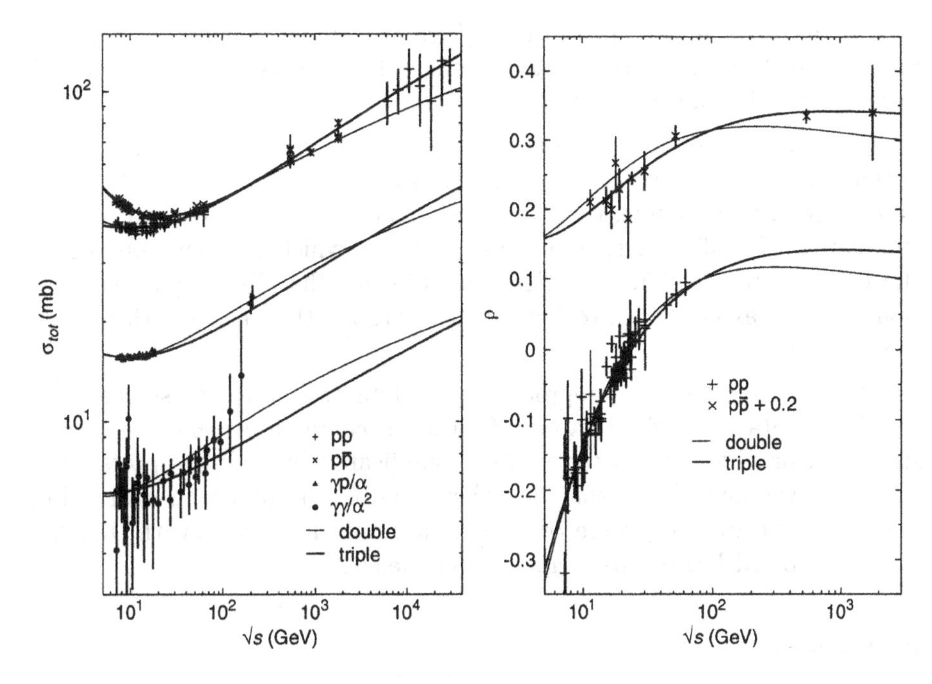

Figure 2. Fits to the total cross-sections and to the ρ parameter. The thick and thin curves correspond respectively to the triple-pole and to double pole cases.

as in ref. [8]. This can be attributed either to too simple a choice for the form factors, or more probably to the onset of perturbative evolution.

Fig. 3 shows the F_2^p fit for some selected Q^2 bins (figures for other Q^2 bins can be found in [5]). As pointed out before, our fits do reproduce the low-Q^2 region quite well, but predict total cross sections on the lower side of the error bands. Hence the extrapolation to $Q^2 = 0$ of DIS data does not require a hard pomeron.

For photon structure functions, one needs to add one singularity at $j = 0$ corresponding to the box diagram [12], but otherwise the $\gamma\gamma$ amplitude is fully specified by the factorization relations. One can see from Fig. 4 that the data on photon structure are well reproduced by both parametrizations.

Even more surprisingly, it is possible to reproduce the $\gamma^*\gamma^*$ cross sections when both photons are off-shell, as shown in Fig. 5. This is the place where BFKL singularities may manifest themselves, but as can be seen such singularities are not needed.

In conclusion, we have shown that t-channel unitarity can be used to map the regions where new singularities can occur, be they of perturbative or non-perturbative origin. Indeed, we have seen that although hadronic singularities must be universal, this is not the case for F_2^p and F_2^γ, as DIS involves off-shell particles. Nevertheless, up to $Q^2 = 150$ GeV^2, the data do not call for the existence of new singularities, except perhaps the box diagram. In the case of total cross sections, this suggests that it is indeed possible to define an S matrix for QED.

For off-shell photons, our fits are rather surprising as the standard claim is that the perturbative evolution sets in quite early. This evolution is indeed allowed by t-channel unitarity constraints: it is possible to have extra singularities in off-shell photon cross sections, which are built on top of the non-perturbative singularities. But it seems that Regge parametrisations can be extended quite high in Q^2 without the need for these new singularities.

Thus it is possible to reproduce soft data (*e.g.* total cross sections) and hard data (*e.g.* F_2 at large Q^2) using a common j-plane singularity structure, provided the latter is more complicated than simple poles. Furthermore, we have shown that it is then possible to predict $\gamma\gamma$ data using t-channel unitarity. How to reconcile such a simple description with DGLAP evolution, or BFKL results, remains a challenge.

References

1. Gribov V.N. and Pomeranchuk I. Ya.: (1962) "Complex Angular Momenta and the Cross Sections of Various Processes at High Energies" *Phys. Rev. Lett.* **Vol. no. 8**, p. 343; Charap J. and Squires E.J.: (1962) "Factorization of the Residues of Regge Poles" *Phys. Rev.* **Vol. no. 127**, p. 1387 .
2. COMPETE collaboration, Cudell J. R. *et al.*: (2002) "Hadronic scattering amplitudes: medium-energy constraints on asymptotic behaviour" *Phys. Rev.* **Vol. no. D65**, p. 074024 [arXiv:hep-ph/0107219]; (2002) " Benchmarks for the Forward Observables at RHIC, the Tevatron-run II and the LHC" *Phys. Rev. Lett.* **Vol. no. 89**, p. 201801 [arXiv:hep-ph/0206172].
3. H1 collaboration: Adloff C.*et al.*: (2001) "Deep Inelastic Inclusive *ep* Scattering at Low x and a Determination of α_s" *Eur. Phys. J.* **Vol. no. C21**, p. 33; ZEUS collaboration: Chekanov S. *et al.*: (2001) "Measurement of the Neutral Current Cross-Section and F(2) Structure Function for Deep Inelastic $e+p$ Scattering at HERA" *Eur. Phys. J.* **Vol. no. C21**, p. 443, and references therein.
4. L3 Collaboration: Acciarri M. *et al.* (2001) "Total Cross Section in gamma gamma Collisions at LEP" *Phys. Lett.* **Vol. no. B519**, p. 33; OPAL Collaboration: G. Abbiendi *et al.* (2002) "Measurement of the Hadronic Photon Structure Function F2gamma at LEP2" *Phys. Lett.* **Vol. no. B533**, p. 207, and references therein.
5. Cudell J. R., Martynov E. and Soyez G. (2002) "t-Channel Unitarity and Photon Cross Sections", arXiv:hep-ph/0207196.
6. Donnachie A. and Landshoff P. V.: (1998) "Small x: Two Pomerons" *Phys. Lett.* **Vol. no. B437**, p. 408; (2002) "Perturbative QCD and Regge Theory: closing the circle" *Phys. Lett.* **B533**, 277 .
7. Desgrolard P. and Martynov E.: (2001) "Regge Models of the Proton Structure Function with and without Hard Pomeron: a Comparative Analysis" *Eur. Phys. J.* **Vol. no. C22**, p. 479 [arXiv:hep-ph/0105277].
8. Cudell J. R and Soyez G.: (2001) "Does F2 Need a Hard Pomeron?" *Phys. Lett.* **Vol. no. B516**, p. 77 [arXiv:hep-ph/0106307].
9. Bloch F. and Nordsieck A.: (1937) "Note on the Radiation Field of the Electron" *Phys. Rev.* **Vol. no. 52**, p. 54.

10. Kulish P. P. and Faddeev L. D.: (1970) "Asymptotic Conditions and Infrared Divergences in Quantum Electrodynamics" *Theor. Math. Phys.* **Vol. no. 4**, p. 745 (1970) *Teor. Mat. Fiz.* **Vol. no. 4**, p. 153 (in Russian);
Zwanziger D.: (1976) "Physical States in Quantum Electrodynamics" *Phys. Rev.* **Vol. no. D14**, p. 2570;
Bagan E., Lavelle M. and McMullan D.: (2000) "Charges from Dressed Matter: Physical Renormalization" *Annals Phys.* **Vol. no. 282**, p. 503 [arXiv:hep-ph/9909262]; Horan R., Lavelle M. and McMullan D.: (2000) "Asymptotic Dynamics in Quantum Field Theory" *J. Math. Phys.* **Vol. no. 41**, p. 4437 [arXiv:hep-th/9909044], and references therein.
11. Desgrolard P., Lengyel A. and Martynov E.: (2002) "Pomeron Effective Intercept, Logarithmic Derivatives of $F_2(x, Q^2)$ in DIS and Regge Models" *JHEP* **Vol. no. 0202**, 029 [arXiv:hep-ph/0110149].
12. Budnev V. M., Ginzburg I. F., Meledin G. V. and Serbo V. G.: (1974) "The Two-Photon Particle Production Mechanism. Physical Problem. Applications. Equivalent Photon Approximation" *Phys. Rep.* **Vol. no. 15**, p. 181.

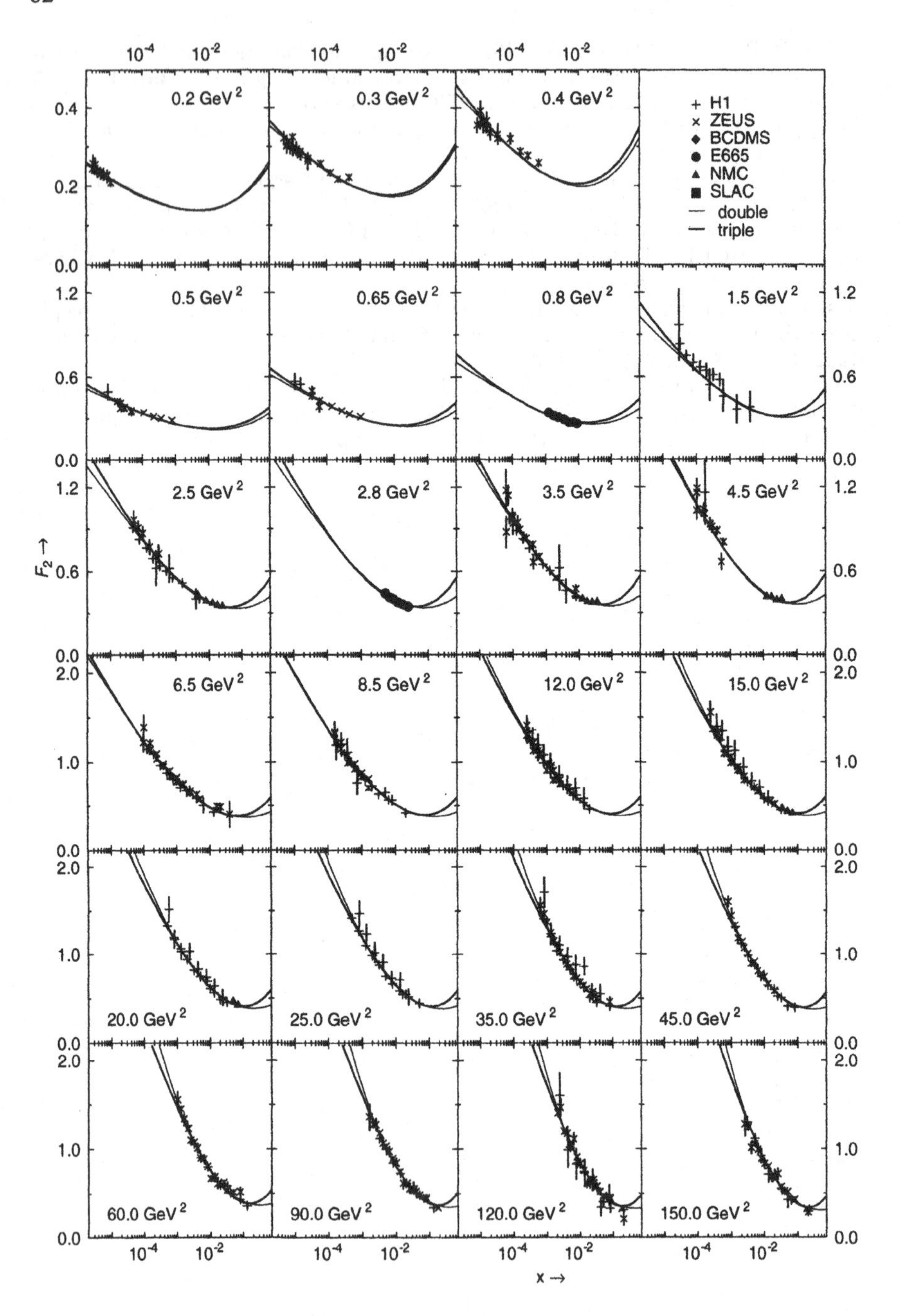

Figure 3. Fits to F_2^p. We show only graphs for which there are more than 6 experimental points, as well as the lowest Q^2 ones. The curves are as in Fig. 2

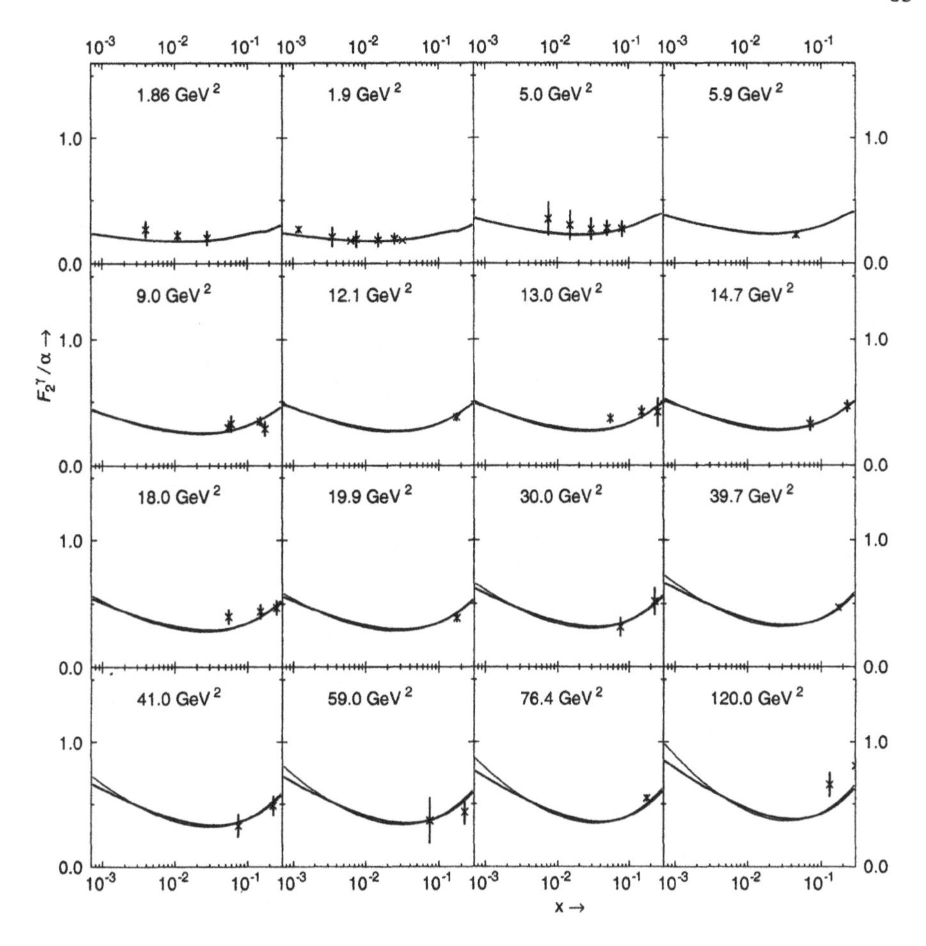

Figure 4. Fits to F_2^γ. The thick and thin curves correspond respectively to the triple-pole and to the double-pole cases. The data are from [4].

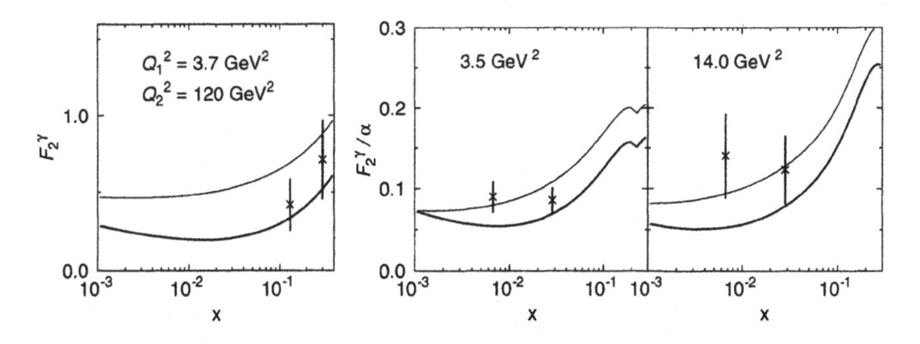

Figure 5. Fits to F_2^γ for nonzero asymmetric values of P^2 and Q^2 and for $P^2 = Q^2$. The curves are as in Fig. 4. The data are from the [4].

DISPERSION RELATIONS AND INCONSISTENCY OF ρ DATA

O.V. SELYUGIN
Institut de Physique, Université de Liège, Belgium
and BLTPh, JINR, Dubna, Russia

Abstract. The high energy elastic nucleon cross section is treated from the viewpoint of the basic principles of local field theory. The connection between the energy dependence of σ_{tot} and the ρ - ratio of the real to imaginary parts of the forward elastic scattering amplitude is examined in the framework of dispersion relations, derivative dispersion relations and crossing symmetry.

1. Introduction

Interest in diffractive processes is now revived due, in most part, to the discovery of rapidity gaps at HERA, i.e. of processes where diffractive interactions contribute a sizeable part to the whole amplitude. The diffraction interaction is defined by the multigluon colorless exchange named pomeron [1]. In the perturbative regime of QCD, it can be considered as a compound system of the two Reggeized gluons [2] in the approximation where one sums the leading ln's in energy, though its nonperturbative structure is basically unknown. Research on the nature of pomeron requires the knowledge of the parameters of purely diffractive processes, such as the total cross section and the phase of the elastic scattering amplitude. These quantities are closely related with the first principles of field theory such as causality, polynomial boundedness, crossing symmetry etc. They are also important for modern nuclear physics at high energies, as these quantities underlie many modern descriptions of nuclear interactions when we study such effects as nuclear shadowing, transparency or broken parity [3].

The energy dependence of the cross section and of the parameter $\rho(s)$ - the ratio of the real to imaginary part of the scattering amplitude in the high energy domain - is a much discussed question which still remains without a definite answer. Quite many efforts were spent to understand

R. Fiore et al. (eds.), Diffraction 2002, 85–95.

the high-energy hadron scattering from such general principles of relativistic quantum field theory as Lorenz invariance, analyticity, unitarity and crossing symmetry. Analyticity, which arises from the principle of causality, leads to dispersion relations which give the real part of the scattering amplitude at $t = 0$ as an energy integral involving σ_{tot} .

The forward dispersion relations for nucleon-nucleon scattering have not been proven, although they were written down a long time ago [4]. Numerous papers have been devoted to calculations of the real parts of the pp and $\bar{p}p$ forward scattering amplitude, using variety of dispersion relations and different representations for the energy dependence of the imaginary part of the scattering amplitude at $t = 0$.

The dispersion relations for scattering amplitudes depend on three properties, analyticity in the energy variable, the optical theorem, and polynomial boundedness. The prediction of the dispersion relation is very important for the discovery of new physical phenomena at superhigh energies. For example, in the case of potential scattering by nonlocal potentials the polynomial boundedness can be broken [5]. With some additional assumptions connected with string models [6], the effect of non-local behavior will decrease the phase, and hence increase $\rho = ReF/ImF$. A different behavior can be obtained in the case of extra internal dimension [7]. So, such a remarkable effect (the deviation of $\rho(s)$ from the prediction of the dispersion relation) could be discovered already at the LHC in the TOTEM experiment.

2. Different connections between the real and the imaginary part of the scattering amplitude

The optical theorem connects the imaginary part of the forward elastic scattering amplitude with the total cross section. $\sigma_{tot}(s) = 4\,\pi ImF(s,t = 0)$. The crossing property of the scattering amplitude is used very often to relate the imaginary part to the real part of the scattering amplitude with the substitution $S \to \tilde{S} = Sexp(-i\pi/2)$. In the case of the maximal behavior, the crossing-even and crossing-odd amplitudes at $t = 0$ are [10]

$$
\begin{aligned}
F_{+}(s,t=0)/is &= F_{+}(0)\ [ln(s\ e^{-i\pi/2}]^2 \\
F_{-}(s,t=0)/is &= iF_{-}(0)\ [ln(s\ e^{-i\pi/2}]^2
\end{aligned}
\tag{1}
$$

On the other hand the integral dispersion relations give us the most powerful connections between σ_{tot} and ρ. Their form depends on the behavior of the crossing-even and crossing-odd parts of the scattering amplitude. In the case where the crossing-even part of the scattering amplitude saturates the Froissart bound, so that the total cross section does not grow faster than a power of $\ln(s)$, the dispersion relation requires a single subtraction. If the crossing-odd part of the scattering amplitude does not grow with energy, so that the difference of the total cross section between hadron-hadron and hadron-antihadron scattering falls with energy, the odd dispersion relation needs no subtraction

$$Ref^{even}(E) = ReF^{even}(0) + P\frac{1}{\pi}\int_m^\infty dE' \frac{2\ E^2}{E'(E'^2 - E^2)} Imf^{even}(E'); \quad (2)$$

$$Ref^{odd}(E) = P\frac{1}{\pi}\int_m^\infty dE' \frac{2\ E}{(E'^2 - E^2)} Imf^{odd}(E'), \quad (3)$$

where P denotes the principal part of the integrals.

On the basis of the integral dispersion relation, one obtained derivative dispersion relations (first used in the context of Regge theory [11] and developed in [12]) which are more suitable for calculation but, of course, have a more narrow region of validity [13, 14]. It is to be noted that in this case we lose the subtraction constant. It means that the derivative dispersion relation can be used only at high energies where this constant is not perceived. The derivative dispersion relations can be written

$$\frac{R(s)}{s^\alpha} = tan[\frac{\pi}{2}(\alpha - 1 + \frac{d}{dln\ s})]\frac{I(s)}{s^\alpha}, \quad (4)$$

where the parameter α is some arbitrary constant modifying the integral. Usually we choose $\alpha = 1$ from the sake of simplicity [14]. In [15] it was shown that if we fit the experimental data, the best value is $\alpha = 1.25$. But in this case, the parameter $\alpha \neq 1$ will change the relation between the imaginary and the real parts of the scattering amplitude in the whole energy region. So, the predictions of the integral dispersion relations and of the derivative dispersion relation will not coincide. That means that one or the other is false. From the viewpoint of basic principles, which lead to the integral dispersion relation, we have to give prefer once to them. Hence, the equality of both representations at high energies requires that α be equal to 1.

3. The experimental data and the check of the integral dispersion relations

Now it is a common belief that the experimental data confirm the validity of the dispersion relations. Of course, there is some ambiguity concerning the energy dependence of the total cross sections. Already the fit by Amaldi et al. [16] and by Amos et al. [17] has given the tendency for $\sigma_{tot}(s)$ and $\rho(s)$ energy to increase. Khuri and Kinoshita [5] have predicted that $\rho(s)_{pp}$ should approach zero from positive values. At the time of this prediction,

$\rho(s)_{pp}$ was known to be a small negative quantity implying, thus, that $\rho(s)_{pp}$ should be zero at some energy, reach a maximum, and then tend to zero at still higher energies.

The comparison of the predicted values of the real part of the pp scattering amplitude with the experimental ones allows one to conclude that, on the whole, the experimental data agree well with the dispersion relation predictions. However if we take into account the whole set of experimental data, without any exception, we find that the total χ^2 is large and that some experimental data contradict others. Some points of one experiment are situated above the theoretical line and some points of other experiment are below the theoretical curve by more than three statistical errors. Moreover, in [8], it was noted that experimental data on $\bar{p}p$-scattering around $p_L = 5$ GeV/c disagreed with the analysis based on the dispersion relations [9].

At low energy, the definition of the subtraction constant is very important. The spin-independent amplitude could be continued into the low-energy region of $\bar{p}p$ scattering and into unphysical region. However, because of the lack of the data for the low energy $\bar{p}p$ scattering, such a continuation (e.g.. by means of the effective range approximation) cannot be carried out. This is why in the dispersion relation calculations the imaginary part of the amplitude in the unphysical and low-energy region is sometimes decomposed into a series in which the coefficients are determined by comparison of the calculated real parts with the experimentally measured ones.

An alternative way of handling these regions is to replace the continuum of states with a set of bound states at fixed energies. Mathematically, this is equivalent to replacing a cut of an analytic function by a sum of poles (resonances). The values of their residues (coupling constants) are either a priori fixed, or are found by comparing the calculated values of the real part with the experimental data. One should note that the replacement of the cut with the poles is, in itself, an arbitrary procedure. Besides, both the exact number of poles-resonances and the values of their coupling constants are unknown.

Therefore, it seems to us that the apparent agreement of theoretical (i.e. by the dispersion relation) calculations with experimental data on the real parts of the pp forward scattering amplitude means only that the parameters in various approximations of unphysical and low-energy region of $\bar{p}p$ scattering can be chosen so as to obtain consistency between the theoretical calculations and the experimental data. Really, the dispersion relations for pp , $\bar{p}p$ forward scattering have been tested at low energy, at best, only qualitatively.

4. The calculation of ρ through different methods

As noted above, we can obtain the value of ρ by different methods. The simplest one is to use the crossing symmetry properties of the scattering amplitude and to obtain the $ReF(s,t)$ by changing $\mathbf{S} \to \mathbf{Sexp}(-\mathbf{i}\pi/\mathbf{2})$ in all energy dependent parameters of a model, see for example [10]. In that case, the real part of the scattering amplitude is obtained straightaway.

Another way to obtain the $ReF(s,t)$ is to use the derivative dispersion relations. The total cross section can be taken a form

$$\sigma_{tot} = Z_{pp} + A\ Ln(s/s_0) + Y_1 s^{-\eta_1} - Y_2 s^{-\eta_2}. \tag{5}$$

Hereafter we suppose, besides the special separate cases, that s has the coefficient $s_1^{-1} = 1\ GeV$. The derivative dispersion relation gives us [18]

$$\rho(s,t)\ \sigma_{tot} =$$
$$\frac{\pi}{2}\ A\ ln(\frac{s}{s_0}) - Y_1\ s^{-\eta_1}[\tan(\frac{1-\eta_1}{2}\pi]^{-1} - Y_2 s^{-\eta_2}[\cot(\frac{1-\eta_2}{2}\pi]^{-1}. \tag{6}$$

To accurately compare the results of the different calculations, let us examine carefully the calculations with the integral dispersion relation. There were many works on the methods of calculating the dispersion integrals [19]. From recent works, one should note the paper [20] where it was shown that for the most common parametrizations of the total cross section, the principal value integrals can be calculated in terms of rapidly convergent series. The advantages are not restricted to a reduction of the computer time.

Let us consider, as in [21], the standard dispersion relations for $pp \to pp\ (+)$ and $\bar{p}p \to \bar{p}p\ (-)$

$$\rho(E)_{\pm}\sigma_{\pm}(E) = \frac{B}{p} + \frac{E}{\pi p}\ P\int_m^\infty dE'p'[\frac{\sigma_{\pm}(E')}{E'(E'-E)} - \frac{\sigma_{\mp}(E')}{E'(E'+E)}], \tag{7}$$

where E and p are the energy and the momentum in the laboratory frame, m is the proton mass and B is a subtraction constant. In terms of the variable $s = 2m(E+m)$, a parametrization of the cross section can be written in the form (5).

In calculating the principal value of the dispersion integral, we face two difficulties. One is the singularity of the integral at the point $E' = E$ and the other is the infinity of the upper bound of the integral which is especially important in the case of a total cross section growing with energy. It is possible to solve both problems if the integral is divided in three parts. Let us cut the integral at the singularity point and select the last part which contains the tail of the integral

$$\int_m^\infty dE'p'[\frac{\sigma_\pm(E')}{E'(E'-E)} - \frac{\sigma_\mp(E')}{E'(E'+E)}] = I_1 + I_2 + I_3, \tag{8}$$

where

$$I_1 = \int_m^{E-\epsilon} dE'p'[\frac{\sigma_\pm(E')}{E'(E'-E)} - \frac{\sigma_\mp(E')}{E'(E'+E)}], \tag{9}$$

$$I_2 = \int_{E+\epsilon}^{kE} dE'p'[\frac{\sigma_\pm(E')}{E'(E'-E)} - \frac{\sigma_\mp(E')}{E'(E'+E)}], \tag{10}$$

$$I_3 = \int_{kE}^{\infty} dE'p'[\frac{\sigma_\pm(E')}{E'(E'-E)} - \frac{\sigma_\mp(E')}{E'(E'+E)}]. \tag{11}$$

Here k is a number, for example, $k = 10$. The third integral can easily be calculated analytically in some of the most important cases, for example, if we take the nonfalling part of the total cross section in the form,

$$\sigma_{tot}(s)^{as} \;=\; Z \;+\; A\; ln^n(s/s_0). \tag{12}$$

Here we take for simplicity $s_0 = 1$ and $n = 1$ or $= 2$. The other cases are slightly more complicated, but do not lead to any principal difficulty.

Let us take $k = 10$. For the constant term the integral is

$$\int_{kE}^{\infty} [\frac{2Z\;E}{E'(E'^2 - E^2)}]dE' = 2Z\;E\;[\frac{1}{2kE}(ln(1+a) - ln(1-a) + ln(-\frac{1}{E}) - ln(\frac{1}{E}))] \;=\; 2.00671\;Z/k \;=\; 0.200671\;Z. \tag{13}$$

For the $n = 1$ the growing term is

$$\int_{kE}^{\infty} [\frac{2E\;A\;lnE'}{E'(E'^2 - E^2)}]dE' \;=\; A\;[0.662285 + 0.20067\;lnE]. \tag{14}$$

In the case of maximal, allowed by analyticity, growth of the total cross section, I_3 is given by

$$\int_{kE}^{\infty} [\frac{2EA\;ln^2E'}{E'(E^2 - E^2)}]dE' \;=\; \frac{A}{12}[28.6337 + lnE\;(15.8948 + 2.40805\;lnE)]. \tag{15}$$

As a result, we can compare our computer calculations with our analytical calculation to find a suitable upper bound of the dispersion integral for

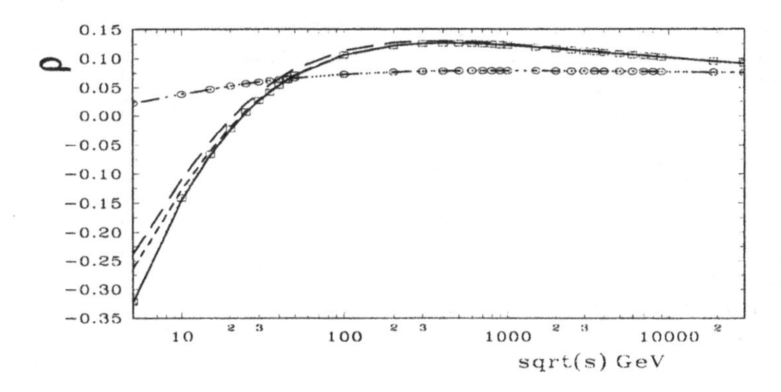

Figure 1. The calculations of $\rho(s)$ for the energy dependence of σ_{tot} given by (5) with $n = 1$ (the solid line - by the derivative dispersion relations (DDR) (6), the short dashed line - by the integral dispersion relations (IDR) with the subtracted constant from work by Amos, the long dashed line - by using the crossing symmetry, the squares - by the integral dispersion relations with fitting subtracting constant). In the upper part of the figure the calculations for the tail part of dispersion integral are presented (the line - by the analytical (14) and the circle - by numerical).

computer calculation. Such a comparison is shown in Figs. 1 and 2 where the line is the analytical calculations, and the box and triangles represent computer calculations. It is clear that both calculations coincide with high accuracy in all energy region.

Now return to the first two integrals. The numerical calculations show that, after removing the tail of the whole integral, the sum of the integrals I_1 and I_2 very fastly tends to its limit and cancels out the divergent parts. Usually, a precision of 1 % is sufficient. As a result, we can easily control the accuracy of our calculations. The calculations of $\rho(s)$ by the dispersion relation are shown in Fig.1 for the case $\sigma_{tot}(s) \approx ln(s)$ and in Fig.2 for the case $\sigma_{tot}(s) \approx ln^2(s)$.

The calculations by the derivative dispersion relation are shown in these figures also. At high energies, both calculations coincide and give the same predictions. If we choose the corresponding value of the subtracting constant B, both calculations coincide with high accuracy at low energy too. The calculation using crossing-symmetry also coincides with both calculations in the high energy region and can coincide with the integral dispersion relations if we choose the corresponding size of the subtraction constant. But it will differ from the subtraction constant required for the derivative dispersion relations.

So, we can conclude that the descriptions at low and high energy of

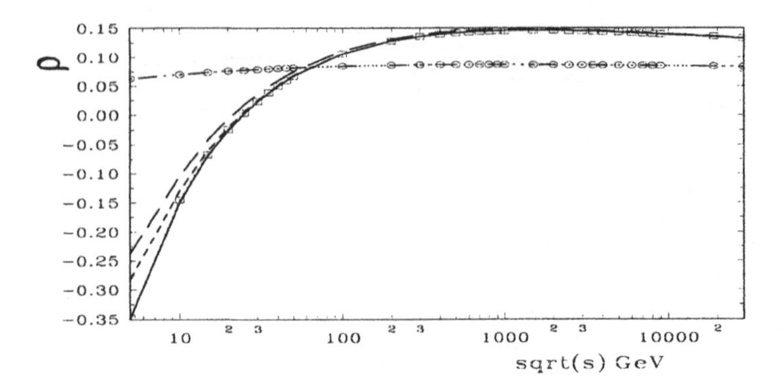

Figure 2. The same as Fig.1 with $n = 2$

$\rho(s)$ are practically independent of each other in the case of the integral dispersion relation. The subtraction constant practically unties these two regions. So, a good description of the data of ρ cannot give valuable information about the validity of our basic principle and weakly impacts on our prediction about the behavior of the total cross section and the parameter $\rho(s)$ at high energies.

5. The experimental data on ρ

Let us now consider very shortly the procedure of extracting parameters of the hadronic scattering amplitude from the experimental data dN/dt. In fact, in experiment we measure dN/dt, which then leads to "experimental" data such as σ_{tot}, B-slope and ρ are extracted from dN/dt with some model assumptions. Some assumptions are also needed to extrapolate the mesured quantities to $t = 0$. Most important in this case is the determination of the normalization of the experimental data. Contributions from the electromagnetic and hadronic interactions exist in the hadron interaction at small angles. We can calculate the Coulomb amplitude with the required accuracy . The contribution of the Coulomb-hadron interference term in dN/dt depends on the parameters of the hadronic amplitudes and the Coulomb-hadron phase also depends on these parameters. After the normalization of the dN/dt data we obtain the differential elastic cross section.

The analysis of the elastic scattering data makes several assumptions. Very often one assumes that the real and imaginary parts of the nuclear amplitude have the same t exponential dependence; the cross-odd part of the scattering amplitude either has the same behavior of momentum transfer as the cross even part, or it is neglected; the contributions of the spin-flip amplitudes are neglected. Note that the value of ρ is heavily correlated with the normalization of $d\sigma/dt$. Its magnitude weakly impacts on the determination of σ_{tot} only in the case where the normalization is known exactly.

There is no experiment which measures the magnitudes of these parameters separately. Sometimes, to reduce experimental errors, the magnitude of some quantities is taken from another experiment. As a result, we can obtain a contradiction between the basic parameters for one energy (for example, if we calculate the imaginary part of the scattering amplitude, we can obtain a nonexponential behavior). It can lead to some errors in the analysis based on the dispersion relations.

For example, let us take one experiment (there are others which are very similar) which was carried out at $\sqrt{s} = 16. \div 27.4$ GeV by the JINR-FNAL Collaboration [22]. To analyze dN/dt in this experiment, the energy dependence of σ_{tot} and B were taken in an analytical form supported by the data of other experiments. But the new fit of the dN/dt data with all free parameters gives values of ρ which are systematically above the original values [23]. Such a conclusion was obtained early in [24].

It means that the theoretical line obtained by the dispersion relation will be higher than in the existing descriptions and close to a simple description with the use of only the crossing symmetric properties of the scattering amplitude (see, Figs.1,2). But it does not change essentially our prediction for high energies, only the value of the subtraction constant will be changed.

6. Conclusions

The magnitudes of σ_{tot}, ρ and slope - B have to be determined in one experiment and determinations of their magnitudes depend on each other. The procedure of extrapolation of the imaginary part of the scattering amplitude is significant for determining σ_{tot}. There can also exist some additional hypotheses which lead to a deviation of differential cross section at very small momentum transfer. For example, the analysis of experimental data shows a possible manifestation of the hadron spin-flip amplitudes at high energies. The research into spin effects will be a crucial stone for different models and will help us understand the interaction and structure of particles, especially at large distances. All this raises the question about the measurement of spin effects in elastic hadron scattering at small angles at future accelerators. Especially, we would like to note the programs at RHIC where the polarizations of both collider beams will be constructed. Additional information on polarization data will help us only if we know sufficiently well the beam polarizations. So, the normalisations of dN/dt and A_N are most important for the determination of the magnitudes of these values. New methods of extracting the magnitudes of these quantities are required.

Acknowledgments. I would like to express my sincere thanks to the organizers for the kind invitation and the financial support; and to J.-R. Cudell, V.V. Ezhela and B. Nicolescu for fruitful discussions.

References

1. Landshoff P.V. (2002) "Soft and hard QCD" hep-ph/0209364.
2. Fadin V.S., Kuraev E.A. and Lipatov L.N., (1975) "On Pomeranchuk singularity in asymptotically free theries" *Phys. Lett.* , **60**, pp.50–52; Lipatov L.N., (1976), *Sov. J. Nucl. Phys.* , **23** p. 642; Kuraev E.A., Lipatov L.N. and Fadin V.S., (1976), *Sov. Phys.JETP* , **44**, p. 45; Balitsky I.I. and Lipatov L.N., "The Pomeranchuk singularity in quantum chromodynamics" (1978), *Sov. J. Nucl. Phys.* , **28**, p. 822–829.
3. Wang E. and Wang Xin-Nian, (2001) "Interplay of soft and hard processes and hadronp_T spectra in pA and AA collision" nucl-th/0104031; Borozan I. and Seymour M.H. (2002) " An eikonal model for multiparticle production in hadron-hadron interections" MC-TH-2002-4 , hep-ph/0207283; .
4. Goldberger M.L., Nambu Y. and Oehme R. (1957) "Dispersion relations for nucleon–nucleon scattering" *Ann. Phys.* **2**, pp. 226–282.
5. Khuri, N. N. (1957) *Thesis, (Princteon University)*; (1967) *Phys. ReV. Lett.*, **18**, p. 1094.
6. Gross D.J., and Mende P. (1988) "String theory beyond the Planck scale" *Nucl. Phys. B* **303**, pp.407–454.
7. Khuri, N. N. and Kinoshita, T. (1985) Real part of the scattering amplitude and behaviour of the total cross section at high energies" *Phys.ReV. D*, **137**, pp. 720–725.
8. Armstrong T.A. et al., Call. E760 (1996) "Precision measurements of antiproton-proton forward elastic scattering parameters" *Phys. Lett.*, **B 385**, pp. 479–483.
9. Kroll P. and Schwenger W. (1989) "Analysis of low-energy amplitude forward scattering" *Nucl. Phys. A* **503**, pp. 865–884.
10. Gauron P., Leader E. and Nicolescu B., (1988) "Similarities and differnces between $\bar{p}p$ and pp scattering at TeV energies and beyond" *Nucl. Phys.* **B 299** , pp. 640–672.
11. Gribov V.N. and Migdal A.B. (1968) "Quasistable Pomeranchuk pole and diffraction scattering at ultra high energies" *Yad. Fiz.* **8**, p. 1213; (1969), *Sov. J. Nucl. Phys.* , **8**, pp. 703–770.
12. Bronzan J.B., (1973), *Argonne Nat. Lab.* , *Report* **ANL/HEP 7327**; Bronzan J.B., Kane G.L., Sukhatme U.P. (1974) "Obtaining real parts of scattering amplitudes directly from cross section data using derivative analiticity relations" *Phys. Lett. B*, **49**, pp. 272-276 .
13. Kolar P. and Fischer J. (1984) "On validity and practical applicability of derivative analyticity relations" J. Math. Phys. **25**, pp. 2538–2544; (1987) *Czech. J. Phys. B* **37**, pp. 297–305.
14. Fischer J. and Kolar P. (2001) *Proc. os IX Blois Workshop, "Elastic and Diffractive Scattering", Průhonice, Czech.Rep., June 9-15, 2001.*, ed. Kundrát V., Závada P., p.305.
15. Martini A.F., Menon M.J., Paes J.T.S. and Silva Nieto M.J., (1998) " Differential dispersion relations and elementary amplitudes in multiple diffraction model" hep-ph/9810498; Avila R.F., Luna E.G.S. amd Menon M.J., (2001) "High-energy proton-

proton forward scattering and derivative analyticity relations" hep-ph/0105065.
16. Amaldi, U., et al. (1985) "The real part of the forward proton-proton scattering amplitude measured at the CERN Intersecting Storage Rings" *Phys.Lett. B*, **66**, pp. 390–394.
17. Amos, N.A. et al. (1985) "Measurement of small-angle antiproton-proton and proton-proton elastic scattering at the CERN Intersecting Storage Rings" *Nucl. Phys. B* **262**, pp. 689–714.
18. Cudell J.R. et al. COMPETE coll. (2002) "Hadronic scattering amplitudes: mediumenergy constrain on asymptotic behaviour" *Phys.ReV. D*, **65**, pp. 074024-57; Cudell J.R. et al. COMPETE coll.(2002) "Bench marks for the forward observables at RHIC, the Tevatron-run II and the LHC" hep/ph/0206172.
19. Hendrick R.E. and Lautrup B. (1975) "Real parts of the forward elastic $\pi^{\pm}p$, $K^{\pm}p$, $\bar{p}p$, and pp scattering amplitudes from 1 to 200 GeV/c" *Phys.ReV. D*, **11**, pp. 529–537.
20. Bertini M., Giffon M., Jenkovszky L. and Paccanoni F., (1996) *Il Nuovo Cim.* **109A** p.257.
21. Söding, P. (1964) "Real part of the proton-proton and proton-antiproton forward scattering amplitude at high energies" *Phys. Lett. B* **8** p. 285–287.
22. D. Gross et al. (1978) The real part of the p-p and p-d forward scattering amplitudes from 50 to 400 GeV *Phys.Rev.Lett.* **41** pp. 217–226.
23. Selyugin O.V. (1992) Amplitude of elastic hadron-hadron scattering in the region of Coulombic momentum transfers *Yad.Fiz.* , **55** pp. 841-853.
24. Shiz A. et al. (1981) "Real part of the forward elastic nuclear amplitude for $\bar{p}p$, pp, π^+p, π^-p, K^+p, and K^-p, scattering between 70 and 200 GeV/c" *Phys. Rev.* **D24**, pp. 46-65.

VECTOR MESON PHOTOPRODUCTION IN THE SOFT DIPOLE POMERON MODEL FRAMEWORK

A. PROKUDIN [1,2], E. MARTYNOV[3,4] AND E. PREDAZZI[1]

[1] *Dipartimento di Fisica Teorica, Università Degli Studi Di Torino, Via Pietro Giuria 1, 10125 Torino, ITALY and Sezione INFN di Torino, ITALY*

[2] *Institute For High Energy Physics, 142281 Protvino, RUSSIA*

[3] *Bogolyubov Institute for Theoretical Physics, National Academy of Sciences of Ukraine, 03143 Kiev-143, Metrologicheskaja 14b, UKRAINE*

[4] *Institut de physique Bat B5-a Université de Liège Sart Tilman B-4000 Liège, BELGIQUE*

Abstract. Exclusive photoproduction of all vector mesons by real and virtual photons is considered in the Soft Dipole Pomeron model. It is emphasized that being the Pomeron in this model a double Regge pole with intercept equal to one, we are led to rising cross-sections but the unitarity bounds are not violated. It is shown that all available data for $\rho, \omega, \varphi, J/\psi$ and Υ in the region of energies $1.7 \leq W \leq 250$ GeV and photon virtualities $0 \leq Q^2 \leq 35$ GeV^2 , including the differential cross-sections in the region of transfer momenta $0 \leq |t| \leq 1.6$ GeV^2, are well described by the model.

1. Introduction

A new precise measurement of J/ψ exclusive photoproduction by ZEUS [1] opens a new window in our understanding of the process and allows us to give more accurate predictions for future experiments.

The key issue of the dataset [1] is the diffractive cone shrinkage observed in J/ψ photoproduction which leads us to consider it a soft rather than pure QCD process so that we can apply the Soft Dipole Pomeron exchange [2] model.

The basic diagram is depicted in Figure 1; s and t are the usual Mandelstam variables, $Q^2 = -q^2$ is the virtuality of the photon.

R. Fiore et al. (eds.), Diffraction 2002, 97–107.

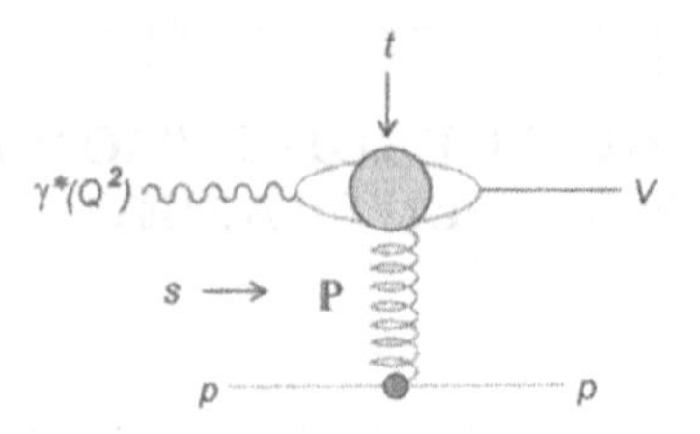

Figure 1. Photoproduction of a vector meson.

It is well known that high-energy representation of the scattering amplitude may be expressed as a sum over the the appropriate Regge poles in the complex j plane [3]

$$A(s,t)_{s\to\infty} \approx \sum_i \eta_i(t)\beta_i(t)(\cos\theta_t)^{\alpha_i(t)}, \tag{1}$$

where $\eta_i(t)$ is the signature factor and θ_t is the scattering angle in the t channel.

In the case of vector meson photoproduction we utilize the variable $z \sim \cos\theta_t$

$$z = \frac{2(W^2 - M_p^2) + t + Q^2 - M_V^2}{\sqrt{(t + Q^2 - M_V^2)^2 + 4M_V^2 Q^2}} \tag{2}$$

where $W^2 = (p+q)^2 \equiv s$, M_V is the vector meson mass, M_p is the proton mass.

Assuming vector meson dominance [4], the relation between the forward cross sections of $\gamma p \to Vp$ and $Vp \to Vp$ is given by

$$\frac{d\sigma}{dt}(t=0)_{\gamma p\to Vp} = \frac{4\pi\alpha}{f_V^2}\frac{d\sigma}{dt}(t=0)_{Vp\to Vp} \tag{3}$$

where the strength of the vector meson coupling $\frac{4\pi}{f_V^2}$ may be found from e^+e^- decay width of vector meson V

$$\Gamma_{V\to e^+e^-} = \frac{\alpha^2}{3}\frac{4\pi}{f_V^2}m_V \tag{4}$$

Using the property $\Gamma_{V\to e^+e^-}/ < Q_j^2 > \simeq const$ we can obtain the following approximate relations

$$m_\rho/f_\rho^2 : m_\omega/f_\omega^2 : m_\varphi/f_\varphi^2 : m_{J/\psi}/f_{J/\psi}^2 = 9:1:2:8 \tag{5}$$

which are in fairly good agreement with experimental measurements of decay widths [5].

We take into account these relations by introducing coefficients N_V (following to [6]) and writing the amplitude as $A_{\gamma p \to Vp} = N_C N_V A_{Vp \to Vp}$, where

$$N_C = 3\,; N_\rho = \frac{1}{\sqrt{2}}\,; N_\omega = \frac{1}{3\sqrt{2}}\,; N_\phi = \frac{1}{3}\,; N_{J/\psi} = \frac{2}{3}\,. \tag{6}$$

The amplitude of the process $Vp \to Vp$ may be written in the following form

$$A(z,t;M_V^2,\tilde{Q}^2) = I\!P(z,t;M_V^2,\tilde{Q}^2) + f(z,t;M_V^2,\tilde{Q}^2) + \ldots\,, \tag{7}$$

where, $\tilde{Q}^2 = Q^2 + M_V^2$.

$I\!P(z,t;M_V^2,\tilde{Q}^2)$ is the Pomeron contribution for which we use the so called dipole Pomeron which gives a very good description of all hadron-hadron total cross sections [7],[8]. Specifically, $I\!P$ is given by [9]

$$\begin{aligned} I\!P(z,t;M_V^2,\tilde{Q}^2) = {} & ig_0(t;M_V^2,\tilde{Q}^2)(-iz)^{\alpha_{I\!P}(t)-1} + \\ & ig_1(t;M_V^2,\tilde{Q}^2)ln(-iz)(-iz)^{\alpha_{I\!P}(t)-1}\,, \end{aligned} \tag{8}$$

where the first term is a single j-pole contribution and the second (with an additional $ln(-iz)$ factor) is the contribution of the double j-pole.

A similar expression applies to the contribution of the f-Reggeon

$$f(z,t;M_V^2,\tilde{Q}^2) = ig_f(t;M_V^2,\tilde{Q}^2)(-iz)^{\alpha_f(t)-1}. \tag{9}$$

It is important to stress that in this model the intercept of the Pomeron trajectory is equal to 1

$$\alpha_{I\!P}(0) = 1. \tag{10}$$

Thus the model does not violate the Froissart-Martin bound [10].

For ρ and φ meson photoproduction we write the scattering amplitude as the sum of a Pomeron and f contribution. According to the Okubo-Zweig rule, the f meson contribution should be suppressed in the production of the φ and J/ψ mesons, but given the present crudeness of the state of the art, we added the f meson contribution in the φ meson case.

For ω meson photoproduction, we include also π meson exchange (see also the discussion in [11]), which is needed in the low energy sector given that we try to describe the data for all energies W. Even though we did not expect it, the model describes well the data down to threshold.

In the integrated elastic cross section

$$\sigma(z,M_V^2,\tilde{Q}^2)_{el}^{\gamma p \to Vp} = 4\pi \int_{t_-}^{t_+} dt |A^{\gamma p \to Vp}(z,t;M_V^2,\tilde{Q}^2)|^2\,, \tag{11}$$

the upper and lower limits

$$2t_{\pm} = \pm\frac{L_1L_2}{W^2} - (W^2 + Q^2 - M_V^2 - 2M_p^2) + \frac{(Q^2 + M_p^2)(M_V^2 - M_p^2)}{W^2}, \quad (12)$$

$$L_1 = \lambda(W^2, -Q^2, M_p^2), \qquad L_2 = \lambda(W^2, M_V^2, M_p^2), \quad (13)$$

$$\lambda^2(x, y, z) = x^2 + y^2 + z^2 - 2xy - 2yz - 2zx, \quad (14)$$

are determined by the kinematical condition $-1 \leq \cos\theta_s \leq 1$ where θ_s is the scattering angle in the s-channel of the process.

For the Pomeron contribution (9) we use a nonlinear trajectory

$$\alpha_{I\!P}(t) = 1 + \gamma(\sqrt{4m_\pi^2} - \sqrt{4m_\pi^2 - t}\,), \quad (15)$$

where m_π is the pion mass. Such a trajectory was utilized for photoproduction amplitudes in [12], [13] and its roots are very old [14].

For the f-meson contribution for the sake of simplicity we use the standard linear Reggeon trajectory

$$\alpha_{I\!R}(t) = \alpha_{I\!R}(0) + \alpha'_{I\!R}(0)\, t\,. \quad (16)$$

In the case of nonzero virtuality of the photon, we have a new variable in play $Q^2 = -q^2$. At the same time, the cross section σ_L is nonzero.

2. The Model

For the Pomeron residues we use the following parametrization

$$g_i(t; M_V^2, \tilde{Q}^2) = \frac{g_i}{Q_i^2 + \tilde{Q}^2} exp(b_i(t; \tilde{Q}^2))\,, \quad (17)$$
$$i = 0, 1\,.$$

The slopes are chosen as

$$b_i(t; \tilde{Q}^2) = \Big(b_{i0} + \frac{b_{i1}}{1 + \tilde{Q}^2/Q_b^2}\Big)(\sqrt{4m_\pi^2} - \sqrt{4m_\pi^2 - t}\,)\,, \quad (18)$$
$$i = 0, 1\,,$$

to comply with the previous choice (15) and analyticity requirements [14]. The Reggeon residue is

$$g_{I\!R}(t; M_V^2, \tilde{Q}^2) = \frac{g_{I\!R} M_p^2}{(Q_{I\!R}^2 + \tilde{Q}^2)\tilde{Q}^2} exp(b_{I\!R}(t; \tilde{Q}^2))\,, \quad (19)$$

where

$$b_{I\!R}(t;\tilde{Q}^2) = \frac{b_{I\!R}}{1+\tilde{Q}^2/Q_b^2}t\,, \tag{20}$$

g_0, g_1, Q_0^2 (GeV^2), Q_1^2 (GeV^2), $Q_{I\!R}^2$ (GeV^2), Q_b^2 (GeV^2), b_{00} (GeV^{-1}), b_{01} (GeV^{-1}), b_{10} (GeV^{-1}), b_{11} (GeV^{-1}), $b_{I\!R}$ (GeV^{-2}) are adjustable parameters. $I\!R = f$ for ρ and φ, $I\!R = f,\pi$ for ω. We use the same slope $b_{I\!R}$ for f and π Reggeon exchanges.

2.1. PHOTOPRODUCTION OF VECTOR MESONS BY REAL PHOTONS ($Q^2 = 0$).

In the fit we use all available data starting from the threshold for each meson. As the new dataset of ZEUS [1] allows us to explore the effects of nonlinearity of the Pomeron trajectory and residues. In the region of non zero Q^2 the combined data of H1 and ZEUS is used.

The whole set of data is composed of 357 experimental points [1] and, with a grand total of 12 parameters, we find χ^2/d.o.f $= 1.49$. The main contribution to χ^2 comes from the low energy region ($W \le 4\ GeV$); had we started fitting from $W_{min} = 4\ GeV$, the resulting χ^2/d.o.f $= 0.85$ for the elastic cross sections would be much better and more appropriate for a high energy model.

In order to get a reliable description and the parameters of the trajectories and residues we use elastic cross sections for each process from threshold up to the highest values of the energy and differential cross sections in the whole t-region where data are available: $0 \le |t| \le 1.6\ GeV^2$.

The parameters are given in [15].

The results are presented in Fig. 2, which shows also the prediction of the model for $\Upsilon(9460)$ photoproduction.

2.2. PHOTOPRODUCTION OF VECTOR MESONS BY VIRTUAL PHOTONS ($Q^2 > 0$).

In (17) and (19) the Q^2- dependence ($\tilde{Q}^2 = Q^2 + M_V^2$) is completely fixed up to an *a priori* arbitrary dimensionless function $f(Q^2)$ such that $f(0) = 1$. Thus, we may introduce a new factor that differentiates virtual from real photoproduction:

$$f(Q^2) = \left(\frac{M_V^2}{\tilde{Q}^2}\right)^n \tag{21}$$

[1]The data are available at
REACTION DATA Database *http://durpdg.dur.ac.uk/hepdata/reac.html*
CROSS SECTIONS PPDS database *http://wwwppds.ihep.su:8001/c1-5A.html*

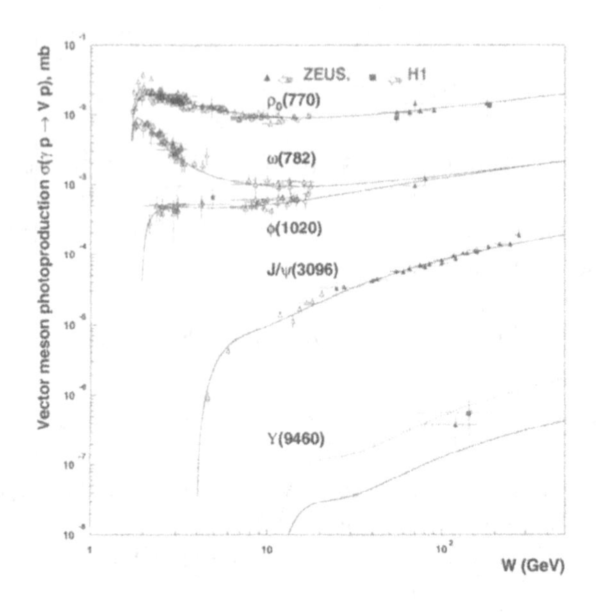

Figure 2. Elastic cross-sections for vector meson photoproduction. The solid curve for $\Upsilon(9460)$ production corresponds to $N_\Upsilon = N_\varphi$, the dotted line to $N_\Upsilon = N_{J/\psi}$.

Accordingly, in the case $Q^2 \neq 0$ we use the following parametrizations for Pomeron couplings (compare with Eq. 17):

$$\hat{g}_i(t; \tilde{Q}^2, M_V^2) = f(Q^2) g_i(t; \tilde{Q}^2, M_V^2), \ i = 0, 1, \tag{22}$$

where, for the sake of completeness, we will examine three different *choices* for the asymptotic Q^2 behaviour of the Pomeron residue

Choice I

$$n = 1, \qquad \sigma_T(Q^2 \to \infty) \sim \frac{1}{Q^8} \,. \tag{23}$$

Choice II

$$n = 0.5, \qquad \sigma_T(Q^2 \to \infty) \sim \frac{1}{Q^6} \,. \tag{24}$$

Choice III

$$n = 0.25, \qquad \sigma_T(Q^2 \to \infty) \sim \frac{1}{Q^5} \,. \tag{25}$$

For the reggeon couplings we have

$$f_{I\!R}(Q^2) = \Big(\frac{c_1 M_V^2}{c_1 M_V^2 + Q^2}\Big)^{n_2} \,, \tag{26}$$

where c_1 is an adjustable parameter and $n_2 = 0.25,\ -0.25,\ -0.5$ for *choice I, II, III.*

Accordingly, in the case $Q^2 \neq 0$ we use the following parametrizations for Reggeons couplings (compare with Eq. 19):

$$\hat{g}_{I\!R}(t;\tilde{Q}^2, M_V^2) = f_{I\!R}(Q^2) g_{I\!R}(t;\tilde{Q}^2, M_V^2)\ . \tag{27}$$

The lack of data on the ratio σ_L/σ_T, especially in the high Q^2 domain, does not allow us to draw definite conclusions about its asymptotic behaviour. There may be several realizations of the model with different asymptotic behaviour of σ_L/σ_T [2]. As a demonstration of such a possibility we use the following (most economical) parametrization for R (which cannot be deduced from the Regge theory)

Choice I, II, III

$$R(Q^2, M_V^2) = \Big(\frac{cM_V^2 + Q^2}{cM_V^2}\Big)^{n_1} - 1 \tag{28}$$

where c and n_1 are adjustable parameters for *choice I, II, III.*

We have, thus, 3 additional adjustable parameters as compared with real photoproduction. In order to obtain the values of the parameters for the case $Q^2 \neq 0$, we fit just the data[2] on ρ_0 meson photoproduction in the region $0 \leq Q^2 \leq 35\ GeV^2$; the parameters for photoproduction by real photons are fixed.

The parameters thus obtained may be found in [15].

The results of the fit are depicted in Fig. 3. In this figure as well as in all following ones the solid lines, dashed lines and dotted lines correspond to the *choice I, II, III* correspondingly.

We can now check the predictions of the model. The $\chi^2/\#point = 0.89$ for J/ψ meson exclusive production follows without any fitting. Both W and Q^2 dependences are reproduced very well. Notice that, so far, the three *choices I, II, III* all give equally acceptable reproduction of the data (see Figs. 4, 5).

We now plot the various ratios σ_L/σ_T (these data were not fitted) corresponding to Eqs (23),(24),(25) (shown with the solid (*choice I*), dashed (*choice II*) and dotted (*choice III*) lines) in Fig. 6, 7, 8. The result shows, indeed, a rapid increase of σ_L/σ_T with increasing Q^2, however one can see that our intermediate *choice II* is preferable to either *I* or *III* on this basis.

Let us examine the obtained dependences. We find that the data prefer

$$R(Q^2 \to \infty) \sim \Big(\frac{Q^2}{M_V^2}\Big)^{n_1}\ , \tag{29}$$

[2]The data are available at
REACTION DATA Database *http://durpdg.dur.ac.uk/hepdata/reac.html*
CROSS SECTIONS PPDS database *http://wwwppds.ihep.su:8001/c1-5A.html*

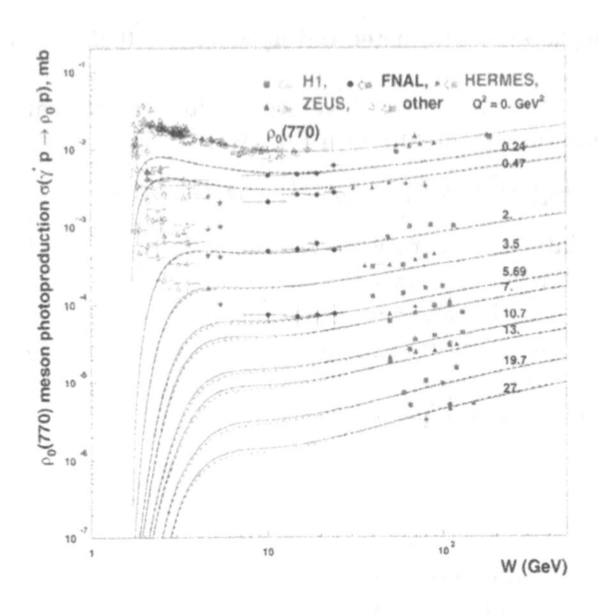

Figure 3. Elastic cross section of exclusive ρ_0 virtual photoproduction as a function of W for different values of Q^2.

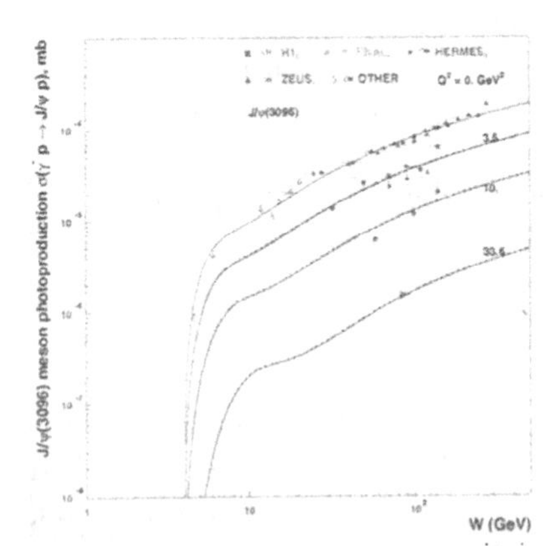

Figure 4. Elastic cross section of exclusive J/ψ virtual photoproduction as a function of W for various Q^2.

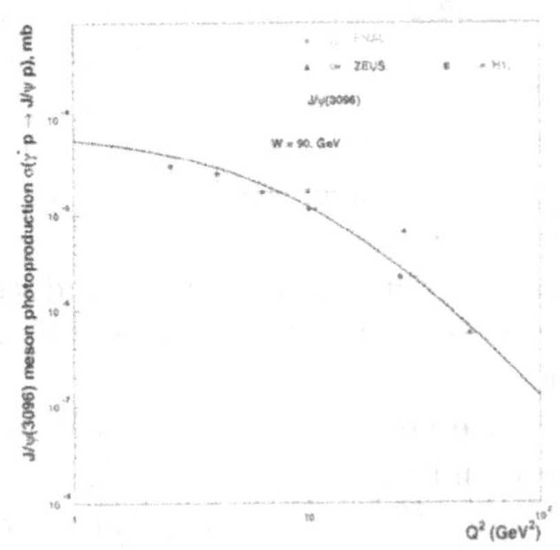

Figure 5. Elastic cross section of exclusive J/ψ virtual photoproduction as a function of Q^2 for $W = 90\ GeV$.

where $n_1 \simeq 2,\ 1,\ 0.3$ in *choice I, II and III.* Our, probably oversimplified, estimates and the data show $0.3 < n_1 < 1$, see Fig. 6, thus $\sigma \sim 1/Q^N$ where $N \in (4, 4.4)$ as $N = 6 - 2n_1$ for the *choice II* and $N = 5 - 2n_1$ for the *choice III.* However it is evident that new more precise data on R are needed.

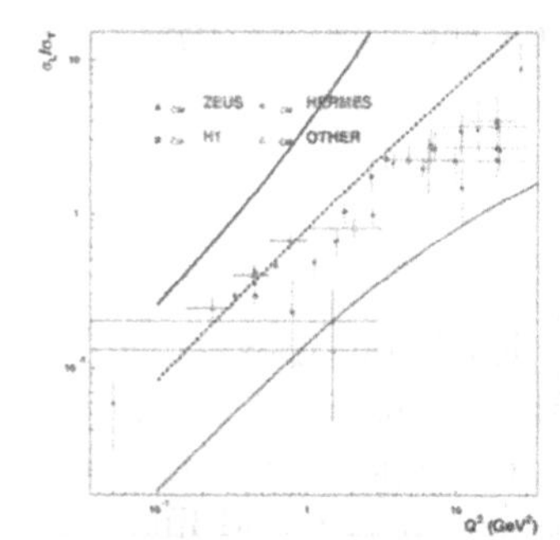

Figure 6. Ratio of σ_L/σ_T for exclusive ρ_0 large Q^2 photoproduction.

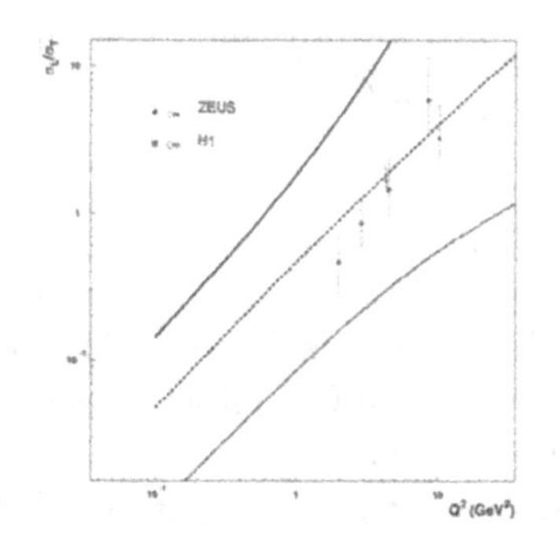

Figure 7. Ratio of σ_L/σ_T for exclusive φ large Q^2 photoproduction.

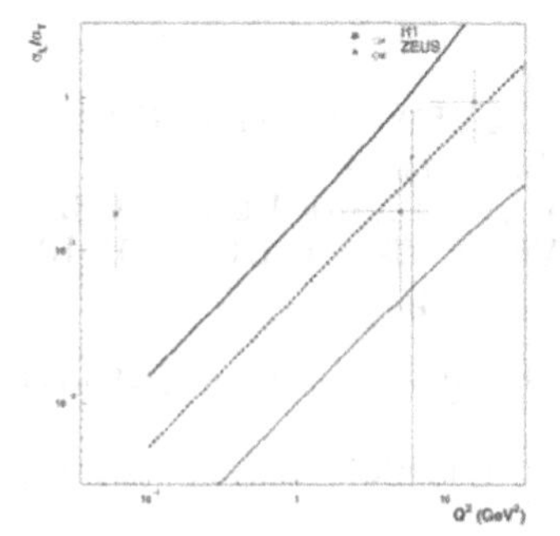

Figure 8. Ratio of σ_L/σ_T for exclusive J/ψ large Q^2 photoproduction.

3. Conclusion

We have shown that the Soft Dipole Pomeron model recently developed [2], [15] for vector meson photoproduction allows us to describe well the new ZEUS data [1] on the differential and integrated cross-sections for $\gamma p \to J/\psi p$. Again, all available data on photoproduction of other vector mesons at $Q^2 = 0$ as well as $Q^2 \neq 0$ are well reproduced.

The nonlinear Pomeron trajectory $\alpha_P(t) = 1 + \gamma\ (\sqrt{4m_\pi^2} - \sqrt{4m_\pi^2 - t})$ turns out to be more suitable for the nonlinearity of the diffractive cone shown by the new ZEUS data.

We would like to emphasize the following important points

1. The Pomeron used in the model is a double pole in the complex j-plane with intercept which is equal to one
2. The new ZEUS data [1] (in contrast to the old ones) quite definitely point towards the nonlinearity of the Pomeron slope and trajectory.

3. Phenomenologically we find that in the region of available Q^2 the ratio $\sigma_L/\sigma_T \sim (Q^2/M_V^2)^{n_1}$, where $0.3 < n_1 < 1$. The definite conclusion can be derived only with new precise data on the ratio σ_L/σ_T, especially for high Q^2.

ACKNOWLEDGEMENT

We would like to thank Michele Arneodo, Alexei Kaidalov, Alexander Borissov, Jean-Rene Cudell and Alessia Bruni for various and fruitful discussions. One of us (E.M.) would like to thank the Department of Theoretical Physics of the University of Torino for its hospitality and financial support during his visit to Turin.

References

1. ZEUS Collaboration: Chekanov S. *et al.*, (2002) EXCLUSIVE PHOTOPRODUCTION OF J/Ψ MESONS AT HERA, *Eur.Phys.J.* C, **24**, pp.345
2. Martynov E., Prokudin A., Predazzi E., (2002) A UNIVERSAL REGGE POLE MODEL FOR ALL VECTOR MESON EXCLUSIVE PHOTOPRODUCTION BY REAL AND VIRTUAL PHOTONS, *Eur.Phys.J.* C, **26**, pp.271-284
3. For a recent update on the Regge theory, see:
   Barone V., Predazzi E. (2002) *"High-Energy Particle Diffraction"*, Springer-Verlag Berlin Heidelberg.
4. Sakurai J.J., (1960) THEORY OF STRONG INTERACTIONS *Ann. Phys. NY* **11**, pp.1;
   Gell-Mann M. and Zachariasen F., (1961) FORM-FACTORS AND VECTOR MESONS, *Phys. Rev.* D, **124**, pp.953-964.
5. Groom D.E.*et al.* (2000) REVIEW OF PARTICLE PHYSICS. PARTICLE DATA GROUP, *Eur.Phys.J.* C, **15**, pp.1.
6. Nemchik J., Nikolaev N. N., Predazzi E. and Zakharov B. G.(1997) COLOR DIPOLE PHENOMENOLOGY OF DIFFRACTIVE ELECTROPRODUCTION OF LIGHT VECTOR MESONS AT HERA, *Z. Phys.* C, **75**, pp.71-78.
7. Desgrolard P., Giffon M., Lengyel A., Martynov E.S., (1994) COMPARATIVE ANALYSIS OF PHENOMENOLOGICAL MODELS FOR POMERON AT $t = 0$, *Nuovo Cimento*, **107**, pp.637-647.
8. Cudell J.R. *et al.*, COMPETE Collaboration, (2002) HADRONIC SCATTERING AMPLITUDES: MEDIUM-ENERGY CONSTRAINTS ON ASYMPTOTIC BEHAVIOR, *Phys. Rev.* D, **65**, pp.074024.
9. Martynov E.S., (1994) UNITARIZATION OF POMERON AND REGGE PHENOMENOLOGY OF DEEP INELASTIC SCATTERING Proceedings of Workshop "Hadrons-94".
   Kiev; (1994) Preprint ITP-94-49E, Kiev;
   Jenkovszky L., Martynov E. and Paccanoni F., (1995) REGGE BEHAVIOUR OF THE NUCLEON STRUCTURE FUNCTION, Padova preprint, DFPD 95/TH/21.
10. Froissart M., (1961) ASYMPTOTIC BEHAVIOR AND SUBTRACTIONS IN THE MANDELSTAM REPRESENTATION, *Phys. Rev.* D, **123**, pp.1053-1057;
    Martin A., (1963) UNITARITY AND HIGH-ENERGY BEHAVIOR OF SCATTERING AMPLITUDES, *Phys. Rev.* D, **129**, pp.1432-1436.
11. Donnachie A., Landshoff P. V., (1987) EXCLUSIVE RHO PRODUCTION IN DEEP INELASTIC SCATTERING, *Phys. Lett.* B, **185**, pp.403;

Donnachie A., Landshoff P. V., (1995) EXCLUSIVE VECTOR MESON PRODUCTION AT HERA, *Phys. Lett.* B, **348**, pp.213-218.

12. Jenkovszky L.L., Martynov E.S., Paccanoni F., (1996) REGGE POLE MODEL FOR VECTOR MESON PHOTOPRODUCTION AT HERA, hep-ph/9608384.
13. Fiore R., Jenkovszky L. L., Paccanoni F. (1999) PHOTOPRODUCTION OF HEAVY VECTOR MESONS AT HERA: A TEST FIELD FOR DIFFRACTION, *Eur.Phys.J.* C, **10**, pp.461-467; Fiore R., Jenkovszky L. L., Paccanoni F., Papa A. (2002) J/Ψ PHOTOPRODUCTION AT HERA, *Phys. Rev.* D, **65**, pp.077505.
14. Predazzi E., (1966) , *Ann. of Phys. (NY)*, **36**, pp.250
15. Martynov E., Prokudin A., Predazzi E., (2002) VECTOR MESON PHOTOPRODUCTION IN THE SOFT DIPOLE POMERON MODEL FRAMEWORK hep-ph/0207272, submitted to *Phys. Rev.* D.

THREE-COMPONENT POMERON IN HIGH ENERGY ELASTIC SCATTERING

YU. ILJIN [1], A. LENGYEL[2] AND Z. TARICS[2]
[1] *Tavria Departament of the International University "Ukraine", Simferopol.*
[2] *Institute of Electron Physics, National Academy of Sciences of Ukraine. Universitetska 21, UA-88016 Uzhgorod, Ukraine.*

Abstract. We assume that the pomeron is a sum of Regge multipoles, each corresponding to a finite gluon ladder. From the fit to the data of $pp-$ and $\bar{p}p-$ scattering at high energy and all available momentum transfer we found that three-term multipole pomeron with different form-factors is significant for description of differential cross section and rise of ratio σ_{el}/σ_{tot} in whole experimental domain.

1. Introduction

The pomeron being an infinite gluon ladder [1] may appear as a finite sum of gluon ladders corresponding to a finite sum of Regge multipoles with increasing multiplicities [2]. The first term in $\ln s$ series contributes to the total cross-section with a constant term and can be associated with a simple pole, the second one (double pole) goes as $\ln s$, the third one (triple pole) as $ln^2 s$, etc. All pomeron poles have unit intercept. Previously the dipole pomeron approach was comprehensively investigated on different applications [3] - [5]. High quality description of $pp-$ and $\bar{p}p-$elastic cross-section was performed by generalized multipole model [6] including the asymptotic amplitude [7]. Recently the same task was performed in three hard pomeron model [8]. This problem is still remain actual and inspired some people to revitalize a well-known old fashion model [9]. In fact, all previously used pomeron models, as a rule, have intercept > 1 which requires the unitarisation, their contributions to the scattering amplitude have complicated structure and the overall number of parameters was fairly high (up to two dozen [6]). Thus, the dipole pomeron model [3] was used to describe $pp-$

R. Fiore et al. (eds.), Diffraction 2002, 109–116.

and $\overline{p}p-$elastic scattering at ISR and Collider energies, however the ratio σ_{el}/σ_{tot} was found [10] to decrease asymptotically. To obtain an increasing function for this ratio there was introduced a factor corresponding to the supercritical pomeron behavior [11]. However, it is obvious that in this case the Froissart bound is broken. To avoid this problem, we consider a model containing a finite series of pomeron terms up to $\ln^2 s$, in accordance with the Froissart constrain. Previously the multipole and multiple pomeron [3] - [11] approaches were comprehensively investigated on different applications. In fact, all used multipole (dipole and tripole) pomeron and multipomeron models, as a rule, have intercept > 1. Recently in paper [12] the contribution of truncated BFKL pomeron series to σ_{tot} of $pp-$ and $\overline{p}p-$scattering was studied and it has been shown that a reliable description can be obtained by using two orders in this series. As a by-product, the elastic differential cross-section was calculated for the diffraction cone at low and high energies. Contrary to [12] we suggest the three-component pomeron model inspired by the finite sum of gluon ladders extended to the whole range of available momentum transfer of high-energy $pp-$ and $\overline{p}p-$elastic scattering and performed a simultaneous fit to the σ_{tot}, ρ and $d\sigma/dt$ data. In fact, at $t \neq 0$ the unitarity in this model is violated (see, for example [13] and references herein). However as seen from calculations presented below, in the energy range under consideration one can neglect this violation. Our goal is to investigate the capabilities of pomeron model as a finite sum of gluon rungs ($\ln s$ power) equivalent to single pole + dipole + tripole pomeron with sufficiently non-linear trajectory for $pp-$ and $\overline{p}p-$ scattering in diffraction cone including the interference of nuclear and coulomb amplitudes.

This paper is organized as follows. In the next section one introduces the main formulae and features of the model. In Sec. 3, we perform the comparison with experiment. In last section the conclusion are drawn up.

2. The model

Our ansatz for the scattering amplitude is:

$$A_{pp}^{\overline{p}p}(s,t) = P(s,t) + R_f(s,t) \pm R_\omega(s,t) \pm O(s,t), \tag{1}$$

where the Pmeron contributiom we introduce in form:

$$P(s,t) = is\left(-i\frac{s}{s_0}\right)^{\alpha_P(t)-1} \sum_{j=0}^{2} g_j \ln^j\left(-i\frac{s}{s_0}\right) e^{\varphi_j(t)} \tag{2}$$

The Pomeron trajectory is:

$$\alpha_P(t) = 1 + \gamma_P\left(\sqrt{t_P} - \sqrt{t_P - t}\right), \tag{3}$$

where the lowest two-pion threshold $t_P = 4m_\pi^2$. The residue functions are [13]:

$$\varphi_j(t) = b_j t + \gamma_j \left(\sqrt{t_P} - \sqrt{t_P - t}\right), \tag{4}$$

In (1), the $R_f(s,t)$, $R_\omega(s,t)$ and $O(s,t)$ contain the subleading reggeon as well as the odderon contributions to the scattering amplitude:

$$R_f\left(s,t\right) = g_f \left(-i\frac{s}{s_0}\right)^{\alpha_f(t)} e^{b_f t}, \tag{5}$$

and

$$R_\omega\left(s,t\right) = i g_\omega \left(-i\frac{s}{s_0}\right)^{\alpha_\omega(t)} e^{b_\omega t}, \tag{6}$$

where

$$\alpha_j\left(t\right) = 1 + \alpha_j^{'} t, \quad j = f, \omega; \quad s_0 = 1\ GeV^2. \tag{7}$$

To describe the different behavior of proton-proton and proton-antiproton differential cross-section in region of deep one needs to include the odderon contribution, which we use in a simple form:

$$O\left(s,t\right) = s\left(-i\frac{s}{s_0}\right)^{\alpha_O(t)-1} \sum_{j=0}^{1} g_j \ln^j \left(-i\frac{s}{s_0}\right) e^{\varphi_O(t)} \tag{8}$$

The odderon trajectory is

$$\alpha_O(t) = 1 + \gamma_O \left(\sqrt{t_O} - \sqrt{t_O - t}\right), \tag{9}$$

where the lowest two-pion threshold $t_O = 4m_\pi^2$. The residue functions are:

$$\varphi_j(t) = b_j t + \delta_j \left(\sqrt{t_O} - \sqrt{t_O - t}\right), \tag{10}$$

3. Comparison with experiment

In order to determine the parameters that control the s-dependence of $A\left(s,0\right)$ we applied a wide energy range $10\ GeV \leq \sqrt{s} \leq 1800\ GeV$, where a good fit is guaranteed [14] and used the available data for total cross sections and ρ (see refs. in paper [14]). A total of 105 experimental points were included for $t = 0$. For the differential cross sections we selected the data at the energies $\sqrt{s} = 19; 23; 31; 44; 53; 62\ GeV$ (for pp–scattering) and $\sqrt{s} = 31; 53; 62; 546; 1800\ GeV$ (for $\bar{p}p$–scattering) [15]. For this energy range the crossover effect is anticipated to be neglected [13]. The squared 4-momentum covers entire available range $0.00065\ (GeV/c)^2 < |t| < 14\ (GeV/c)^2$.

TABLE 1. Parameters of the fit.

Pomeron		γ_P (GeV^{-1})	g_0	g_1	g_2
		0.118	4.61	$-0,0551$	$0,0526$
b_0 (GeV^{-2})	b_1 (GeV^{-2})	b_2 (GeV^{-2})	β_0 (GeV^{-1})	β_1 (GeV^{-1})	β_2 (GeV^{-1})
$3,59$	0	$5,28$	0.228	2.52	0.895
odderon	a_0 (GeV^{-1})	a_1 (GeV^{-1})	γ_O (GeV^{-1})	δ_0 (GeV^{-1})	δ_1 (GeV^{-1})
	$-2,06$	$0,00715$	$0,0824$	3.19	1.99
$f-reggeon$	g_f	$\alpha_f(0)$	b_f (GeV^{-2})		
	$8,74$	$0,78$	0.862		
$\omega-reggeon$	g_ω	$\alpha_\omega(0)$	b_ω (GeV^{-2})		
	4.40	0.881	3,72		
	$\chi^2_{d.o.f.}$		3.05		

The grand total number of 1650 experimental points were used in overall fit.

In the calculations we use the following normalization for the dimensionless amplitude:

$$\sigma_{tot} = \frac{4\pi}{s} ImA\,(s,t=0)\,, \quad \frac{d\sigma}{dt} = \frac{\pi}{s^2}\left|A\,(s,t) + A_c\,(s,t)\right|^2, \tag{11}$$

where $A_c(s,t)$ is the Coulomb amplitude [16]. Strictly speaking, in our case more general approach is required (see, e.g.[17]). However for a global fit performed in this work we have restricted ourselves to a simple Coulomb phase that appeared quite acceptable.

The resulting fits for σ_t, ρ, $\frac{d\sigma}{dt}$ are shown in Figs. 1,2 with the values of the fitted parameters quoted in Table 1. From these figures we conclude that the multipole pomeron model corresponding to a sum of gluon ladders up to two rungs fits the data well in a wide energy and momentum transfer regions. As result, the model gives a good behavior for the ratio σ_{el}/σ_{tot} for $\overline{p}p-$scattering (see Fig. 1).

The rapid decrease of the coefficients of the first three terms in the pomeron series in (2) provides the fast convergence of the series and ensures the applicability of this approximation at still much higher energies.

In this paper, we have explored only the simplest extension of the dipole pomeron to the tripole one. In fact, the scattering amplitude is much more complicated than just a simple power series in $\ln s$. Earlier a comprehensive analysis of the $pp-$ and $\bar{p}p-$diffraction cone scattering using the dipole and supercritical pomeron models was done in [13]. On the one hand, though we used just a simplified t-dependence in the model, reasonably good results were obtained. Because the slopes of secondary reggeons do not influence the fit sufficiently, we have fixed them at $\alpha'_f = 0.84$ (GeV^{-2}) and $\alpha'_\omega = 0.93$ (GeV^{-2}), which correspond to the values of Chew-Frautschi plot. On the other hand, we included the curvature of the pomeron and odderon trajectories that cannot be negligible. The quality of our fits $\chi^2/dof = 3.05$ is comparable with that of the best fit of [6],[8].

4. Conclusions

We have approved the dipole pomeron model [10] adding a term not violating the Froissart bound, each corresponding to a finite gluon ladder instead of a factor corresponding to the supercritical pomeron and conserving other useful properties of the model. This corresponds to the finite sum of gluon ladders with up to two rungs or alternatively up to the tripole pomeron contribution. We have obtained very good description of pp and $\bar{p}p$ hadron scattering data at intermediate and high energies and, as a by-product, the right behavior of ratio σ_{el}/σ_{tot}.We conclude that the nonfactorisable form of the pomeron and odderon amplitudes as well as the nonlinearity of its trajectory and the residue function is strongly suggested by data at all available momentum transfer.

Acknowledgments

We are grateful to the Organizing Committee, in particular to L. Jenkovszky, for the possibility to participate in nice Workshop. We thank V.Kundrat and E.Martynof for discussions. This work was supported by the Arany Janos Foundation.

References

1. Forshaw J., Ross D.A.(1997) *Quantum chromodynamics and the pomeron*, Cambridge Univ. Press, and references therein.
2. Fiore R. *et al.* (2001) Finite sum of gluon ladders and high energy cross sections, *Phys. Rev.* **D63**, 056010.

3. Jenkovszky L.L. (1986) Phenomenology of Elastic Hadron Diffraction, *Fortschr. phys.* **34**, 791-816.
4. Vall A.N., Jenkovszky L.L, Struminsky B.V. (1988) High energy hadron interactions, *Particles and nuclei* **19**, 180-223.
5. Desgrolard P., Giffon M, Jenkovszky L. (1993) Phenomenological implications of a "perturbative" pomeron beyond the dipole approximation, *Phys. Atom. Nucl.* **56**, 226-237.
6. Desgrolard P, Giffon M., Predazzi E. (1994) High quality description of elastic high energy data and prediction on new phenomena, *Z. Phys. C* **63**, 241-252.
7. Gauron P., Nicolescu B., Leader E. (1988) Similarities and differences between pp and $p\bar{p}$ scattering at TeV energies and beyond, *Nucl. Phys.* **B299**, 640-671.
8. Petrov V.A. and Prokudin A.V. (2002) The first three pomerons, in *Proceedings of the IXth Blois Workshop on Elastic and Diffractive Scattering*, Pruhonice near Prague, 9-15 June 2001, eds. V.Kundrat and P.Zavada, Prague, Czech Republic, pp 257-264.
9. Bourrely C,, Soffer J. and Wu N.N. (2002) Impact-Picture Phenomenology for $\pi^{\pm}p, K^{\pm}p$ and $p\bar{p}$ Elastic Scattering at High Energies, *arXiv* **HEP-PH/0210264**.
10. Vall A.N., Jenkovszky L.L. and Struminsky B.V. (1987) Where is asimptotic of hadron interactions? *Yad. Fiz.* **40**, 1519-1524.
11. Donnachie A., Landshoff P. (1992) Total cross sections, *Phys. Lett. B* **296**, 227-239.
12. Gay Ducati M.B., Machado M.V.T. (2001) Truncated BFKL Series and Hadronic Collisions, *Phys. Rev. D* **63**, 094018.
13. Desgrolard P., Lengyel A.I., Martynov E.S. (1997) Nonlinearities and pomeron non-factorizability in conventional diffraction , *Nuovo Cimento* **110A**, 251-266.
14. Cudell J.R. et al. (2000) High-Energy Forward Scattering and the Pomeron: Simple Pole versus Unitsrized Models, *Phys. Rev. D* **61**, 034019.
15. Carter M.K., Collins P.D.B and Whalley M.R. (1986) *A compilation of nucleon-nucleon and nucleon-antinucleon elastic scattering data*, **RAL-87-107**
16. Nikitin V.A. (1970) Investigation of elastic scattering of protons by nucleons at energy 1 - 70 GeV, *Particles and nuclei* **1**, 7-70.
17. Kundtat V., Lokajicek M. (1996) Description of high-energy elastic hadron scattering
in both the coulomb and hadronic domains, *Mod. Phys. Lett.* **11**, 2241- 2250.

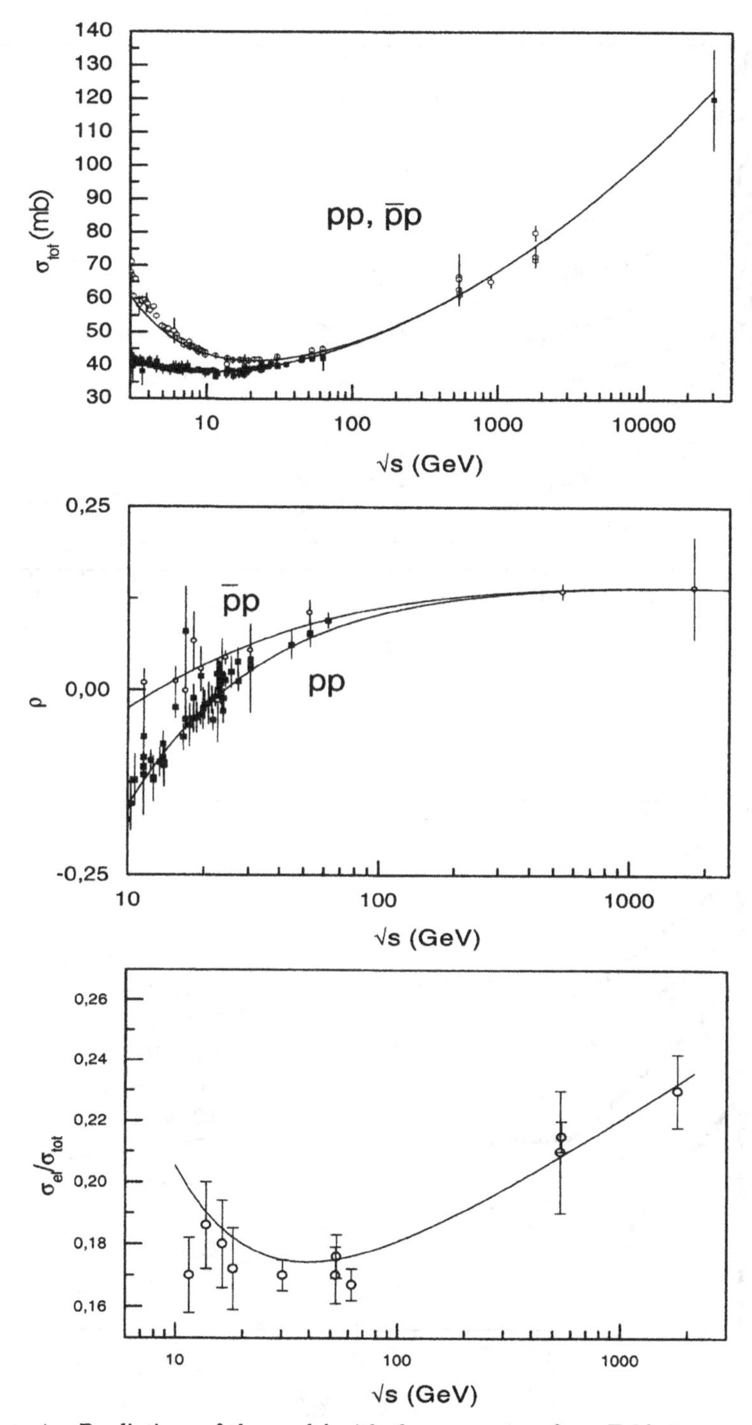

Figure 1. Predictions of the model with the parameters from Table 1 compared to the experimental data on σ_{tot}, ρ for $pp-$ and $\overline{p}p-$ scattering and σ_{el}/σ_{tot} for $\overline{p}p-$ scattering.

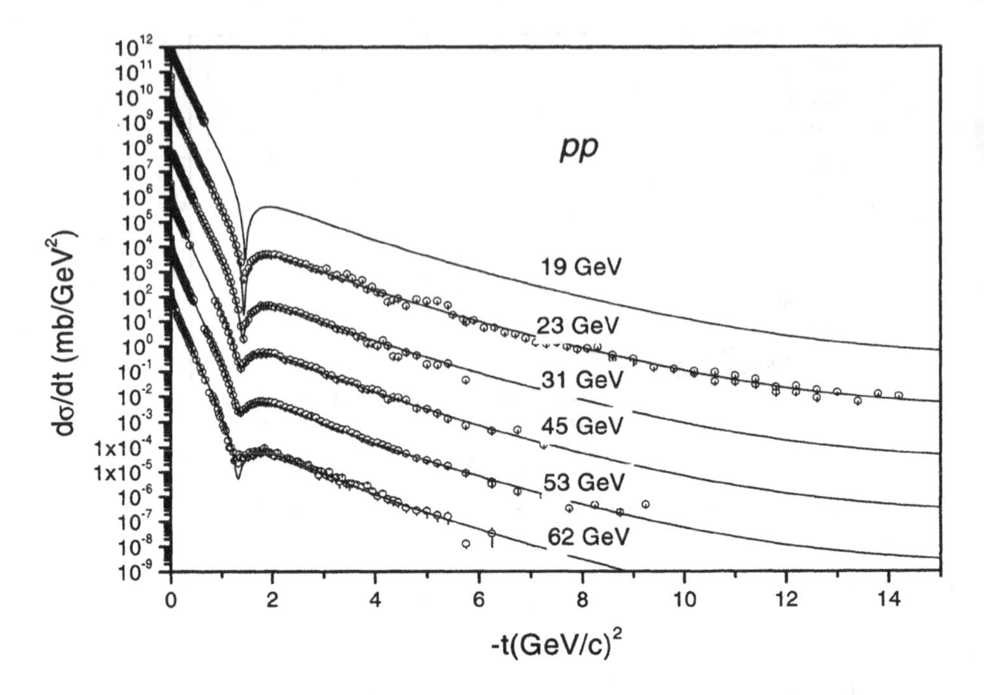

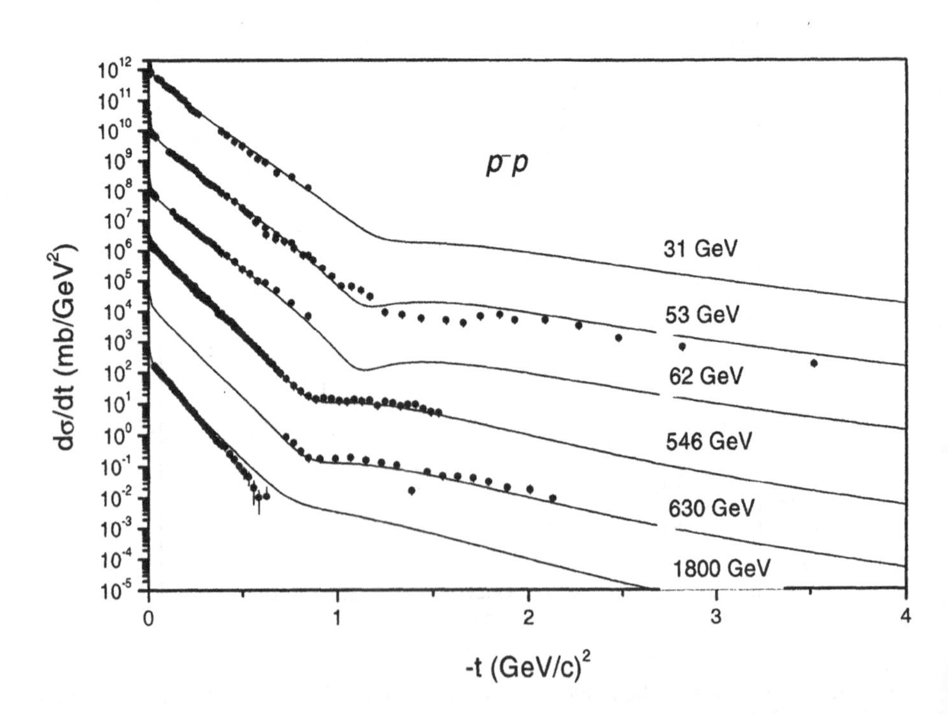

Figure 2. Differential cross-section for $pp-$ and $\overline{p}p-$ scattering (a factor of 10^2 between successive curves is present). The solid curves are the fits to the model.

LOW-ENERGY DIFFRACTION; A DIRECT-CHANNEL POINT OF VIEW: THE BACKGROUND

L.L. JENKOVSZKY
N.N. Bogolyubov Institute for Theoretical Physics
Academy of Sciences of Ukraine
Metrolohichna 14b, 03143 Kiev, Ukraine

S.YU. KONONENKO
Physics Department, T.H. Shevchenko National State University
Academician Glushkov avenue 6, 03022 Kiev, Ukraine

AND

V.K. MAGAS
N.N. Bogolyubov Institute for Theoretical Physics
Academy of Sciences of Ukraine
Metrolohichna 14b, 03143 Kiev, Ukraine &
Center for Physics of Fundamental Interactions (CFIF)
Physics Department, Instituto Superior Tecnico
Av. Rovisco Pais, 1049-001 Lisbon, Portugal

Abstract. We argue that at low-energies, typical of the resonance region, the contribution from direct-channel exotic trajectories replaces the Pomeron exchange, typical of high energies. A dual model realizing this idea is suggested. While at high energies it matches the Regge pole behavior, dominated by a Pomeron exchange, at low energies it produces a smooth, structureless behavior of the total cross section determined by a direct-channel nonlinear exotic trajectory, dual to the Pomeron exchange.

In this paper we investigate the role of the low-energy background in diffractive processes. We follow the ideas of two-component duality [1], [2], by which the high-energy Pomeron exchange is dual to the low-energy background. In describing this background we use a dual amplitude with Mandelstam analyticity (DAMA) [3], where, contrary to narrow resonance dual models, nonlinear trajectories are not only allowed, but even required by their general properties. The asymptotic rise of the trajectories in DAMA

R. Fiore et al. (eds.), Diffraction 2002, 117–122.

is limited by the condition $|\frac{\alpha(s)}{\sqrt{s}\ln s}| \leq const, \quad s \to \infty$. The most popular trajectories, satisfying above condition, are square root type of trajectories [3] , used also in this work. Such trajectories have an upper bound on the real part of these trajectories, which results in the termination of the resonances on it. The maximal value of this real part is determined by the free parameters of the trajectories, that can be fitted to the resonance spectra.

An extreme case is that of an exotic trajectory, whose real part does not reach any resonance. Relevant examples are presented in this talk.

The (s, t)-term of DAMA is

$$D(s,t) = \int_0^1 dz \left(\frac{z}{g}\right)^{-\alpha(s')-1} \left(\frac{1-z}{g}\right)^{-\alpha(t')-1}, \tag{1}$$

where $s' = s(1-z)$, $t' = tz$, g is a parameter, $g > 1$, and s, t are the Mandelstam variables.

For $s \to \infty$ and $t = 0$ it has the following Regge asymptotic behavior

$$D(s,t) \approx \sqrt{\frac{2\pi}{\alpha_t(0)}} g^{1+a+ib} \left(\frac{s\alpha'(0) g \ln g}{\alpha_t(0)}\right)^{\alpha_t(0)-1}, \tag{2}$$

where $a = Re\ \alpha\left(\frac{\alpha_t(0)}{\alpha'(0)\ln g}\right)$ and $b = Im\ \alpha\left(\frac{\alpha_t(0)}{\alpha'(0)\ln g}\right)$.

The pole structure of DAMA is similar to that of the Veneziano model except that multiple poles may appear at daughter levels. The presence of these multipoles does not contradict the theoretical postulates. On the other hand, they can be removed without any harm to the dual model by means the so-called Van der Corput neutraliser. The procedure [3] is to multiply the integrand of (1) by a function $\phi(x)$ with the properties:

$$\phi(0) = 0, \quad \phi(1) = 1, \quad \phi^n(=0), \quad n = 1, 2, 3, ...$$

The function

$$\phi(x) = 1 - exp\left(\frac{-x}{1-x}\right),$$

for example, satisfies the above conditions and results [3] in a standard, "Veneziano-like" pole structure:

$$D(s,t) = \sum_n g^{n+\alpha_t(0)} \frac{C_n}{n - \alpha(s)}, \tag{3}$$

where C_n are the residues, whose form is fixed by the dual amplitude [3]:

$$C_n = \frac{\alpha_t(0)\left(\alpha_t(0)+1\right)...\left(\alpha_t(0)+n+1\right)}{n!}.$$

The pole term in DAMA is a generalization of the Breit-Wigner formula, comprising a whole sequence of resonances lying on a complex trajectory $\alpha(s)$. Such a "reggeized" Breit-Wigner formula has little practical use in the case of linear trajectories, resulting in an infinite sequence of poles, but it becomes a powerful tool if complex trajectories with a limited real part and hence a restricted number of resonances are used. If $Re\ \alpha(s) > 0$, equation (3) produces a sequence of Breit-Wigner resonances lying on the trajectory $\alpha(s)$.

Near the threshold, $s \to s_0$

$$D(s,t) \simeq \frac{g^2}{\alpha(s_0)} \left(\frac{1-\frac{s_0}{s}}{g}\right)^{-\alpha(s_0)} Im\ \alpha(s) \cdot$$

$$\cdot \left[\ln\left(\frac{x_1\left(1-\frac{s_0}{s}\right)}{g}\right) \frac{Im\ \alpha(s_1)}{Im\ \alpha(s)} + \left(\frac{1}{\alpha(s_0)-\ln x_1}\right)\right] , \tag{4}$$

where $0 < x_1 < 1$ and $s_1 = s_0 + (s - s_0)(1 - x_1)$.

A simple model of trajectories satisfying the threshold and asymptotic constraints is a sum of square roots [3]

$$\alpha(s) = \alpha_0 + \sum_i \gamma_i \sqrt{s_i - s}. \tag{5}$$

The number of thresholds included depends on the model; while the lightest threshold gives the main contribution to the imaginary part, the heaviest one promotes the rise of the real part (terminating at the heaviest threshold).

A particular case of the model eq. (5) is that with a single threshold. Imposing an upper bound on the real part of this trajectory, $Re\ \alpha(s) < 0$, and inserting it to eqs. (1) or (3), we get an amplitude that does not produce resonances, since the real part of the trajectory does not reach $n = 0$ where the fist pole could appear. Its imaginary part instead rises indefinitely, contributing to the total cross section with a sooth background.

By using the dual model (1), we now calculate numerically the total cross sections

$$\sigma_t(s) = Im\ A(s, t = 0) , \tag{6}$$

where

$$A(s,t) = c\,(D(s,t) + D(u,t)) \tag{7}$$

and c is a normalization coefficient.

In this work we propose the following simplified model.[1] In the $t-$ channel we use a Pomeron trajectory of the form

$$\alpha_P(t) = 1.1 + 0.2t + 0.02(\sqrt{4m_\pi^2} - \sqrt{4m_\pi^2 - t}), \tag{8}$$

while the exotic $s-$ channel trajectory is

$$\alpha_E(s) = \alpha_0 + \alpha_1(\sqrt{s_0} - \sqrt{s_0 - s}). \tag{9}$$

In the Pomeron trajectory (8) the linear term mimics high-mass thresholds, important for the nearly exponential shape of the cone, while the lowest, $2m_\pi$ threshold is manifest as a "break" in the diffraction cone near $0.1\ GeV^2$.

By definition, the real part of the exotic trajectory should not pass through resonances; we tentatively set

$$Max\{Re\ \alpha_E(s)\} = \alpha_E(s_0) = \alpha_0 + \alpha_1\sqrt{s_0} = -1\ , \tag{10}$$

so that it does not reach $n = 0$, where summation in eq. (3) starts. For the lowest threshold in the exotic trajectory we choose $s_0 = (m_p + m_{J/\psi})^2$, with J/ψ photoproduction (or $J/\psi p$ scattering) in mind.

With this exotic trajectory the threshold behaviour of our amplitude is:

$$D(s,t) \sim (s - s_0)^{1/2 - \alpha_E(s_0)} [const + \ln(1 - s_0/s)] =$$

$$(s - s_0)^{3/2} [const + \ln(1 - s_0/s)]\ . \tag{11}$$

The only remaining free parameters are g, $g > 1$ and the slope of the exotic trajectory α_1. For illustrative purposes we set $g = 10$ and $\alpha_1 = 0.2\ GeV^{-2}$.

With these inputs in hand, now we can calculate the total cross section for all values of s, from the threshold to the highest asymptotic values. By inserting the trajectories (8) (setting $t = 0$) and (9) into (1), we calculate the integrals numerically.

To see more clearly the contribution from the background, we have divided the cross section by its Regge asymptotics $\sim s^{0.1}$. The result, in arbitrary units, is shown in Fig. 1. One can clearly see a cusp between the flat asymptotic behavior and steep rise from the threshold. This is the contribution from the background, modeled by the contribution from the direct-channel exotic trajectory, amounting to about $10 - 12\ \%$.

The background plays an important role in the resonance region. Of particular practical interest is the correct account for the background in the

[1] Our earlier attempts to model the background (as well as resonance contributions) can be found in Refs. [4].

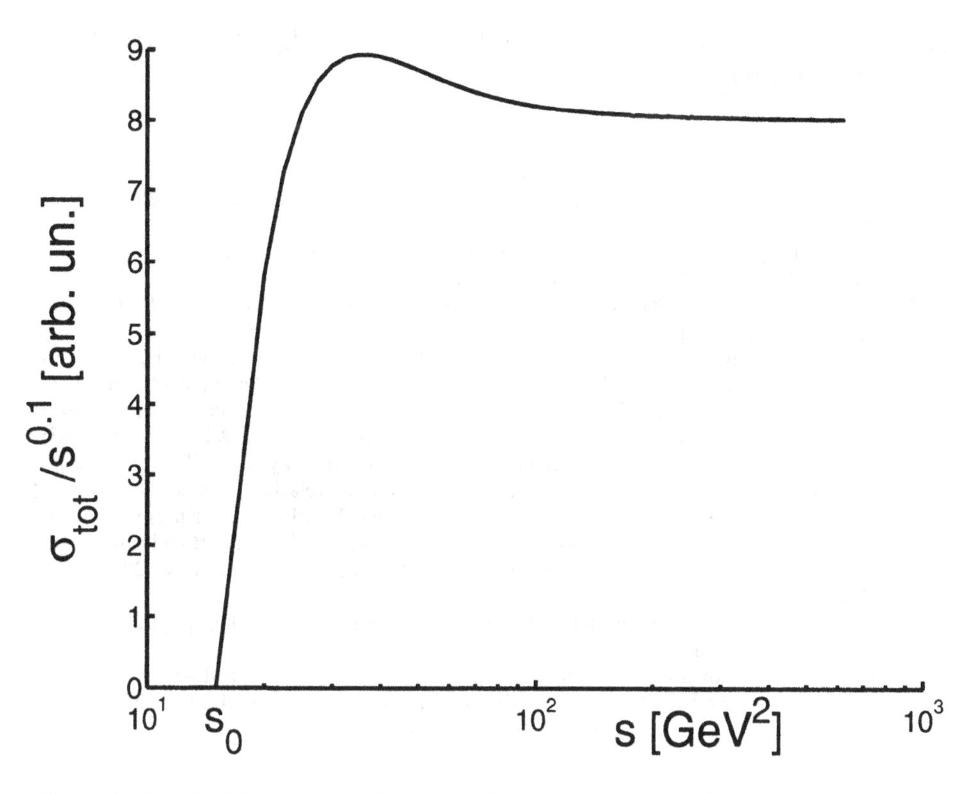

Figure 1. Purely diffractive contribution to the total cross section (in arbitrary units) calculated from DAMA, eq. (1), scaled to its Regge asymptotic behavior. The direct-channel exotic trajectory gives at low energies non-neglectable contribution (background), which could be seen as a cusp of $\sim 10 - 12$ %.

analysis of the JLab data on the electroproduction of nucleon resonances (see [5]).

The best testing field for this model is J/ψ scattering, where the exchange of secondary trajectories is forbidden and consequently no resonances are expected in the low energies region, dominated by the background, or the contribution of the direct-channel exotic trajectory. Since J/ψ total cross section is not measured directly, the relevant reaction is J/ψ photoproduction. In most of the papers on this subject, analyzing the HERA data on photoproduction (see e.g. [6] and refernces therein), the amplitude is assumed to be Regge behaved, eventually corrected by a threshold factor, but in any case ignoring the contribution from the background. The evaluation of the elastic cross section from (1) however involves also integration in t, thus making the calculations more complicated. This will be done in a forthcoming paper.

Acknowledgment. We thank A.I. Bugrij for fruitful discussions on the subject of this paper. L.L.J. and V.K.M. acknowledge the support from INTAS, grant 00-00366.

References

1. Freund, P. (1968) Finite energy sum rules and bootstraps, *Phys. Rev. Lett.*, **20**, pp. 235-237.
2. Harari, H. (1968) Pomeranchuk trajectory and its relation to low-energy scattering amplitudes, *Phys. Rev. Lett.*, **20**, pp. 1395-1398.
3. Bugrij, A.I. et al. (1973) Dual Amplitudes with Mandelstam Analyticity, *Fortschritte der Physik*, **21**, p. 427.
4. Fiore, R., Jenkovszky, L.L., Magas, V. (2001) Connection between lepton- and hadron-induced diffraction phenomena, *Nucl. Phys. Proc. Suppl.*, **99A**, pp. 131-138; Jenkovszky, L.L., Magas, V.K., Predazzi, E. (2001) Resonance-reggeon and parton-hadron duality in strong interactions, *Eur. Phys. J.*, **A12**, pp. 361-367; Duality in strong interactions, nucl-th/0110085; Jenkovszky, L.L., TKorzhinskaja, T., Kuvshinov, V.I. and Magas, V.K. (2001) Duality relation between small- and large-x structure functions, Proceedings of the New Trend in High-Energy Physics, Yalta, Crimea, Ukraine, September 22-29, 2001, edited by P.N. Bogolyubov and L.L. Jenkovszky (Bogolyubov Institute for Theoretical Physics, Kiev, 2001), pp. 178-185.
5. Fiore, R. et al. Explicit model realizing parton-hadron duality, hep-ph/0206027, to appear in Eur. Phys. J A.
6. Fiore, R., Jenkovszky, L.L. and Paccanoni, F. (1999) Photoproduction of heavy vector mesons at HERA: a test for diffraction, *Eur.Phys. J.*, **C10**, 461-467.

GENERALIZED DISTRIBUTION AMPLITUDES: NEW TOOLS TO STUDY HADRONS' STRUCTURE AND INTERACTION

R. FIORE
Dipartimento di Fisica, Università della Calabria &
INFN-Cosenza, I-87036 Arcavacata di Rende, Cosenza, Italy

A. FLACHI
IFAE, Universidad Autònoma de Barcelona
08193 Bellaterra, Barcelona, Spain

L.L. JENKOVSZKY
N.N. Bogolyubov Institute for Theoretical Physics
Academy of Sciences of Ukraine
Metrogohichna 14b, 03143 Kiev, Ukraine

A. LENGYEL
Institute of Electron Physics
Universitetska 21, UA-88000 Uzhgorod, Ukraine

AND

V.K. MAGAS
N.N. Bogolyubov Institute for Theoretical Physics
Academy of Sciences of Ukraine
Metrogohichna 14b, 03143 Kiev, Ukraine &
Center for Physics of Fundamental Interactions (CFIF)
Physics Department, Instituto Superior Tecnico
Av. Rovisco Pais, 1049-001 Lisbon, Portugal

Abstract. A non-perturbative approach to Generalized Parton Distributions, and to Deeply Virtual Compton Scattering in particular, based on off mass shell extension of dual amplitudes with Mandelstam analyticity (DAMA) is developed with the spin and helicity structure as well as the threshold behavior accounted for. The model is tested against the data on deep inelastic electron-proton scattering from JLab.

R. Fiore et al. (eds.), Diffraction 2002, 123–134.

1. Introduction

Parton distributions measure the probability that a quark or gluon carry a fraction x of the hadron momentum. Relevant structure functions (SF) are related by unitarity to the imaginary part of the forward Compton scattering amplitude. Generalized parton distributions (GPD) [1, 2, 3] represent the interference of different wave functions: one, where a parton carries momentum fraction $x+\xi$, and the other one, with momentum fraction $x-\xi$, correspondingly; here ξ is skewedness and can be determined by the external momenta in a deeply virtual Compton scattering (DVCS) experiment. Basically, the DVCS amplitude can be viewed as a binary hadronic scattering amplitude continued off mass shell. DVCS is related to the GPD in the same way as elastic Compton scattering is related to the ordinary SFs.

Apart from the longitudinal momentum fraction of gluons and quarks, GPDs contain also information about their transverse location, given by a Fourier transformation over the t-dependent GPD. Real-space images of the target can thus be obtained in a completely new way [4]. Spacial resolution is determined by the virtuality of the incoming photon. Quantum photographs of the nucleons and nuclei with resolutions on the scale of a fraction of femtometer are thus feasible.

One of the first experimental observations of DVCS was based on the recent analysis of the JLab data from the CLAS collaboration. New measurements at higher energies are currently being analyzed, and dedicated experiments are planned [5].

On the theoretical side, much progress has been achieved [1, 2, 3] in treating GPD in the framework of quantum chromodynamics (QCD) and the light-cone technique. On the other hand, non-nonpertubative effects (resonance production, the background, low-Q^2 effects) dominating the kinematical region of present measurements and the underlying dynamics still leave much ambiguity in the above-mentioned field-theoretical approach. Therefore, as an alternative or complementary approach we have suggested [6, 7, 8] to use dual amplitudes with Mandelstam analyticity (DAMA) as a model for GPD in general and DVCS in particular. We remind that DAMA realizes duality between direct-channel resonances and high-energy Regge behavior ("Veneziano-duality"). By introducing Q^2-dependence in DAMA, we have extended the model off mass shell and have shown [6, 7] how parton-hadron (or "Bloom-Gilman") duality is realized in this way. With the above specification, DAMA can serve as and explicit model valid for all values of the Mandelstam variables s, t and u as well as any Q^2, thus realizing the ideas of DVCS and related GPDs. The historical and logical connections between different kinametical regions and relevant models are depicted in Fig. 1.

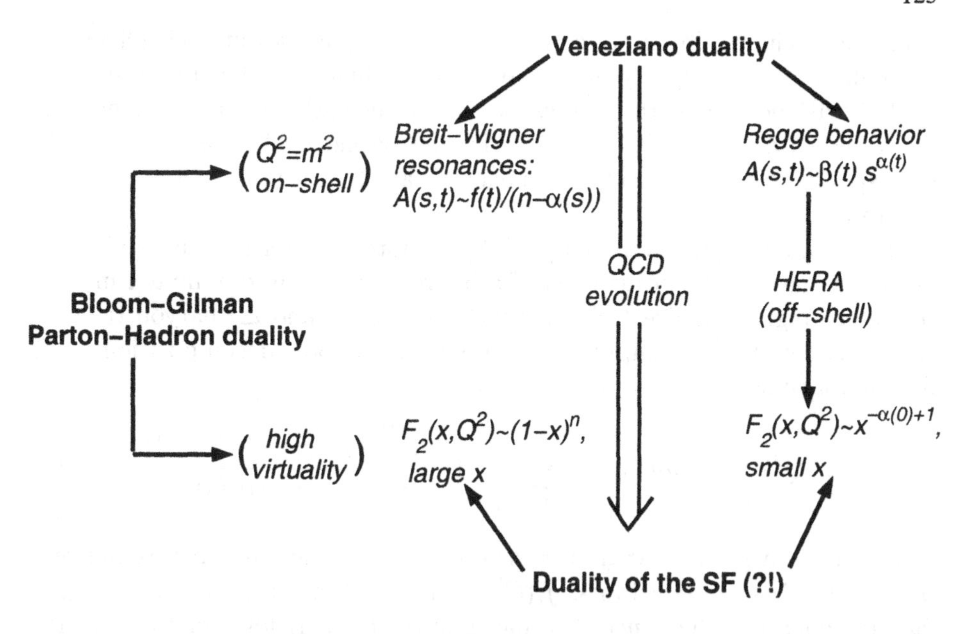

Figure 1. Generalized parton distribution amplitudes. A road map.

In this work we concentrate on the very delicate and disputable problem of the off mass shell continuation (introduction of the variable Q^2) in the dual model, starting with inclusive electron-nucleon scattering both at high energies, typical of HERA, and low energies, with the JLab data in mind (see ref. [8] for more details).

2. Simplified model

The central object of the present study is the nucleon SF, uniquely related to the photoproduction cross section by

$$F_2(x,Q^2) = \frac{Q^2(1-x)}{4\pi\alpha(1+\frac{4m^2x^2}{Q^2})}\sigma_t^{\gamma^*p}(s,Q^2)\ , \tag{1}$$

where the total cross section, $\sigma_t^{\gamma^*p}$, is the imaginary part of the forward Compton scattering amplitude, $A(s,Q^2)$, $\sigma_t^{\gamma^*p}(s) = \mathcal{I}m\ A(s,Q^2)$; m is the nucleon mass, α is the fine structure constant. The center of mass energy of the γ^*p system, the negative squared photon virtuality Q^2 and the Bjorken variable x are related by $s = Q^2(1-x)/x + m^2$.

We adopt the two-component picture of strong interactions [9], according to which direct-channel resonances are dual to cross-channel Regge exchanges and the smooth background in the $s-$channel is dual to the

Pomeron exchange in the $t-$channel. This nice idea, formulated [9] more than three decades ago, was first realized explicitly in the framework of DAMA with nonlinear trajectories (see [11, 6] and earlier references therein).

As explained in Refs. [6] and [11], the background in a dual model corresponds to a pole term with an exotic trajectory that does not produce any resonance.

In the dual-Regge approach [6, 7, 8] Compton scattering can be viewed as an off mass shell continuation of a hadronic reaction, dominated in the resonance region by direct-channel non-strange (N and Δ) baryon trajectories. The scattering amplitude follows from the pole decomposition of a dual amplitude [6]

$$A(s,Q^2)\Big|_{t=0} = norm \sum_{i=N_1^*,N_2^*,\Delta,E} A_i \sum_{n=n_i^{min}}^{n_i^{max}} \frac{f_i(Q^2)^{2(n-n_i^{min}+1)}}{n-\alpha_i(s)}, \quad (2)$$

where i runs over all the trajectories allowed by quantum number exchange, $norm$ and A_i's are constants, $f_i(Q^2)$'s are the form factors. These form factors generalize the concept of inelastic (transition) form factors to the case of continuous spin, represented by the direct-channel trajectories. The n_i^{min} refers to the spin of the first resonance on the corresponding trajectory i (it is convenient to shift the trajectories by 1/2, therefore we use $\alpha_i = \alpha_i^{phys} - 1/2$, which due to the semi-integer values of the baryon spin leaves n in Eq. (2) integer). The sum over n runs with step 2 (in order to conserve parity).

It follows from Eq. (2) that

$$\mathcal{I}m\, A(s,Q^2) = norm \sum_{i=N_1^*,N_2^*,\Delta,E} A_i \sum_{n=n_i^{min}}^{n_i^{max}} \frac{[f_i(Q^2)]^{2(n-n_i^{min}+1)}\mathcal{I}m\ \alpha_i(s)}{(n-\mathcal{R}e\ \alpha_i(s))^2+(\mathcal{I}m\ \alpha_i(s))^2}. \quad (3)$$

The first three terms in (3) are the non-singlet, or Reggeon contributions with the N^* and Δ trajectories in the s-channel, dual to the exchange of an effective bosonic trajectory (essentially, f) in the t-channel, and the fourth term is the contribution from the smooth background, modeled by a non-resonance pole term with an exotic trajectory $\alpha_E(s)$, dual to the Pomeron (see Ref. [6]). As argued in Ref. [6], only a limited number, $\mathcal{N}$, of resonances appear on the trajectories, for which reason we tentatively set $\mathcal{N} = 3$ - one resonance on each trajectories (N_1^*, N_2^*, Δ), i.e. $n_i^{max} = n_i^{min}$. Our analyses [8] shows that $\mathcal{N} = 3$ is a reasonable approximation. The limited (small) number of resonances contributing to the cross section results not only from the termination of resonances on a trajectory but even more due to the strong suppression coming from the numerator (increasing powers of the form factors).

TABLE 1. Values of the fitted parameters. In the first column we show the result of the fit when the parameters of the baryonic trajectories are fixed. The second column contains the result of the fit when the parameters of the trajectories are varied. † denotes the parameters of the physical baryon trajectories from ref. [8]. * The coefficient *norm* is chosen in such a way as to keep $A_{N_1^*} = 1$ in order to see the interplay between different resonances. ◇ Using intercepts and thresholds as a free parameters does not improve the fit, but they may get values far from original.

N_1^*	α_0	-0.8377 (fixed)†	-0.8070	-0.8377 (fixed)◇
	α_1	0.95 (fixed)†	0.9632	0.9842
	α_2	0.1473 (fixed)†	0.1387	0.1387
	$A_{N_1^*}$	1 (fixed)*	1 (fixed)*	–
	$Q^2_{N_1^*}$, GeV^2	2.4617	2.6066	–
N_2^*	α_0	-0.37(fixed)†	-0.3640	-0.37(fixed)◇
	α_1	0.95 (fixed)†	0.9531	0.9374
	α_2	0.1471 (fixed)†	0.1239	0.1811
	$A_{N_2^*}$	0.5399	0.6086	–
	$Q^2_{N_2^*}$, GeV^2	2.9727	2.6614	–
Δ	α_0	0.0038 (fixed)†	-0.0065	0.0038 (fixed)◇
	α_1	0.85 (fixed)†	0.8355	0.8578
	α_2	0.1969 (fixed)†	0.2320	0.2079
	A_Δ	4.2225	4.7279	–
	Q^2_Δ, GeV^2	1.5722	1.4828	–
	s_0, GeV^2	1.14 (fixed)†	1.2871	1.14 (fixed)◇
E	α_0	0.5645	0.5484	7.576
	α_2	0.1126	0.1373	0.0276
	s_E, GeV^2	1.3086	1.3139	1.311 (fixed)◇
	A_{exot}	19.2694	14.7267	–
	G_{exot}	–	–	24.16
	Q^2_{exot}, GeV^2	4.5259	4.6041	4.910
DS	Q_0^2	–	–	2.691
	${Q'_0}^2$	–	–	0.4114
	norm	0.021	0.0207	0.0659
	$\chi^2_{d.o.f.}$	28.29	11.60	18.1

We use Regge trajectories with a threshold singularity and nonvanishing imaginary part in the form:

$$\alpha(s) = \alpha_0 + \alpha_1 s + \alpha_2(\sqrt{s_0} - \sqrt{s_0 - s}), \tag{4}$$

where s_0 is the lightest threshold, $s_0 = (m_\pi + m_p)^2 = 1.14$ GeV^2 in our case, and the linear term approximates the contribution from heavy thresholds [6, 7, 8].

For the exotic trajectory we also keep only one term in the sum [1]. n_E^{min} is the first integer larger then $Max(\mathcal{R}e\ \alpha_E)$ – to make sure that there are no resonances on the exotic trajectory. The exotic trajectory is taken in the form

$$\alpha_E(s) = \alpha_E(0) + \alpha_{1E}(\sqrt{s_E} - \sqrt{s_E - s})\ , \tag{5}$$

where $\alpha_E(0)$, α_{1E} and the effective exotic threshold s_E are free parameters.

To start with, we use the simplest, dipole model for the form factors, disregarding the spin structure of the amplitude and the difference between electric and magnetic form factors:

$$f_i(Q^2) = \left(1 + Q^2/Q_{0,i}^2\right)^{-2}\ , \tag{6}$$

where $Q_{0,i}^2$ are scaling parameters, determining the relative growth of the resonance peaks and background.

We test our model against the experimental data from SLAC [12] and JLab [15][2]. This set of the experimental data is not homogeneous, i.e. points at low s (high x) are given with very small experimental errors, thus "weighting" the fitting procedure not uniformly. Consequently, a pre-selection of the date involved in the fitting procedure was made, although we present all the experimental points in the Figure. The results of our fits and the values of the fitted parameters are presented in Figs. 2 and Table 1. More details can be found in ref. [8].

For the first fit - first column of Table 1 and dashed-dotted lines in Fig. 2 - we fix the parameters of the Δ and N^* trajectories to the physical ones, reproducing the correct masses and widths of the resonances, leaving the four scaling constants Q_i^2, four factors A_i and the parameters of the exotic trajectories to be fitted to the data. Then, to improve the model we effectively account for the large number of overlapping resonances (about 20) present in the energy range under investigation. For this reason we consider the dominant resonances (N_1^*, N_2^* and Δ) as "effective" contributions to the SF. In other words, we require that they mimic the contribution of

[1]In Ref. [11] the whole DAMA integral was calculated numerical and it has been shown that in the resonance region the direct-channel exotic trajectory gives a non-negligible contribution, amounting to about 10-12%.

[2]We are grateful to M.I. Niculescu for making her data compilation available to us.

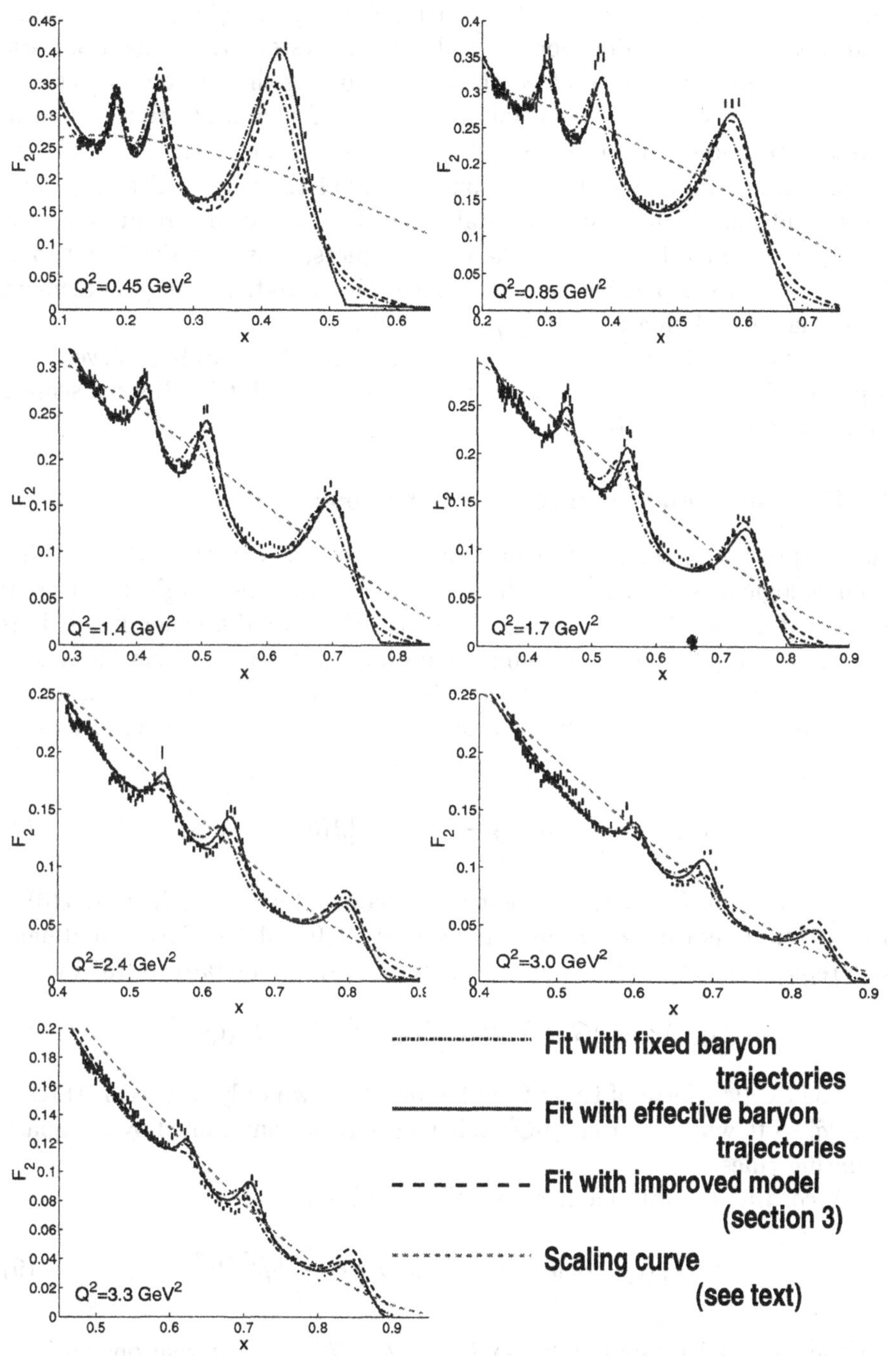

Figure 2. F_2 as a function of x for $Q^2 = 0.45 - 3.3$ GeV^2.

the dominant resonances plus the large number of subleading contributions, which, together, fully describe the real physical system. In the light of these considerations, we have refitted the data, allowing the baryon trajectories parameters to vary. The resulting parameters of such a fit are reported in Table 1 (second column). It is worth noting that although the range of variation was not restricted, the new parameters of the trajectories stay close to their physical values, showing stability of the fit and thus reinforcing our previous considerations. From the relevant plots, shown in Fig. 2 with full lines, one can see that the improvement is significant, although agreement is still far from being perfect ($\chi^2_{d.o.f.} = 11.6$).

The smooth dashed lines in Fig. (2) correspond to a "scaling curves", i.e. a phenomenological parameterizations of the SF exhibiting Bjorken scaling and fitting the data [10].

3. Spin and helicity; threshold behaviour

In the preceding section the nucleon form factor was treated in a phenomenological way, fitted to the data, neglecting the longitudinal cross section σ_L, and the $Q^2 \to 0$ limit of the SFs. On the other hand, it is known [14, 13] that with account for both σ_L and σ_T the form factor can be presented as sum o three terms: $G_+(Q^2)$, $G_0(Q^2)$ and $G_0(Q^2)$, corresponding to $\gamma^* N \to R$ helicity transition amplitudes in the rest frame of the resonance R:

$$G_{\lambda_\gamma} = \frac{1}{m} < R, \lambda_R = \lambda_N - \lambda_\gamma | J(0) | N, \lambda_N >; \tag{7}$$

here λ_R, λ_N and λ_γ are the resonance, nucleon and photon helicities, $J(0)$ is the current operator; λ_γ assumes the values $-1, 0$ and $+1$. Correspondingly, one replaces the Q^2−dependent expression in the numerator of (3) by

$$f(Q^2)^N \to |G_+(Q^2)|^2 + 2|G_0(Q^2)|^2 + |G_-(Q^2)|^2. \tag{8}$$

The explicit form of these form factors is known only near their thresholds $|\vec{q}| \to 0$, while their large-Q^2 behavior may be constrained by the quark counting rules.

According to [16], one has near the threshold

$$G_\pm(Q^2) \sim |\vec{q}|^{J-3/2}, \quad G_0(Q^2) \sim \frac{q_0}{|\vec{q}|} |\vec{q}|^{J-1/2} \tag{9}$$

for the so-called normal $(1/2^+ \to 3/2^-, 5/2^+, 7/2^-, ...)$ transitions and

$$G_\pm(Q^2 \sim |\vec{q}|^{J-1/2}, \quad G_0(Q^2) \sim \frac{q_0}{|\vec{q}|} |\vec{q}|^{J+1/2} \tag{10}$$

for the anomalous $(1/2^+ \to 1/2^-, 3/2^+, 5/2^-, ...)$ transitions, where

$$|\vec{q}| = \frac{\sqrt{(M^2 - m^2 - Q^2)^2 + 4M^2Q^2}}{2M}, \quad |\vec{q}|_{Q=0} = \frac{M^2 - m^2}{2M}, \tag{11}$$

$$q_0 = \frac{M^2 - m^2 - Q^2}{2M}, \tag{12}$$

M is a resonance mass [3].

Following the quark counting rules, in refs. [13] (for a recent treatment see [14]), the large-Q^2 behavior of G's was assumed to be

$$G_+(Q^2) \sim Q^{-3}, \quad G_0(Q^2) \sim Q^{-4}, \quad G_-(Q^2) \sim Q^{-5}. \tag{13}$$

Let us note that while this is reasonable (modulo logarithmic factors) for elastic form factors, it may not be true any more for inelastic (transition) form factors. For example, dual models (see Eq. (1) and ref. [6]) predict powers of the form factors to increase with increasing excitation (resonance spin). This discrepancy can be resolved only experimentally, although a model-independent analysis of the Q^2-dependence for various nuclear excitations is biased by the (unknown) background.

In ref. [14] the following expressions for the G's, combining the above threshold- (9), (10) with the asymptotic behavior (13), was suggested:

$$|G_\pm|^2 = |G_\pm(0)|^2 \left(\frac{|\vec{q}|}{|\vec{q}|_{Q=0}} \frac{Q_0'^2}{Q^2 + Q_0'^2} \right)^{2J-3} \left(\frac{Q_0'^2}{Q^2 + Q_0'^2} \right)^{m_\pm} \tag{14}$$

$$|G_0|^2 = C^2 \left(\frac{Q_0^2}{Q^2 + Q_0^2} \right)^{2a} \frac{q_0^2}{|\vec{q}|^2} \left(\frac{|\vec{q}|}{|\vec{q}|_{Q=0}} \frac{Q_0'^2}{Q^2 + Q_0'^2} \right)^{2J-1} \left(\frac{Q_0^2}{Q^2 + Q_0^2} \right)^{m_0} \tag{15}$$

for the normal transitions and

$$|G_\pm|^2 = |G_\pm(0)|^2 \left(\frac{|\vec{q}|}{|\vec{q}|_{Q=0}} \frac{Q_0'^2}{Q^2 + Q_0'^2} \right)^{2J-1} \left(\frac{Q_0'^2}{Q^2 + Q_0'^2} \right)^{m_\pm} \tag{16}$$

$$|G_0|^2 = C^2 \left(\frac{Q_0^2}{Q^2 + Q_0^2} \right)^{2a} \left(\frac{q_0^2}{|\vec{q}|^2} \frac{Q_0'^2}{Q^2 + Q_0'^2} \right)^{2J+1} \left(\frac{Q_0^2}{Q^2 + Q_0^2} \right)^{m_0} \tag{17}$$

[3]In our approach M is defined by trajectory: $\mathcal{R}e\alpha(M^2) = J$, where J is a spin of the resonance; and therefore for the varying baryon trajectory M might differ from the mass of physical resonance.

for the anomalous ones, where $m_+ = 3$, $m_0 = 4$, $m_- = 5$ count the quarks, C and a are free parameters. The form factors at $Q^2 = 0$ are related to the known (measurable) helicity photoproduction amplitudes $A_{1/2}$ and $A_{3/2}$ by

$$|G_{+,-}(0)| = \frac{1}{\sqrt{4\pi\alpha}}\sqrt{\frac{M}{M-m}}|A_{1/2,3/2}|. \tag{18}$$

The values of the helicity amplitudes are quoted by experimentalists [17] (those relevant to the present discussion are compiled also in [14]).

We have fitted the above model to the JLab data [15] again by keeping the contribution from three prominent resonances, namely $\Delta(1232)$, $N^*(1520)$ and $N^*(1680)$. For the sake of simplicity we neglected the cross term containing G_0 (and the coefficient C and parameter a), since it is small relative to the other two terms in Eq. (8) [14].

We use the same background term as in the previous section, but with the normalization coefficient G_E:

$$background = \frac{f_E I_E}{(n_E^{min} - R_E)^2 + I_E^2}\ , \quad f_E = G_E\left(\frac{Q_E^2}{Q^2 + Q_E^2}\right)^4 .$$

To be specific, we write explicitly the three resonance terms, to be fitted to the data, (cf. with Eq. (3) of the previous section)[4]:

$$\mathcal{I}m\ A(s, Q^2) = norm\cdot$$

$$\frac{f_\Delta I_\Delta}{(1-R_\Delta)^2 + I_\Delta^2} + \frac{f_{N^-} I_{N^-}}{(1-R_{N^-})^2 + I_{N^-}^2} + \frac{f_{N^+} I_{N^+}}{(2-R_{N^+})^2 + I_{N^+}^2} + background, \tag{19}$$

where e.g. f_Δ is calculated according to Eqs. (8,14,16):

$$f_\Delta = \left(\frac{|\vec{q}|}{|\vec{q}|_{Q=0}}\frac{Q_0'^2}{Q^2 + Q_0'^2}\right)^{2J-1=2}\left(|G_+(0)|^2\left(\frac{Q_0^2}{Q^2+Q_0^2}\right)^3 + |G_-(0)|^2\left(\frac{Q_0^2}{Q^2+Q_0^2}\right)^5\right); \tag{20}$$

here R and I denote the real and the imaginary parts of the relevant trajectory, specified by the subscript. Similar expressions can be easily cast for f_{N^+} and for f_{N^-} as well.

The form factors at $Q^2 = 0$ can be simply calculated from Eq. (18) by inserting the known [17] (see also Table 1 in [14]) values of the relevant photoproduction amplitudes:

[4]Remember that our trajectories are shifted by 1/2 from the physical trajectories $\alpha_i = \alpha_i^{phys} - 1/2$.

$A_{\Delta(1232)}(1/2, 3/2) = (-0.141\ GeV^{-1/2}, -0.258\ GeV^{-1/2});$
$A_{N(1520))}(1/2, 3/2) = (-0.022\ GeV^{-1/2}, 0.167\ GeV^{-1/2});$
$A_{N(1680)}(1/2, 3/2) = (-0.017\ GeV^{-1/2}, 0.127\ GeV^{-1/2}).$

Note that in this way the relative normalization of the resonance terms is fixed (no A_i's appearing in previous section), leaving only two adjustable parameters, Q_0^2 and $Q_0'^2$ (provided C is set zero, then a also disappears), which means that this version of the model (with the cross term, containing G_0, neglected!) is very restrictive.

The resulting fits to the SLAC and JLab data are presented in Fig. 2 (dashed lines) and Table 1 (third column). The fit is not so good ($\chi^2_{d.o.f.} =$ 18.1), what probably tells us not to neglect the cross term. We hope to account for this and to improve the fit in the forthcoming works.

4. Future Prospects

In this work we have presented new results on the extension of a dual model (Sec. 2) to include the spin and helicity structure of the amplitudes (SFs) as well as its threshold behavior as $Q^2 \to 0$. Let us remind that the lowest threshold (in the direct channel) of a hadronic reaction (if Compton scattering is considered in analogy with πN scattering) is $s_0 = (m_\pi + m_N)^2$. Hence its imaginary part and the relevant structure function starts at this value and vanishes below s_0. The "gap" between the lowest electromagnetic m_p^2 and the above-mentioned hadronic thresholds is filled by the arguments and formulas of Sec. 3. The $Q^2 \to 0$ is also important as the transition point between elastic and inelastic dynamics.

The important step performed in our work after Ref. [14] is the "Reggization" of the Breit-Wigner pole terms (19), i.e. single resonance terms in Ref. [14] are replaced by those including relevant baryon trajectories[5]. The form of these trajectories, constrained by analyticity, unitarity and by the experimental data is crucial for the dynamic. The use of baryon trajectories instead of individual resonances not only makes the model economic (several resonances are replaced by one trajectory) but also helps in classifying the resonances, by including the "right" ones and eliminating those nonexistent. The construction of feasible models of baryon trajectories, fitting the data on resonances and yet satisfying the theoretical bounds, is in progress.

Note also that the dual model constrains the form factors: as seen from Eq. (3), the powers of the form factors rise with increasing spin of the excited state. This property of the model can and should be tested experimentally.

[5]Another important step is adding the background into consideration.

The Q^2 dependence, introduced in this model via the form factors can be studied and constrained also by means of the QCD evolution equations.

To summarize, the model presented in this paper, can be used as a laboratory for testing various ideas of the analytical S-matrix and quantum chromodynamics. Its virtue is a simple explicit form, to be elaborated and constrained further both by theory and experimental tests.

Acknowledgment. L.J., A.L. and V.M. acknowledge the support by INTAS, grants 00-00366 and 97-31726.

References

1. Müller, D. et al. (1994) Wave Functions, Evolution Equations and Evolution Kernels from Light-Ray Operators of QCD, *Fortschr. Phys.* **42** pp. 101-142.
2. Ji, X. (1997) Gauge invariant decomposition of nucleon spin and its spin-off, *Phys. Rev. Lett.* **78** pp. 610-613.
3. Radyushkin, A.V. (1996) Scaling limit of deeply virtual Compton scattering, *Phys. Lett.* **B380** pp. 417-425; (1997) Nonforward parton distributions, *Phys. Rev.* **D56** pp. 5524-5557.
4. Pire, B. (2002) Generalized Parton Distributions and Generalized Distribution Amplitudes : New Tools for Hadronic Physics, hep-ph/0211093.
5. Elouadrhirs, L., (2002) Deeply Virtual Compton Scattering at Jefferson Lab, Results and Prospects, hep-ph/0210341.
6. Jenkovszky, L.L., Magas, V.K., Predazzi, E. (2001) Resonance-reggeon and parton-hadron duality in strong interactions, *Eur. Phys. J.*, **A12**, pp. 361-367;
7. Jenkovszky, L.L., Magas, V.K., Predazzi, E. Duality in strong interactions, nucl-th/0110085; Jenkovszky, L.L., Magas, V.K., Dual Properties of the Structure Functions, hep-ph/0111398, Proceedings of the 31st International Symposium On Multiparticle Dynamics (ISMD 2001) , September 1-7, 2001, Datong, China, pp. 74-77;
8. Fiore, R. et al. (2002) Explicit model realizing parton-hadron duality, hep-ph/0206027, to appear in Eur. Phys. J A.
9. Freund, P. (1968) Finite energy sum rules and bootstraps, *Phys. Rev. Lett.*, **20**, pp. 235-237; Harari, H. (1968) Pomeranchuk trajectory and its relation to low-energy scattering amplitudes, *Phys. Rev. Lett.*, **20**, pp. 1395-1398.
10. Csernai, L.P. et al. (2002) From Regge Behavior to DGLAP Evolution *Eur. Phys. J.* **C24** (2002) pp. 205-211.
11. Jenkovszky, L.L., Kononenko, S.Yu., Magas, V.K (2002) Diffraction from the direct-channel point of view: the background, hep-ph/0211158.
12. Stoler, P. (1991) Form-factors of excited baryons at high Q**2 and the transition to perturbative QCD, *Phys. Rev. Lett.* **66** pp. 1003-1006; (1991) Form-factors of excited baryons at high Q**2 and the transition to perturbative QCD. 2., *Phys. Rev.* **D44** pp. 73-80.
13. Carlson, C.E., Mukhopadhyay, N.C. (1995) Leading log effects in the resonance electroweak form-factors, *Phys. Rev. Lett.* **74** pp. 1288-1291; (1998) Bloom-Gilman duality in the resonance spin structure functions, *Phys. Rev.* **D58** 094029.
14. Davidovsky, V.V. and Struminsky, B.V. (2002) The Behavior of Form Factors of Nucleon Resonances and Quark-Hadron Duality, hep-ph/0205130 and in these Proceedings.
15. http://www.jlab.org/resdata.
16. Bjorken, J.D. and Walecka, J.D., (1966) Electroproduction of nucleon resonances *Ann. Physics* **38** pp. 35-62.
17. Particle Data Group (1998) *Eur. Phys. J* **C3** pp.1-794.

INVESTIGATION OF QUARK-HADRON DUALITY: THE MODEL FOR NUCLEON RESONANCE FORM FACTORS

V.V. DAVYDOVSKYY
Institute for Nuclear Research
Nauki Ave., 47, 03680 Kyiv, Ukraine

AND

B.V. STRUMINSKY
Bogolyubov Institute for Theoretical Physics
Metrolohichna 14b, 03143 Kyiv, Ukraine

Abstract. The model for resonance production form factors, dependent on photon virtuality Q^2, which have correct threshold behavior and take into account the available data on resonance decays, is presented. The behavior of nucleon structure function F_2 in the resonance region is investigated. Qualitative description of the quark-hadron duality is achieved.

1. Introduction

Two types of duality are known in particle physics. They are the reggeon-resonance and quark-hadron duality.

Veneziano amplitude was the first example of successful implementation of resonance-reggeon duality. Another example is a Dual Amplitude with Mandelstam Analyticity (DAMA) [4, 10] , in which the complex non-linear Regge trajectories with correct thresholds can be introduced.

The second type of the duality, quark-hadron, was discovered by Bloom and Gilman [2] (we refer the reader to this paper for introduction to this matter).

The exact mathematical formulation of the duality (both resonance-reggeon and quark-hadron) was given on the basis of the finite-energy sum rules, formulated independently in the works [9, 11, 7].

R. Fiore et al. (eds.), Diffraction 2002, 135–142.

Formally, the quark-hadron duality is exact, but practically the necessity to cut the expansion of any Fock state leads to different manifestations of the duality under various kinematic conditions and in various reactions.

For theoretical description of the duality it's necessary to construct the structure functions in the resonance region (threshold energies and small photon virtualities). For this purpose the dependence of resonance production form factors $\gamma^* N \to R$ on photon virtuality Q^2 must be known. Below we construct the form factors and study the quark-hadron duality.

2. Resonance form factors

Lepton-hadron scattering in the lowest order on electromagnetic coupling constant is described as an exchange of virtual photon with virtuality $Q^2 = -q^2$ (q is four-dimensional momentum of photon) and energy (in nucleon rest frame) $\nu = (pq)/m$ (p is four-dimensional momentum of nucleon, m is nucleon mass). In this case q^2 serves the role of squared momentum transferred from lepton to nucleon.

The mechanism of virtual photon and nucleon interaction depends on the quantity of transferred momentum Q^2 and photon energy. At high Q^2 and ν noncoherent scattering of virtual photon on nucleon quarks takes place, and huge amount of hadrons is produced.

At moderate Q^2 ($\sim$ 1 GeV^2) and ν internal nucleon spin states are excited, which leads to the resonance production.

Below we introduce main notations and define necessary quantities.

Four-dimensional momentum of nucleon, photon and resonance are denoted as p, q and P respectively. In the resonance rest frame components of this vectors read:

$$p = (\frac{M^2 + m^2 + Q^2}{2M}; 0, 0, -\frac{\sqrt{(M^2 - m^2 - Q^2)^2 + 4M^2 Q^2}}{2M}), \tag{1}$$

$$q = (\frac{M^2 - m^2 - Q^2}{2M}; 0, 0, \frac{\sqrt{(M^2 - m^2 - Q^2)^2 + 4M^2 Q^2}}{2M}), \tag{2}$$

$$P = (M; 0, 0, 0), \tag{3}$$

where $p^2 = m^2$, $P^2 = M^2$ and M is the mass of a resonance.

The vertex of virtual photon absorption by nucleon $\gamma^* N \to R$ is described by three independent form factors $G_{\pm,0}(Q^2)$ (or by two, when a spin of resonance equals to 1/2, and $G_-(Q^2) \equiv 0$), which are (in the resonance rest frame) helicity amplitudes of the transition $\gamma^* N \to R$:

$$G_{\lambda_\gamma} = \frac{1}{2m} < R, \lambda_R = \lambda_N - \lambda_\gamma | J(0) | N, \lambda_N >, \tag{4}$$

where λ_R, λ_N and λ_γ are the helicity of resonance, nucleon and photon respectively; $J(0)$ is current operator; λ_γ takes values of -1, 0, $+1$.

Nucleon structure functions can be expressed in terms of form factors (4) [5]. For F_2 we write:

$$F_2(x, Q^2) = \frac{m\,\nu\, Q^2}{\nu^2 + Q^2}\, \delta(W^2 - M^2)\, \left[|G_+(Q^2)|^2 + 2|G_0(Q^2)|^2 + |G_-(Q^2)|^2\right], \tag{5}$$

where $W^2 = (p+q)^2$ is the square of total energy of photon and nucleon in the c.m. frame; x is Bjorken variable.

Formula (5) determines a contribution of one infinitely narrow resonance into nucleon structure function F_2. For a resonance with width Γ in the expression (5) we change delta-function $\delta(W^2 - M^2)$ to

$$\frac{1}{\pi}\frac{M\Gamma}{(W^2 - M^2)^2 + M^2\Gamma^2}. \tag{6}$$

The expression (6) originates from the propagator of a resonance. But, in principle, other shapes could be used.

The basic idea of this work is to take contributions of all resonances, which data is published in literature [8], into account. If we let F_2^R to denote the contribution of resonance R into spin-independent structure function, then the contribution of only resonances to the structure function will be written in the form of the sum:

$$F_2 = \sum_R F_2^R. \tag{7}$$

To calculate resonance contribution to the structure function one must construct form factors of resonance production as functions of photon virtuality Q^2. Notice, that experimentally the dependence of form factors of known resonances [8] on Q^2 practically is not studied. In tables [8] only their values at $Q^2 = 0$ are listed.

The transitions $\gamma^* N \to R$ by the parity may be of two types: normal, i.e.

$$1/2^+ \to 3/2^-, 5/2^+, 7/2^-, \ldots \tag{8}$$

and abnormal:

$$1/2^+ \to 1/2^-, 3/2^+, 5/2^-, \ldots \tag{9}$$

About corresponding form factors $G_{\pm,0}(Q^2)$ it's known: a) form factor threshold behavior at $|\vec{q}| \to 0$ [1], b) form factor asymptotic behavior at high Q^2, c) form factor value at $Q^2 = 0$ [8].

So, as it was shown in [1], form factors of production of resonance with spin J in case of normal parity transition $\gamma^* N \to R$ (8) have the following threshold behavior:

$$G_\pm(Q^2) \sim |\vec{q}|^{J-3/2}, \tag{10}$$

$$G_0(Q^2) \sim \frac{q_0}{|\vec{q}|} |\vec{q}|^{J-1/2}. \tag{11}$$

In case of abnormal parity transitions (9):

$$G_\pm(Q^2) \sim |\vec{q}|^{J-1/2}, \tag{12}$$

$$G_0(Q^2) \sim \frac{q_0}{|\vec{q}|} |\vec{q}|^{J+1/2}. \tag{13}$$

Special case is transitions $1/2^+ \to 1/2^+$, which are determined by only two form factors G_+ and G_0 (G_- corresponds to resonance helicity 3/2 and thus is absent for resonances with spin 1/2). Their threshold behavior for the transition $1/2^+ \to 1/2^+$ is as follows:

$$G_+(Q^2) \sim |\vec{q}|, \tag{14}$$

$$G_0(Q^2) \sim \frac{q_0}{|\vec{q}|} |\vec{q}|^2. \tag{15}$$

The form factors of transition $1/2^+ \to 1/2^-$ are determined by the expression (12) at $J = 1/2$, i.e.

$$G_+(Q^2) \sim Const, \tag{16}$$

$$G_0(Q^2) \sim \frac{q_0}{|\vec{q}|} |\vec{q}|. \tag{17}$$

The behavior of form factors at high Q^2 is determined by quark counting rules [3, 12], according to which

$$G_+(Q^2) \sim Q^{-3}, \quad G_0(Q^2) \sim Q^{-4}, \quad G_-(Q^2) \sim Q^{-5}. \tag{18}$$

So, we suggest the expressions for form factors, possessing all the above-mentioned properties, to be written in the following form:

$$\left|G_\pm(Q^2)\right|^2 = |G_\pm(0)|^2 \left(\frac{|\vec{q}|}{|\vec{q}|_{Q=0}} \frac{Q_0'^2}{Q^2 + Q_0'^2}\right)^{2J-3} \left(\frac{Q_0^2}{Q^2 + Q_0^2}\right)^{m_\pm}, \tag{19}$$

$$\left|G_0(Q^2)\right|^2 = C^2 \left(\frac{Q^2}{Q^2 + Q_0''^2}\right)^{2a} \frac{q_0^2}{|\vec{q}|^2} \left(\frac{|\vec{q}|}{|\vec{q}|_{Q=0}} \frac{Q_0'^2}{Q^2 + Q_0'^2}\right)^{2J-1} \left(\frac{Q_0^2}{Q^2 + Q_0^2}\right)^{m_0} \tag{20}$$

for normal parity transitions and

$$\left|G_\pm(Q^2)\right|^2 = |G_\pm(0)|^2 \left(\frac{|\vec{q}|}{|\vec{q}|_{Q=0}} \frac{Q_0'^2}{Q^2 + Q_0'^2}\right)^{2J-1} \left(\frac{Q_0^2}{Q^2 + Q_0^2}\right)^{m_\pm}, \tag{21}$$

$$\left|G_0(Q^2)\right|^2 = C^2 \left(\frac{Q^2}{Q^2 + Q_0''^2}\right)^{2a} \frac{q_0^2}{|\vec{q}|^2} \left(\frac{|\vec{q}|}{|\vec{q}|_{Q=0}} \frac{Q_0'^2}{Q^2 + Q_0'^2}\right)^{2J+1} \left(\frac{Q_0^2}{Q^2 + Q_0^2}\right)^{m_0} \tag{22}$$

for abnormal parity transitions, where

$$|\vec{q}| = \frac{\sqrt{(M^2 - m^2 - Q^2)^2 + 4M^2Q^2}}{2M}, \quad |\vec{q}|_{Q=0} = \frac{M^2 - m^2}{2M} \tag{23}$$

and $m_+ = 3$, $m_- = 5$, $m_0 = 4$.

Form factors of transition $1/2^+ \to 1/2^+$ are written as:

$$\left|G_+(Q^2)\right|^2 = |G_+(0)|^2 \left(\frac{|\vec{q}|}{|\vec{q}|_{Q=0}} \frac{Q_0'^2}{Q^2 + Q_0'^2}\right)^2 \left(\frac{Q_0^2}{Q^2 + Q_0^2}\right)^{m_+}, \tag{24}$$

$$\left|G_0(Q^2)\right|^2 = C^2 \left(\frac{Q^2}{Q^2 + Q_0''^2}\right)^{2a} \frac{q_0^2}{|\vec{q}|^2} \left(\frac{|\vec{q}|}{|\vec{q}|_{Q=0}} \frac{Q_0'^2}{Q^2 + Q_0'^2}\right)^4 \left(\frac{Q_0^2}{Q^2 + Q_0^2}\right)^{m_0}. \tag{25}$$

In the expressions (19)-(22) the quantities Q_0^2, $Q_0'^2$, $Q_0''^2$, a and C are free parameters, which could be determined by fitting to experimental data, and they are the same for all resonances.

The values of form factors at $Q^2 = 0$ are related to helicity amplitudes of photoproduction $A_{1/2}$ and $A_{3/2}$, listed in [8], as follows [5]:

$$|G_{+,-}(0)| = e^{-1} \sqrt{\frac{M^2 - m^2}{m}} \left|A_{1/2,3/2}\right|, \tag{26}$$

where $e = \sqrt{4\pi/137}$ is electron charge. Notice, that longitudinal form factor at $Q^2 = 0$ turns to zero: $G_0(0) = 0$.

Substituting expression (5), written for each particular resonance, taking into account parity of the transition and using proper expressions for form factors, to (7), we get the structure function in the resonance region:

$$\begin{aligned} F_2(x, Q^2) = \frac{1}{\pi} \sum_R \frac{2m^2 x}{1 + 4m^2x^2/Q^2} \frac{M\Gamma}{(m^2 + Q^2(1/x - 1) - M^2)^2 + M^2\Gamma^2} \times \\ \times \left[\left(\frac{|\vec{q}|}{|\vec{q}|_{Q=0}} \frac{Q_0'^2}{Q^2 + Q_0'^2}\right)^{n_+} \times \right. \\ \times \left(|G_+(0)|^2 \left(\frac{Q_0^2}{Q^2 + Q_0^2}\right)^{m_+} + |G_-(0)|^2 \left(\frac{Q_0^2}{Q^2 + Q_0^2}\right)^{m_-}\right) + \\ \left. + 2C^2 \left(\frac{Q^2}{Q^2 + Q_0''^2}\right)^{2a} \frac{q_0^2}{|\vec{q}|^2} \left(\frac{|\vec{q}|}{|\vec{q}|_{Q=0}} \frac{Q_0'^2}{Q^2 + Q_0'^2}\right)^{n_0} \left(\frac{Q_0^2}{Q^2 + Q_0^2}\right)^{m_0}\right] \end{aligned} \tag{27}$$

where $n_+ = 2J - 3$, $n_0 = 2J - 1$ for normal parity transitions and $n_+ = 2J - 1$, $n_0 = 2J + 1$ for abnormal parity transitions, and the sum is over the resonances. We take into account the contributions of the following resonances: $N(1440)$, $N(1520)$, $N(1535)$, $N(1650)$, $N(1675)$, $N(1680)$, $N(1700)$, $N(1710)$, $N(1720)$, $N(1990)$, $\Delta(1232)$, $\Delta(1550)$, $\Delta(1600)$, $\Delta(1620)$, $\Delta(1700)$, $\Delta(1900)$, $\Delta(1905)$, $\Delta(1910)$, $\Delta(1920)$, $\Delta(1930)$, $\Delta(1950)$.

The above expressions determine the contribution of resonances into nucleon structure function F_2. It's obvious that the production of resonances in electron-nucleon scattering is not the only process, contributing to structure functions. The production of mesons and other hadrons forms non-resonant background, which also must be taken into account. As far as background is not studied thoroughly yet, we don't take it into account, rejecting various existing parametrisations.

It's shown in [6] how the expressions for structure function F_2 containing resonance terms, which have strong dependence on Q^2 (because of form factors), change over in the limit of high Q^2 to scaling expressions, that weakly depend on Q^2.

3. Results

The dependence of resonance part of the structure function F_2 on Nahtmann variable $\xi = 2x/(1 + \sqrt{1 + 4m^2x^2/Q^2})$ at various values of Q^2 is shown on fig. 1. Specific values of Q_0, $Q_0^{'}$ and $Q_0^{''}$, used to plot fig. 1, are

$$Q_0 = 1.6\,GeV;\; Q_0' = 0.8\,GeV;\; Q_0'' = 1.0\,GeV;\; a = 1/4;\; C = 0.2\,. \quad (28)$$

Solid line corresponds to the fit to deep inelastic data for $F_2(\xi)$ [13].

One can see, that $\Delta(1232)$-isobar gives the largest contribution to the structure function. Non-resonant background in this region appears to be relatively small, unlike the regions corresponding to more massive resonances. One can also see that the $\Delta(1232)$ peak exactly follows the scaling curve as Q^2 changes, which is a manifestation of the quark-hadron duality.

For the comparison with experimental data the parametrization of non-resonant background contribution must be carried out. However now this problem has no unique solution.

Taking the non-resonant background into account must lead to the rise of the structure function in the region of higher resonances the way it follows the scaling curve (in varying Q^2) in broad region of ξ.

Expressions for form factors of nucleon resonance production, obtained in this work, taking into account correct threshold behavior and experimental data on resonance decays, may serve as starting ones for the investigation of nucleon properties in the language of hadronic degrees of freedom.

Obtained expressions for structure functions allow to show qualitatively the manifestation of the quark-hadron duality in the behavior of structure function F_2.

Acknowledgements

I would like to thank Laszlo Jenkovszky for the support of my participation in "Diffracton 2002". *(V.V.D.)*

References

1. Bjorken, J. and Walecka, J. (1966) Electroproduction of nucleon resonances, *Ann. Phys.*, **38**, pp. 35-62.
2. Bloom, E. and Gilman, F. (1971) Scaling and the behavior of nucleon resonances in inelastic electron-nucleon scattering, *Phys. Rev.*, **D4**(9), pp. 2901-2916.
3. Brodsky, S. and Farrar, G. (1973) Scaling laws at large transverse momentum, *Phys. Rev. Lett.*, **31**, pp. 1153-1156.
4. Bugrij, A., Jenkovszky, L. and Kobylinsky, N. (1971) Dual models with mandelstam analyticity, *Nucl. Phys.*, **B35**, pp. 120-132.
5. Carlson, C. and Mukhopadhyay, N. (1998) Bloom-Gilman duality in the resonance spin structure functions, *Phys. Rev.*, **D58**, p. 094029.
6. Davidovsky, V. and Struminsky, B. (2003) Structure functions of nucleon, form factors of resonances and duality, *Phys. Atom. Nucl.*, **66**(5), to be published.
7. Dolen, R., Horn, D. and Schmid, C. (1968) Finite energy sum rules and their application to πN charge exchange, *Phys. Rev.*, **166**, pp. 1768-1781.
8. Particle Data Group (1988) *Phys. Lett.*, **B204**, p. 379.
9. Igi, K. and Matsuda, S. (1967) New sum rules and singularities in the complex J plane, *Phys. Rev. Lett.*, **18**, pp. 625-627.
10. Jenkovszky, L., Magas, V. and Predazzi, E. (2001) Resonance-reggeon and parton-hadron duality in strong interactions, *Eur. Phys. J.*, **A12**, pp. 361-367.
11. Logunov, A., Soloviev, L. and Tavkhelidze, A. (1967) Dispersion sum rules and high energy scattering, *Phys. Lett.*, **B24**, pp. 181-182.
12. Matveev, V., Muradian, R. and Tavkhelidze, A. (1973) Automodellism in the large - angle elastic scattering and structure of hadrons, *Lett. Nuovo Cimento*, **7**, pp. 719-723.
13. Niculescu, I., Armstrong, C., Arrington, J. and *et. al.* (2000) Experimental verification of the quark-hadron duality, *Phys. Rev. Lett.*, **85**(6), pp. 1186-1189.

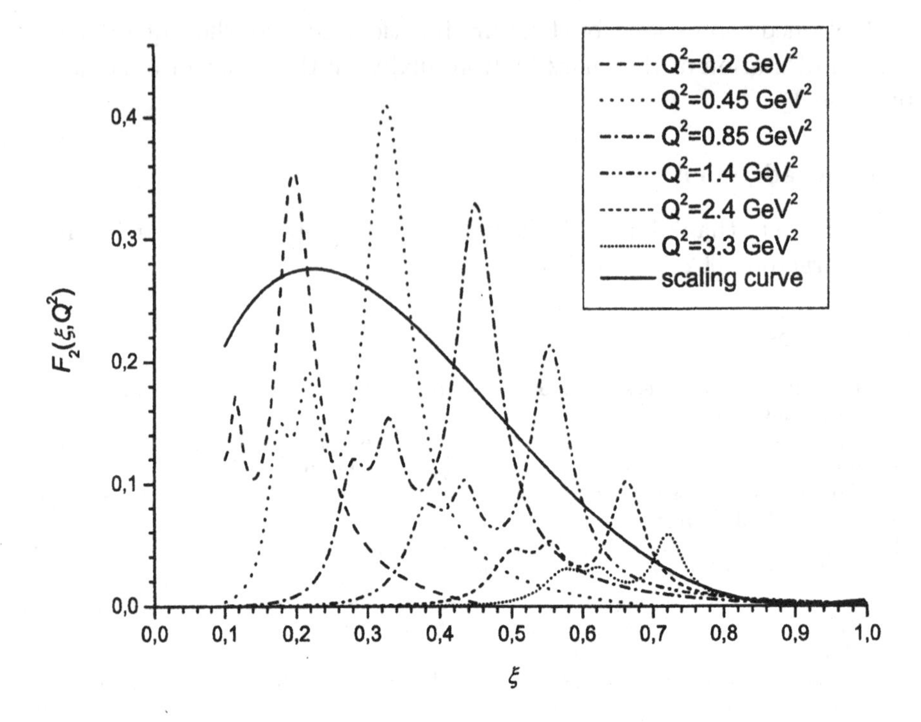

Figure 1. Resonance contribution to the structure function $F_2(\xi, Q^2)$ in resonance region.

QCD HYDRODYNAMICS FOR LHC AND RHIC

L.P. CSERNAI
Department of Physics, University of Bergen
Allegaten 55, N-5007 Bergen, Norway &
KFKI Research Institute for Particle and Nuclear Physics
P.O.Box 49, H-1525 Budapest 114, Hungary

M.I. GORENSTEIN
University of Frankfurt
Robert-Mayer St. 8-10, D-60054 Frankfurt am Main, Germany
& Bogolyubov Institute for Theoretical Physics
Metrologichna str. 14-b, 03143 Kiev, Ukraine

V.K. MAGAS
CFIF, Physics Department, Instituto Superior Tecnico
Av. Rovisco Pais, 1049-001 Lisbon, Portugal

AND

D.D. STROTTMAN
Theory Division, Los Alamos National Laboratory,
Los Alamos, NM 87545, USA

Abstract. The realistic and detailed description of an energetic heavy ion reaction requires a Multi Module Model, where the different stages of the reaction are each described with a suitable theoretical approach. One fluid dynamical models provide an adequate and accurate description of the middle stages of the reaction. In addition, fluid dynamical calculations require initial and freeze out conditions. In this work we concentrate on the modeling of the initial stages of the reaction, before the local thermal equilibrium is achieved, and on the freeze out process. We discuss the possibility of the fast simultaneous hadronization and chemical freeze out of supercooled QGP, as a possible solution of the HBT "puzzle".

Fluid dynamical models are widely used to describe heavy ion collisions. Their advantage is that one can vary flexibly the Equation of State (EoS) of the matter and test its consequences on the reaction dynamics and the

R. Fiore et al. (eds.), Diffraction 2002, 143–154.

outcome. This makes fluid dynamical (FD) models a very powerful tool to study possible phase transitions in heavy ion collisions – such as the nuclear matter liquid-gas phase transition in medium energy heavy ion collisions or the Quark-Gluon Plasma (QGP) formation in high energy collisions. For example, the only models that can handle the supercooled QGP (sQGP) are hydrodynamic models with a corresponding EoS.

For highest energies achieved nowadays at RHIC hydrodynamic models give a good description of the observed radial and elliptic flows [1, 2, 3, 4], in contrast to microscopic models, like HIJING [5] and UrQMD [6]. We would like to stress here that "hydrodynamic model" means, apart of well known hydrodynamic equations, the particular choice of the initial conditions (usually fitted to reproduce the data), the particular choice of the EoS and the way freeze out (FO) is realized. In this work we are going to concentrate on the modeling of initial conditions for RHIC and LHC energies and on the FO.

Recent data from RHIC give a good test for all the models in ultra-relativistic heavy ion collisions business, namely the HBT radii. Two-particle interferometry has become a powerful tool for studying the size and duration of particle production from elementary collisions to heavy ions like Au+Au at RHIC or Pb+Pb at SPS. For the case of nuclear collisions, the interest mainly focuses on the possible transient formation of a deconfined state of matter. This could affect the size of the region from where the hadrons (mostly pions) are emitted as well as the time for particle production. Comparing recent data [7] from RHIC with SPS data one finds a "puzzle" [8]: all the HBT radii are pretty similar although the center of mass energy is changed by an order of magnitude. Discussions at "Quark Matter 2002" [9] lead to the conclusion that the duration of particle emission, as well as the lifetime of the system before Freeze Out, appear to be shorter than the predictions of most models at the physics market. Although, there is a set of models based on very fast 3D Bjorken hydrodynamic expansion and hadronization from sQGP [10], which are in much better agreement with experimental data than all the others. We will discuss this possibility in the section 2.

1. Multi Module Model

A realistic and detailed description of an energetic heavy ion reaction requires a Multi Module Model, where the different stages of the reaction are each described with a suitable theoretical approach [30, 31, 32, 33]. It is important that these modules are coupled to each other correctly: on the interface, all conservation laws should be satisfied, and the entropy should not decrease. This last feature sometimes not given proper attention.

The nondecreasing entropy condition should be satisfied at each "stage" or "Module" of our collision model, as well as, at the transition between stages or modules.

In energetic collisions of large heavy ions, especially if a QGP is formed in the collision, one-fluid dynamics is a valid and good description for the intermediate stages of the reaction. Here, interactions are strong and frequent, so that other models, (e.g. transport models, string models, etc., that assume binary collisions, with free propagation of constituents between collisions) have limited validity. On the other hand, the initial and final, FO, stages of reaction are outside the domain of the applicability of the fluid dynamical model.

After hadronization and FO, the matter is already dilute and can be described well with kinetic models. Note that even after the chemical FO the hadron species are not completely frozen – elastic particle collisions and resonance decays have to be taken into account.

The modeling of the initial stages is the most problematic. None of the theoretical appraches can unambiguosly describe the initial stages. Therefore the phenomenological models have to be used.

For the ultra-relativistic energies the boost invariant Bjorken model is frequently used (it requires only 2D hydro calculations in transverse plane), with the initial conditions fitted to the data. Usually the energy density profile in the plane transverse to beam direction is assumed to be proportional to the number of binary collisions or to number of participants, and the central energy is used as a free parameter to fit the data (see [11] for recent overview). The preliminary experimental results from RHIC show for most particle rapidity spectra the plateau-like behavior around mid-rapidity [12], and a strong elliptic flow (v_2 flow component) with a clear peak around mid rapidity [13]. To build such a strong elliptic flow, strong stopping and momentum equilibration are required. Also the $\bar{p}/p$ ratio at mid-rapidity measured at RHIC [14] (preliminary) is still far from one, which tells us that the middle region is not baryonfree.

In Refs. [4, 15] the two module model has been proposed. Authors combine hydro and UrQMD – replacing the hadronic phase of hydrodynamics with hadronic transport model to describe properly chemical and thermal FO. We would like to underline the following: A) the initial state for hydro was not modeled, but fitted to the data; B) in this approach at some "point", i.e. 3D hypersurface in the space-time, we have to transfer our global hydro quantity into a distribution of different hadron species. This problem B) is absolutely similar to the FO problem. One has to specify the condition for this transfer and define the corresponding hypersurface. Again this transfer hypersurface may have both timelike and space-like parts, where the modification of the Cooper-Frye formula is needed

[16, 17, 18, 19]. The nondecreasing entropy condition should be satisfied. In our opinion the realization of hydro-UrQMD transfer is even a more difficult and less certain problem than FO modeling. Noninteracting hadrons can be used as a good and valid approximation for the post FO stage whereas the hydro-UrQMD transfer happens at much higher energy densities where the properties of interacting hadron system are not well known. For the recent treatment of this problem see Ref. [20].

Perturbative Quantum Chromo Dynamics (pQCD) would be, in principle, the proper model for describing our systems at very high energies. Unfortunately, pQCD itself is not applicable for heavy ion reactions at RHIC energies. Nevertheless, the pQCD calculations with some extra non-perturbative, phenomenological assumptions, like saturation of a gluon plasma, can be performed. Different models following this scenario have been proposed - see for example [21, 22]. In Ref. [22] authors combine Gyulassy-Levai-Vitev (GLV) non-abelian energy loss formalism with Bjorken 1+1D expansion, and successfully reproduced $v_2(p_T)$ STAR Minimum Bias Data up to $p_T \approx 2 - 2.5\ GeV$.

1.1. EFFECTIVE STRING ROPE MODEL

One important conclusion of heavy ion research in the last decade is that standard 'hadronic' string models fail to describe heavy ion experiments. All string models had to introduce new, energetic objects: string ropes [23, 24], quark clusters [25], or fused strings [26], in order to describe the abundant formation of massive particles like strange antibaryons. Based on this, we describe the initial moments of the reaction in the framework of classical, or coherent, Yang-Mills theory, following the Bjorken model with baryon recoil [27] and assuming a larger field strength (string tension) than in ordinary hadron-hadron collisions. For example, calculations both in the Quark Gluon String Model [28, 29] and in the Monte Carlo string fusion model [26] indicate that the energy density of strings reaches $8 - 10\ GeV/fm$ already in SPS reactions. This is nearly 10 times more than the tension used in standard, 'hadronic', string models where $\sigma \approx 1\ GeV/fm$, what allows us to talk about string ropes. We also take care about exact satisfaction of all the conservation laws. Thus, in the framework of Effective String Rope Model (ESRM) for the first time the initial transparency/stopping and energy deposited into strings and string ropes was determined consistently with each other [30, 31].

We describe the hadronization of our string ropes in an effective way (see Refs. [30, 31] for more details). Thus, as the output of ESRM we get energy and baryon density distributions and flow velocity field of the locally thermalized QGP. Then, the hydrodynamic evolution can be started for the

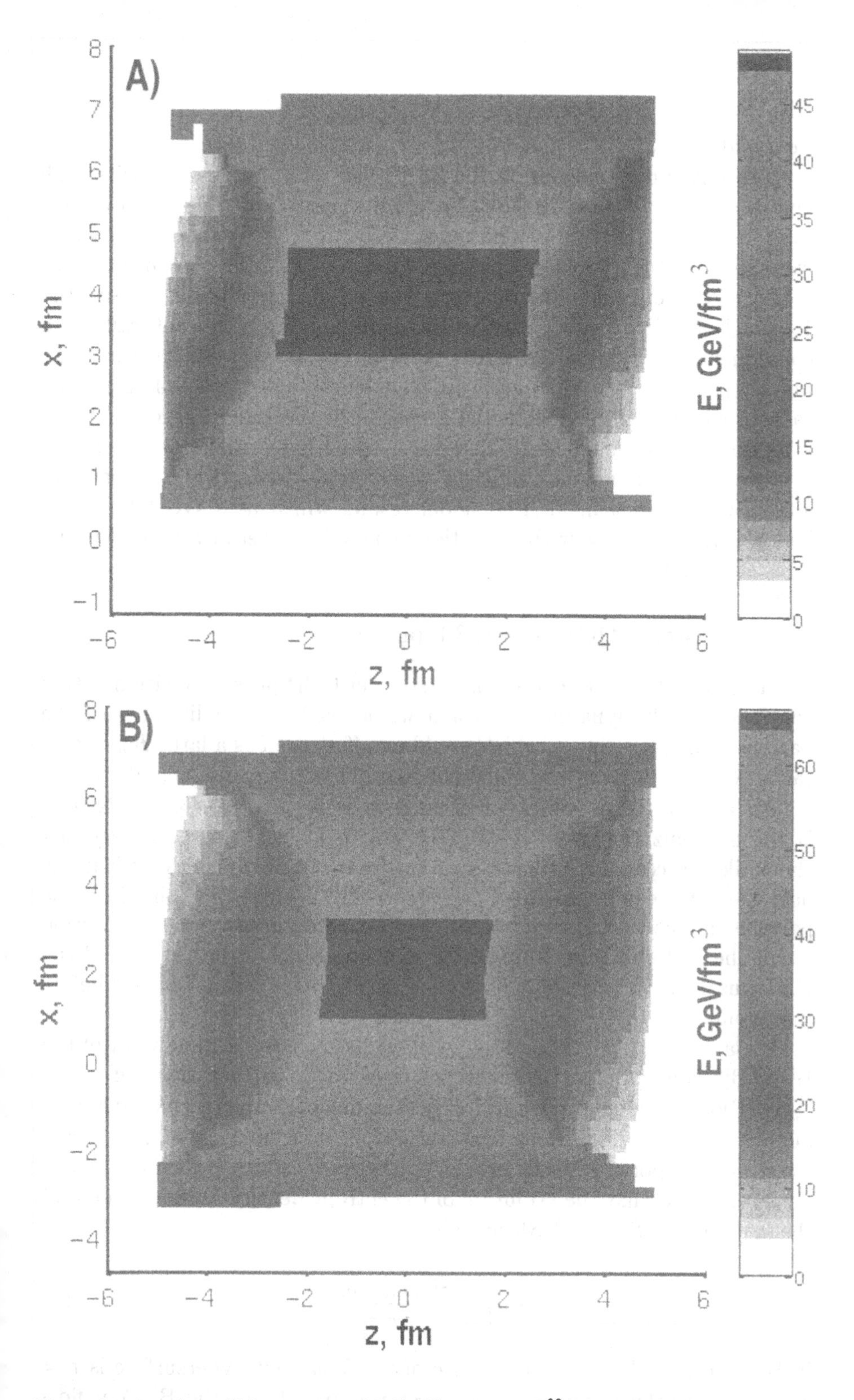

Figure 1. Au+Au collision at $\varepsilon_0 = 100\ GeV/nucl$, $E = T^{00}$ is presented in the reaction plane as a function of x and z for $t_h = 5\ fm/c$. Subplot A) ($b = 0.5 \cdot 2\ R_{Au}$), subplot B) ($b = 0.25 \cdot 2\ R_{Au}$). The QGP volume has a shape of a tilted disk and may produce a third flow component.

intermediate state.

The initial dynamics using the ESRM led to an initial state with hi energy density, reaching 20 GeV/fm^3 on the average and $60-80$ GeV/fn peak values in the middle. The model led to a QGP configuration whi was similar to the initial state described in Landau's fluid dynamical mode with the difference that the relatively flat initial state was tilted – see Fi 1. In a heavy ion reaction such a tilted initial state arises naturally as consequence of the streak by streak momentum conservation, since at fini impact parameters streaks with different length and mass collide again each other, finally moving in the direction of the heavier streak (this a base of firestreak model [37]). Such a tilted flat initial state results a "third flow component" which is observed at the highest SPS energi and it was present in all FD model results which used QGP EoS [3 Unfortunately the flow within reaction plane was not yet measured at RH – see section 3.

2. Supercooled QGP and HBT puzzle

It was demonstrated that a strong first-order QCD phase transition with continuous hydrodynamic expansion would lead to long lifetimes of t particle source [38, 39, 4], which would manifest itself as a large R_{out}/R_{si} ratio. Now this type of hadronization is excluded by experimental data.

An alternative possibility, discussed in Refs. [41, 42, 43, 34, 44, 4 is the hadronization from the sQGP. This is expected to be a very fa shock-like process. If the hadronization from sQGP coincides with free out, then this could explain a part of the HBT puzzle, i.e. the flash-li particle emission ($R_{out}/R_{side} \approx 1$). Now we would like to ask the questic – can the hadronization from sQGP explain also the another part of t HBT puzzle, i.e. a very short ($\sim 6[10] - 10[9]$ fm/c) expansion time befo freeze out?

For a study of the expanding QGP we have chosen a framework of t 1+1D Bjorken model (actually our principal results will not change if we u 3+1D Bjorken model). Within the Bjorken model all the thermodynamic quantities are constant along constant proper time curves, $\tau = \sqrt{t^2 - z^2}$ *const*. The important result of Bjorken hydrodynamics (which assumes perfect fluid) is that the evolution of the entropy density, is independent the Equation of State (EoS), namely

$$s(\tau) = \frac{s(\tau_{init})\tau_{init}}{\tau} . \quad ($$

In Bjorken model the natural choice of the freeze out hypersurface is τ *const* hypersurface, where normal vector is parallel to the Bjorken flo

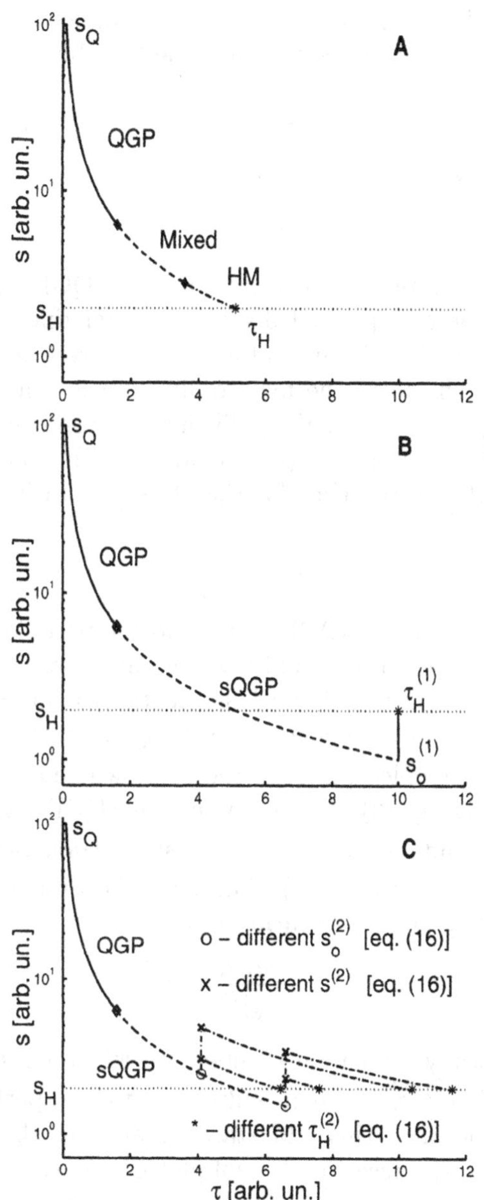

Figure 2. Different ways for a system to go from Q state (s_Q) to H state (s_H) are presented on $\{s, \tau\}$ plane. Subplot A shows continuous expansion, which takes time τ_H, eq. (4). Subplot B presents flash-like particle emission, i.e. simultaneous hadronization and freeze out; which takes time $\tau_H^{(1)}$, eq. (5). Subplot C shows several possibilities according to scenario 2 with shock-like hadronization into superheated HM. Time $\tau_H^{(2)}$ (7) can be smaller or larger than $\tau_H^{(1)}$, depending on details of the EoS, but always larger than τ_H.

velocity, $v = z/t$. Thus, $d\sigma^\nu = (1, 0)$ in the rest frames of each fluid element. This leads to the simple solution of the time like shock equations (see Ref. [45]):

$$\tilde{v}^2 = \tilde{v}_o^2 = 0\ , \quad \epsilon = \epsilon_o,\ n = n_o,\ p \neq p_o\ . \tag{2}$$

The entropy condition is reduced to

$$s \geq s_o\ . \tag{3}$$

Now let us try to answer the question – can QGP expansion with t.l. shock hadronization of supercooled state be faster than the hadronization through the mixed phase? The initial state is given at the proper time $\tau_{init} \equiv \tau_Q$, when the local thermal equilibrium is achieved in the QGP state $Q \equiv (\epsilon_Q, p_Q, s_Q)$. The final equilibrium hadron state is also fixed, by experiment or otherwise, as $H \equiv (\epsilon_H, p_H, s_H)$. For the continuous expansion given by eq. (1) the proper time for the $Q \rightarrow H$ transition is (see Fig. 2 - subplot A):

$$\tau_H \ = \ \frac{s_Q \tau_Q}{s_H}\ . \tag{4}$$

If our system enters the sQGP phase and the particle emission is flash-like, i.e. the system hadronizes and freezes out at the same time, then eq. (1) is also valid all the time with final t.l. shock transition to the same H state. We call this as a scenario number one (see Fig. 2 - subplot A). Our system should go into supercooled phase to the point where $\epsilon_o^{(1)} = \epsilon_H, n_o^{(1)} = n_H$, as it is required by eq. (2). At this point our sQGP has entropy density $s_o^{(1)}$. It's value depend on the EoS, but t.l. shock transition is only possible if $s_o^{(1)} \leq s_H$ according to eq. (3). Thus, for the proper time of $Q \rightarrow H$ transition according to first scenario we have:

$$\tau_H^{(1)} \ = \ \frac{s_Q\ \tau_Q}{s_o^{(1)}} \ \geq \ \tau_H\ . \tag{5}$$

We can also study a scenario number two when our system supercools to the state $(\epsilon_o^{(2)}, p_o^{(2)}, s_o^{(2)})$, then hadronizes to a superheated HM state $(\epsilon^{(2)}, p^{(2)}, s^{(2)})$, and then this HM state expands to the same freeze out state $H \equiv (\epsilon_H, p_H, s_H)$ (see Fig. 2 - subplot C). At the point of the shock transition one has:

$$\tau_o^{(2)} \ = \ \frac{s_Q\ \tau_Q}{s_o^{(2)}}\ , \tag{6}$$

Then we have a t.l. shock transition satisfying eq. (2), and following the HM branch of the hydrodynamic expansion we find:

$$\tau_H^{(2)} \ = \ \frac{s^{(2)}\cdot \tau_o^{(2)}}{s_H} \ = \ \frac{s_Q \tau_Q}{s_H}\ \frac{s^{(2)}}{s_o^{(2)}} \ \geq \ \tau_H\ , \tag{7}$$

since $s^{(2)} \geq s_o^{(2)}$ due to non-decreasing entropy condition (3). In this second scenario the value of the entropy density $s_o^{(2)}$ of sQGP can be both smaller and larger than HM final value s_H. Depending on details of the EoS the proper time $\tau_H^{(2)}$ (7) of the $Q \rightarrow H$ transition can also be smaller as well as larger than $\tau_H^{(1)}$ (5), but always larger than τ_H.

One can assume that for late stages of the reaction Bjorken assumptions become unapplicable and therefore we have more freedom in, for example, choosing FO hupersurface. This possibility was discussed in Ref. [34], where the FO and hadronization of sQGP happened at a constant time (Lab frame time) hupersurface after Bjorken hydro expansion. Our results show that such a process is very sensitive on the properties of the EoS (contrary to the one dicussed above) and the nondecreasing entropy conditions works very restrictively. Also, fast simultaneous hadronization and chemical freeze out of sQGP leads to re-heating of the expanding matter and to a change in a collective flow profile.

The conclusion of the above analysis seems to be a rather general one: the system's evolution through a supercooled phase and time-like shock hadronization can not be shorter than a continuous expansion within the perfect fluid hydrodynamics independently of the details of EoS and the parameter values of the initial, Q, and final, H, states. Although in such a way we may achieve flash-like particle emission, supported by the HBT data, the expansion time becomes longer, making it harder to reproduce the experimental HBT radii.

3. Determining the reaction plane at RHIC

At RHIC the directed transverse flow was not measured up to now. Different experimental difficulties were mentioned as the reason. In particular that the reaction plane could not be identified, although the so called v_2 parameter was measured.

This problem is partly due to the method used to determine v_2, which did not explicitly required the parallel determination of the reaction plane. It is, however, clear that the existence of strong elliptic flow indicates a strong collective correlation among the particles in every event: particles are preferably emitted upwards (+x) and downwards (-x) in the reaction plane and much less to the sides. Thus, if this correlation is detected and strong, the plane must be identifiable also! It is relatively easy to design methods, which work well statistically for very low event multiplicities.

Of course, to tell which is the target and projectile side of the plane may be difficult. If in a symmetric, A+A reaction one assumes, the idealized Bjorken picture with infinite length in the [z]-direction, and the lens shape source in the transverse plane, the target and projectile side cannot be distinguished in principle.

On the other hand, if one takes a more realistic three dimensional model, one will get both azimuthal asymmetry and forward/backward asymmetry. This can be used then to determine which side of the plane is which.

As v_2 is measured, the reaction plane, $[x, z]$, is known, just the target/projectile side should be selected. This is not done due to the prejudice that the distribution of emitted particles is mirror-symmetric in Central Mass Frame (CM): $f_{CM}(p_x, p_y, p_z) = f_{CM}(p_x, p_y, -p_z)$. However, at finite impact parameters $(2 - 15\%)$ there is a forward/backward asymmetry, which we can utilize the following way to determine the reaction plane.

The Q-vector (a la Danielewicz, and Odyniecz (1985), see in section 7.7 of ref. [35]) $Q_k = \sum_{ik} y_{CM,i} p_{x,i}$ has to be calculated, event by event, for all particles, i, of type k. Only the sign is relevant, as the plane is known already. This Q-vector will select the same side (e.g. projectile) in each event.

4. Conclusions

In conclusion we can say that the hydrodynamic modeling of heavy ion reactions is alive and is better than ever. Clear hydrodynamic effects are seen everywhere, and from early on.

The recent results from RHIC show that we are in the regime where collective matter type of behavior is dominant. We hope to gain more and more detailed information on QGP and its dynamical properties. Continued hard work is needed to exploit all possibilities, and the task of theoretical modeling and analysis is vital in future progress of the field.

Acknowledgment. Three of us (M.G., V.M. and D.S.) acknowledge the support of the Bergen Computational Physics Laboratory in the framework of the European Community - Access to Research Infrastructure action of the Improving Human Potential Programme.

References

1. Kolb, P.F., Sollfrank, J., Heinz, U. (2000) Anisotropic transverse flow and the quark-hadron phase transition, *Phys. Rev.* **C62**, 054909.
2. Kolb, P.F., Huovinen, P., Heinz, U., Heiselberg, H. (2001) Elliptic flow at SPS and RHIC: from kinetic transport to hydrodynamics, *Phys. Lett* **B500** pp. 232-240.
3. Huovinen, P., Kolb, P.F., Heinz, U., Heiselberg, H. (2001) Radial and elliptic flow at RHIC: further predictions, *Phys. Lett* **B503** pp. 58-64.
4. Teaney, V, Lauret, V, and Shuryak, E.V. (2001) Flow at the SPS and RHIC as a Quark Gluon Plasma Signature, *Phys. Rev. Lett.* **86** pp. 4783-4786; A Hydrodynamic Description of Heavy Ion Collisions at the SPS and RHIC, nucl-th/0110037.

5. Molnar, D. and Gyulassy, M. (2002) Saturation of Elliptic Flow and the Transport Opacity of the Gluon Plasma at RHIC, *Nucl. Phys.* **A697** pp. 495-520.
6. Bleicher, M. and Stöcker, H. (2002) Anisotropic flow in ultra-relativistic heavy ion collisions *Phys. Lett.* **B526** pp. 309-314.
7. RHIC data: STAR Collaboration (2001) Pion Interferometry of $\sqrt{s_{NN}} = 130$ GeV Au+Au Collisions at RHIC, *Phys. Rev. Lett.* **87** 082301; PHENIX Collaboration (2002) Transverse mass dependence of two-pion correlations in Au+Au collisions at $\sqrt{s_{NN}} = 130$ GeV, *Phys. Rev. Lett.* **88** 192302;
SPS data: NA49 Collaboration (2002) New Results From Na49, *Nucl. Phys.* **A698** pp. 104-; Appelshäuser, H., Pion interferometry: Recent results from SPS, hep-ph/0204159.
8. Gyulassy, M., Lectures on the theory of high energy A + A at RHIC, nucl-th/0106072.
9. Pratt, S., talk at the Quark Matter 2002, July, Nantes, France.
10. Csörgő, T., Ster, A., The reconstructed final state of Au+Au collisions from PHENIX and STAR data at $\sqrt{s} = 130$ AGeV - indication for quark deconfinement at RHIC, nucl-th/0207016.
11. Kolb, P.F., Heinz, U., Huovinen, P., Eskola, K.J., Tuominen, K. (2001) Centrality dependence of multiplicity, transverse energy, and elliptic flow from hydrodynamics, *Nucl. Phys.* **A696** pp. 197-215.
12. STAR Collaboration (2001) Multiplicity distribution and spectra of negatively charged hadrons in Au+Au collisions at $\sqrt{s_{NN}} = 130$ GeV, *Phys. Rev. Lett.* **87** 112303.
13. PHENIX Collaboration (2002) Elliptic Flow Measurements with the PHENIX Detector, *Nucl. Phys.* **A698** pp. 559-563; PHOBOS Collaboration (2002) Charged particle flow measurement for $|\eta| < 5.3$ with the PHOBOS detector, *Nucl. Phys.* **A698** pp. 564-567; STAR Collaboration (2002) Elliptic flow in Au+Au collisions at $\sqrt{s_{NN}} = 130$ GeV, *Nucl. Phys.* **A698** pp. 193-198.
14. STAR Collaboration (2001) Mid-rapidity anti-proton to proton ratio from Au+Au collisions at $\sqrt{s_{NN}} = 130$ GeV, *Phys. Rev. Lett.* **86** pp. 4778-4782.
15. Bass, S., Dumitru, A. (2000) Dynamics of Hot Bulk QCD Matter: from the Quark-Gluon Plasma to Hadronic Freeze-Out, *Phys. Rev.* **C61** 064909.
16. Bugaev, K.A. (1996) Shock-like Freezeout in Relativistic Hydrodynamics, *Nucl. Phys.* **A606** pp. 559-567.
17. Anderlik, Cs. et al. (1999) Nonideal particle distributions from kinetic freeze-out models, *Phys. Rev.* **C59** pp. 388-394,.
18. Anderlik, Cs. et al. (1999) Freeze out in hydrodynamic models, *Phys. Rev.* **C59** pp. 3309-3316 ; (1999) Large p_t enhancement from freeze out, *Phys. Lett.* **B459** pp. 33-36.
19. Magas, V.K. et al. (1999) Kinetic freeze out models , *Heavy Ion Phys.* **9** pp. 193-216 ; (1999) Freeze-out in hydrodynamical models in relativistic heavy ion collisions, *Nucl. Phys.* **A661** pp. 596-599 .
20. Bugaev, K.A. (2002) Relativistic Kinetic Equations for Finite Domains and Freeze-Out Problem, nucl-th/0210087.
21. Eskola, K.J., Ruuskanen, P.V., Rasanen, S.S., Tuominen, K. (2001) Multiplicities and Transverse Energies in Central AA Collisions at RHIC and LHC from pQCD, Saturation and Hydrodynamics, *Nucl. Phys.* **A696** pp. 715-728.
22. Gyulassy, M., Vitev, I., Wang, X.-N. (2001) High p_T Azimuthal Asymmetry in Non-central A+A at RHIC, *Phys. Rev. Lett.* **86** pp. 2537-2540.
23. Biró, T.S., Nielsen, H.B., Knoll, J. (1984) Color rope model for extreme relativistic heavy ion collisions, *Nucl. Phys.* **B245** pp. 449-468.
24. Sorge, H. (1995) Flavor Production in Pb(160AGeV) on Pb Collisions: Effect of Color Ropes and Hadronic Rescattering, *Phys. Rev.* **C52** pp. 3291-3314.
25. Werner, K., Aichelin, J. (1996) Linking Dynamical and Thermal Models of Ultra-

relativistic Nuclear Scattering, *Phys. Rev. Lett.* **76** pp. 1027-1030.

26. Amelin, N.S., Braun, M.A., Pajares, C. (1993) Multiple Production In The Monte Carlo String Fusion Model, *Phys. Lett.* **B306** pp. 312-318 ; (1994) String Fusion And Particle Production At High-Energies: Monte Carlo String Fusion Model, *Z. Phys.* **C63** pp. 507-516.
27. Gyulassy, M., Csernai, L.P. (1986) Baryon Recoil And The Fragmentation Regions In Ultrarelativistic Nuclear Collisions, *Nucl. Phys.* **A460** pp. 723-754.
28. Amelin, N.S. et al.(1991) Collectivity, Energy Density And Baryon Density In Pb On Pb Collisions, *Phys. Lett.* **B261** pp. 352-356; (1991) Transverse Flow And Collectivity In Ultrarelativistic Heavy Ion Collisions, *Phys. Rev. Lett.* **67** pp. 1523-1526.
29. Amelin, N.S., Csernai, L.P., Staubo, E.F., and Strottman, D. (1992) Collectivity in ultrarelativistic heavy ion collisions, *Nucl. Phys.* **A544** pp. 463-466.
30. Magas, V.K., Csernai, L.P., Strottman, D. (2001) The initial state of ultrarelativistic heavy ion collision, *Phys. Rev.* **C64** 014901.
31. Magas, V.K., Csernai, L.P., Strottman, D. (2002) Effective String Rope Model for the initial stages of Ultra-Relativistic Heavy Ion Collisions, *Nucl. Phys.* **A712** pp.167-204.
32. Magas, V.K., Csernai, L.P., Strottman, D., The source of the Elliptic Flow and Initial Conditions for Hydrodynamical Calculations, nucl-th/0009049; The source of the "third flow component", hep-ph/0101125; Multi Module Model for Relativistic Heavy Ion Collisions, hep-ph/0110347; Initial conditions for RHIC collisions, hep-ph/0202118.
33. Csernai, L.P. , Anderlik, Cs., Magas, V.K., Phase transitions in high energy heavy ion collisions within fluid dynamics, nucl-th/0010023.
34. Keränen, A., Manninen, J., Csernai, L.P. and Magas, V.K., Statistical Hadronization of Supercooled Quark-Gluon Plasma, nucl-th/0205019.
35. Csernai, L.P. (1994) *Introduction to Relativistic Heavy Ion Collisions*, Willey.
36. Csernai, L.P., Röhrich, D. (1999) Third flow component as QGP signal, *Phys. Lett.* **B458** pp.454-465.
37. Myers, D. (1978) A Model For High-Energy Heavy Ion Collisions, *Nucl. Phys.* **A296** pp. 177-188; Gosset, J., Kapusta, J.I., Westfall, G.D. (1978) Calculations With The Nuclear Firestreak Model, *Phys. Rev.* **C18** pp. 844-855.
38. Pratt, S. (1984) Pion Interferometry For Exploding Sources, *Phys. Rev. Lett.* **53** pp. 1219-1221; (1986) Pion Interferometry Of Quark - Gluon Plasma, *Phys. Rev.* **D33** pp. 1314-1327; Bertsch, G., Gong, M., Tohyama, M. (1988) Pion Interferometry In Ultrarelativistic Heavy Ion Collisions, *Phys. Rev.* **C37** pp. 1896-1900; Bertsch, G. (1989) ion Interferometry As A Probe Of The Plasma, *Nucl. Phys.* **A498** pp. 173-180.
39. Rischke, D., Gyulassy, M. (1996) The Time-Delay Signature of Quark-Gluon-Plasma Formation in Relativistic Nuclear Collisions, *Nucl. Phys.* **A608** pp. 479-512.
40. Csernai, L.P., Kapusta, J.I. (1992) Dynamics Of The QCD Phase Transition, *Phys. Rev. Lett.* **69** pp. 737-740; (1992) Nucleation of relativistic first order phase transitions, *Phys. Rev.* **D46** pp. 1379-1390.
41. Csörgő, T., Csernai, L.P. (1994) Quark-Gluon Plasma Freeze-Out from a Supercooled State?, *Phys. Lett.* **B333** pp. 494-499.
42. Csernai, L.P., Mishustin, I.N. (1995) Fast Hadronization Of Supercooled Quark - Gluon Plasma, *Phys. Rev. Lett.* **74** pp. 5005-5008.
43. Gorenstein, M.I., Miller, H.G., Quick, R.M., Ritchie, R.A. (1994) Time-Like Shock Hadronization of a Supercooled Quark-Gluon Plasma, *Phys. Lett.* **B340** pp. 109-114.
44. Jenkovszky, L.L., Kampfer, B., Sysoev, V.M. (1990) On The Expansion Of The Universe During The Confinement Transition, *Z. Phys.* **C48** pp. 147-150.
45. Csernai, L.P., Gorenstein, M.I., Jenkovszky, L.L., Lovas, I., Magas, V.K., Can supercooling explain the HBT puzzle?, hep-ph/0210297.

NUCLEAR COLLISION DESCRIPTION IN TERMS OF THE GINZBURG-LANDAU MODEL

L. F. BABICHEV[1], V. I. KUVSHINOV[2] AND V. A. SHAPARAU[3]
B.I. Stepanov Institute of Physics
National Academy of Sciences of Belarus
F. Skaryna avenue 68, 220072 Minsk, Belarus
E-mail:kuvshino@dragon.bas-net.by

AND

A. A. BUKACH[4]
Belarusian State University,
F. Skaryna avenue 4, 220050 Minsk, Belarus

Abstract. We use the generalized squeezed state description in the framework of the Ginzburg-Landau theory. Multiplicity distributions of the squeezed states are studied at second-order phase transition for different squeeze factors. It is shown that the normalized factorial moments exhibit a specific behaviour as functions of the resolution scale. We obtain the values of the scaling exponent which coincides with experimental data at small squeeze factor.

1. Introduction

With increasing of the collision energy in $e^+e^-, p\bar{p}, ep$, heavy-ion experiments the role of the multiparticle production and the collective effects of particle interaction becomes more significant. Large progress has been achieved in study of the considered processes within perturbative Quantum Chromodynamics (QCD) [1]-[4]. Perturbative theory is not able to reproduce all consequences inherent to corresponding Lagrangian of the interaction, in particular, the collective aspects of the behaviour of considered systems as a whole can have fundamental significance for description of the features of the confinement and hadronization. Here methods of the statistical physics are fruitful since a mathematical apparatus used at investigation of the multiplicity distribution and particle correlations is common

R. Fiore et al. (eds.), Diffraction 2002, 155–164.

both for the statistical physics and for the processes of the multiparticle production [5].

For statistical systems the fluctuations are large near critical points. Therefore the multiplicity fluctuations of hadrons produced in high-energy heavy-ion collisions can be used as a measure of whether a quark-gluon system has undergone a phase transition [6]. Until today the question concerning order of the parton-hadron PT in high-energy collisions is opened. Lattice gauge calculations indicate that for two flavors the PT is most likely of the second order [7]. When strange quarks are included, it may become a weak first-order PT.

At present accelerator energies the average number of produced particles is large enough and therefore the scattering operator and other operators are inconvenient to consider in terms of number states. At the same time in such systems as lasers where the average number of photons is large it is conveniently to use the coherent state representation (P-representation) [8, 9]. There are a number of efforts to apply P-representation for investigation of the multiplicity fluctuations as a phenomenological manifestation of quark-hadron PT in the framework of the Ginzburg-Landau (GL) formalism both for the second-order [10, 11] and first-order [12]-[14] PT. In both cases it was supposed that multiplicity distribution of the hadrons without PT is a Poissonian and scaling behaviour of the factorial moments has been found. Moreover the scaling exponent ν is equal 1.305 for second-order PT [11] and $1.32 < \nu < 1.33$ for the generalized GL model with first-order PT [14]. There was the disagreement between experiment and theory. Indeed, NA22 data on particle production in hadronic collisions give $\nu = 1.45 \pm 0.04$ [15], heavy-ion experiments $\nu = 1.55 \pm 0.12$ [10] and $\nu = 1.459 \pm 0.021$ [16].

Idea of applying stochastic methods developed for studying photon-counting statistics of light beams to particle production processes was used for explanation of the experimentally observed properties hadrons, in particular, the multiplicity distributions, factorial and cumulant moments. It was shown that the most general distribution that characterizes $e^+e^-, p\bar{p}$, neutrino-induced collisions is a k-mode squeezed state distribution [17, 18]. Squeezed states (SS) involve coherent states as the specific case and posses uncommon properties: they display a specific behaviour of the factorial and cumulant moments [19] and can have both sub-Poissonian and super-Poissonian statistics corresponding to antibunching and bunching of photons [8], [20]-[22]. Although the SS are constructed in quantum optics (QO) their relevance to hadron production in high-energy collisions was recognized long ago [23]. In particular, the multiplicity distribution of the pions has been explained by formalism of the squeezed isospin states [24]. In addition, study of the evolution of gluon states at the non-perturbative stage of jet development has obtained the new squeezed gluon states [25]-[28] which could be necessary element of hadronization and, in particular, QGP$\rightarrow$hadrons. Then using the Local parton hadron duality it is easy to show that in this case behaviour of hadron multiplicity distribution in jet events is differentiated from the negative binomial one. Such specific behaviour of the multiplicity distribution is confirmed by experiments for $pp, p\bar{p}$-collisions [29, 30].

Considering multiplicity distribution in different collisions without PT as squeezed one we generalize the coherent state representation taking into account the squeezed states within GL-approach for investigation of the multiplicity fluctuations at phase transition in QGP. Since the multiplicity fluctuations exhibit intermittency behaviour which is observed in a large number of experiments, we investigate conditions of appearance of this effect in depending on the parameters of GL model.

2. Multiplicity distribution of the squeezed states at second-order phase transition

It is convenient to start description of the squeezed-state formalism in GL theory with definition of the photon SS. Two basic kinds of single-mode ideal SS are used in QO: coherent squeezed state (CSS) and scaling SS (SSS) [20] defined as

$$\left.\begin{aligned} |\psi,\eta\rangle &= D(\psi)S(\eta)|0\rangle \qquad (CSS), \\ |\psi,\eta\rangle &= S(\eta)D(\psi)|0\rangle \qquad (SSS), \end{aligned}\right\} \tag{1}$$

where $D(\psi) = \exp\{\psi a^{+} - \psi^{*} a\}$ is a displacement operator, $S(\eta) = \exp\Big\{\frac{\eta^{*}}{2}a^{2} - \frac{\eta}{2}(a^{+})^{2}\Big\}$ is a squeeze operator, $\psi = |\psi|e^{i\gamma}$ is an eigenvalue of non-Hermitian annihilation operator a, $|\psi|$ and γ are an amplitude and a phase of the coherent state correspondingly, $\eta = re^{i\theta}$ is an arbitrary complex number, r is a squeeze factor, phase θ defines the direction of squeezing maximum [21]. Using general formula for two-photon coherent state distribution [31] and the formulas for the average multiplicity for CSS and SSS

$$\left.\begin{aligned} \langle n\rangle &= \int\limits_V |\psi(z)|^2 dz + \sinh^2(r), \\ \langle n\rangle &= \Bigg(\int\limits_V |\psi(z)|^2 dz\Bigg)\Big[\cosh(2r) - \sinh(2r)\cos(2\gamma-\theta)\Big] + \sinh^2(r). \end{aligned}\right\} \tag{2}$$

we can write the corresponding expression for the probability density of finding n particles in SS $|\langle n|\psi(z),\eta\rangle|^2 = P_n^0$ in the form

$$\begin{aligned} P_n^0 &= \frac{1}{\cosh(r)n!}\left(\frac{\tanh(r)}{2}\right)^n \left| H_n\left(\left[\int\limits_V |\psi(z)|^2 dz\right]^{\frac{1}{2}} F_1(r,\gamma,\theta)\right)\right|^2 \\ &\quad \times \exp\Bigg\{\int\limits_V |\psi(z)|^2 dz\ F_2(r,\gamma,\theta)\Bigg\}, \end{aligned} \tag{3}$$

where $F_1(r,\gamma,\theta), F_2(r,\gamma,\theta)$ are functions of the parameters r,γ,θ and in case of CSS are equal to

$$\left.\begin{aligned} F_1(r,\gamma,\theta) &= \frac{\cosh(r)e^{i(\gamma-\theta/2)}+\sinh(r)e^{-i(\gamma-\theta/2)}}{\sqrt{\sinh(2r)}}, \\ F_2(r,\gamma,\theta) &= \cosh(2r)[\tanh(r)\cos(2\gamma-\theta)-1] \\ &\qquad\qquad +\sinh(2r)[\tanh(r)-\cos(2\gamma-\theta)] \end{aligned}\right\} \tag{4}$$

and for SSS

$$F_1(r,\gamma,\theta) = \frac{e^{i(\gamma-\theta/2)}}{\sqrt{\sinh(2r)}}, \qquad F_2(r,\gamma,\theta) = \tanh(r)\cos(2\gamma-\theta)-1. \tag{5}$$

At $\theta = 0$ we have a sub-Poissonian distribution and at $\theta = \pi$ — super-Poissonian one. If the squeeze factor is more than one we have oscillations of given distribution. Obviously that at $r \to 0$ (the squeezing effect is absent) the probability density of finding n particles is Poissonian

$$P_n^0 = \frac{1}{n!}\exp\Big\{-\int_V |\psi(z)|^2 dz\Big\}\left(\int_V |\psi(z)|^2 dz\right)^n. \tag{6}$$

Within standard GL model the free energy of the system is

$$F[\psi] = \int dz\{a|\psi(z)|^2 + b|\psi(z)|^4 + c|\partial\psi/\partial z|^2\}, \tag{7}$$

where $\psi(z)$ is introduced to serve as a complex order parameter. Then the hadron multiplicity distribution can be given by the functional integral of the type [5]

$$P_n = Z^{-1}\int D\psi P_n^0 e^{-F[\psi]}, \tag{8}$$

here $Z = \int D\psi e^{-F[\psi]}$. Thus the probability of having a large n in volume V is controled by deviation of ψ from ψ_0 (minimum of the GL potential) as specified by the thermodynamical factor $e^{-F[\psi]}$. Taking into account the explicit form of the probability density of finding n particles in SS P_n^0 (3) we obtain the next expression for the hadron multiplicity distribution

$$\begin{aligned} P_n &= \frac{1}{2\,Z\cosh(r)}\int D\psi\,\frac{\tanh^n(r)}{2^n n!} \\ &\quad\times\exp\left\{-F[\psi]+\int_V|\psi(z)|^2 dz\,F_2(r,\gamma,\theta)\right\} \\ &\quad\times\left|H_n\left(\left[\int_V|\psi(z)|^2dz\right]^{\frac{1}{2}}F_1(r,\gamma,\theta)\right)\right|^2. \end{aligned} \tag{9}$$

To investigate obtained expression we identify $V = \delta^d$ (d is a dimension) and regard that $|\psi(z)|$ is constant in every bin width δ. Then multiplicity distribution after phase transition is

$$P_n = \frac{D_{-1}^{-1}\left(-|a|\sqrt{\frac{\delta^d}{2b}}\right)}{2\,\pi\cosh(r)} \int_0^{2\pi} d\gamma\ \tanh^n(r)\exp\left\{\frac{\delta^d F_2(F_2+2|a|)}{8b}\right\}$$

$$\times \sum_{k=0}^{n/2}\sum_{l=0}^{n/2}(-1)^{k+l}\frac{(2k-1)!!\,(2l-1)!!\,n!\,(n-k-l)!}{(2k)!\,(2l)!\,(n-2k)!\,(n-2l)!}\left(\frac{2\delta^d}{b}\right)^{\frac{1}{2}(n-k-l)}$$

$$\times F_1^{n-2k}(F_1^*)^{n-2l}\,D_{-(n-k-l+1)}\left(-\left[|a|+F_2\right]\sqrt{\frac{\delta^d}{2b}}\right), \tag{10}$$

where $D_{-f}(w)$ is a function of the parabolic cylinder. Obviously, this expression for P_n is not depended on the phase which defines the direction of squeezing maximum since integrand is a harmonic function of this squeeze parameter.

This expressions for P_n (9), (10) will be essential at analysis phenomenon of intermittency.

3. Intermittency

One of the effective way to manifest the nature of the multiplicity fluctuations in high-energy collisions is to examine the dependence of the normalized factorial moments F_q [15, 32]

$$F_q = \frac{\langle n(n-1)\cdots(n-q+1)\rangle}{\langle n\rangle^q} = \frac{f_q}{f_1^q} \tag{11}$$

on the bin width δ in rapidity. Here $f_q = \langle n\,(n\ -\ 1\,)\cdot\ \cdot\ \cdot\ (n-q+1)\,\rangle$, n is the number of hadrons detected in δ in an event, and the average are taken over all events. The multiplicity fluctuations can exhibit intermittency behaviour which is manifested by power-law behaviour of F_q on δ [15]

$$F_q \propto \delta^{-\varphi_q}, \tag{12}$$

where φ_q is referred to as the intermittency index. Indeed, apart from collision energy and nuclear size we can vary only the size of a cell δ in phase space that is just the central theme of intermittency. This effect has been observed in a large number of experiments: e^+e^-, μp, pp, pA and AA collisions [15].

Therefore the intermittency analysis is used to explore universal characteristics of quark-hadron PT in the GL model. In this section we examine whether (12) is valid under taking into account PT. Since

$$f_q = \sum_{n=q}^{\infty} \frac{n!}{(n-q)!} P_n, \tag{13}$$

using (9) and (7) we obtain the next explicit form of f_q

$$f_q = \frac{1}{2\,Z\cosh(r)} \int D\psi\, e^{-F[\psi]} \exp\left\{ \int_V |\psi(z)|^2 dz\, F_2(r,\gamma,\theta) \right\}$$
$$\times \sum_{n=q}^{\infty} \frac{1}{(n-q)!} \left(\frac{\tanh(r)}{2}\right)^n \left| H_n\left(\left[\int_V |\psi(z)|^2 dz \right]^{\frac{1}{2}} F_1(r,\gamma,\theta) \right) \right|^2, \tag{14}$$

where V is the volume of the cell in which the factorial moment is measured. Taking into account the formula [33]

$$\sum_{k=0}^{\infty} \frac{t^k}{k!} H_{k+m}(x) H_{k+n}(y) = (1-4t^2)^{-(m+n+1)/2}$$
$$\times \exp\left[\frac{4xyt - 4t^2(x^2+y^2)}{1-4t^2} \right]$$
$$\times \sum_{k=0}^{min(m,n)} 2^{2k} k! \binom{m}{k} \binom{n}{k} t^k \tag{15}$$
$$\times H_{m-k}\left(\frac{x-2ty}{\sqrt{1-4t^2}} \right) H_{n-k}\left(\frac{y-2tx}{\sqrt{1-4t^2}} \right)$$

and identifying $V = \delta^d$, regarding that $|\psi(z)|$ is constant in every bin width δ, we rewrite the expression (14) taking into account the explicit form of $F_1(r,\gamma,\theta), F_2(r,\gamma,\theta)$ for CSS (4) and SSS (5) correspondingly in the next form

(CSS)

$$f_q = \frac{\sinh^{2q}(r)}{2\,Z} \int_0^{2\pi} d\gamma \int_0^{\infty} d|\psi|^2 e^{-F[\psi]} \sum_{n=0}^{q} \left(\frac{q!}{n!}\right)^2 \frac{1}{(q-n)!(2\tanh(r))^n} \tag{16}$$
$$\times \left| H_n\left(\sqrt{\frac{|\psi|^2 \delta^d}{\sinh(2r)}}\, e^{i(\gamma-\theta/2)} \right) \right|^2,$$

(SSS)

$$f_q = \frac{\sinh^{2q}(r)}{2\,Z} \int_0^{2\pi} d\gamma \int_0^{\infty} d|\psi|^2 e^{-F[\psi]} \sum_{n=0}^{q} \left(\frac{q!}{n!}\right)^2 \frac{1}{(q-n)!(2\tanh(r))^n}$$
$$\times \left| H_n\left(\sqrt{\frac{|\psi|^2\delta^d}{\sinh(2r)}} \left[\cosh(r)e^{i(\gamma-\frac{\theta}{2})} - \sinh(r)e^{-i(\gamma-\frac{\theta}{2})}\right]\right)\right|^2. \tag{17}$$

Integrating obtained expressions we can represent their as

$$f_q = \frac{J_q}{J_0}, \tag{18}$$

where in case CSS

$$J_q = \frac{\pi \sinh^{2q}(r)}{\sqrt{2b\delta^d}} \exp\left\{\frac{|a|^2\delta^d}{8b}\right\} \sum_{n=0}^{q} \frac{(q!)^2}{(q-n)!} \tanh^{-n}(r) \sum_{k=0}^{n/2} \left(\frac{(2k-1)!!}{(2k)!}\right)^2$$
$$\times \frac{(\sinh(2r))^{2k-n}}{(n-2k)!} \left(\frac{2\delta^d}{b}\right)^{\frac{1}{2}(n-2k)} D_{-(n-2k+1)}\left(-|a|\sqrt{\frac{\delta^d}{2b}}\right) \tag{19}$$

and for SSS

$$J_q = \frac{\pi \sinh^{2q}(r)}{\sqrt{2b\delta^d}} \exp\left\{\frac{|a|^2\delta^d}{8b}\right\} \sum_{n=0}^{q} \frac{(q!)^2}{(q-n)!} \tanh^{-n}(r) \sum_{k=0}^{n/2}\sum_{l=0}^{n/2} \frac{(2k-1)!!}{(2k)!\,(2l)!}$$
$$\times (2l-1)!!(n-k-l)!(\sinh(2r))^{k+l-n} \left(\frac{2\delta^d}{b}\right)^{\frac{1}{2}(n-k-l)}$$
$$\times D_{-(n-k-l+1)}\left(-|a|\sqrt{\frac{\delta^d}{2b}}\right)$$
$$\times \sum_{j=0}^{n-2l} \frac{(\sinh(r))^{l-k+2j}\,(\cosh(r))^{2n-k-2l-2j}}{j!\,(l-k+j)!\,(n-k-l-j)!\,(n-2l-j)!}. \tag{20}$$

Then according to (11), (18) the normalized factorial moments F_q have the next form

$$F_q = J_q J_1^{-q} J_0^{q-1}. \tag{21}$$

If the local slope of $\ln F_q$ vs $\ln F_2$ is approximately constant then we would have the scaling behaviour (Ochs-Wosiek scaling law) [34]

$$F_q \propto {F_2}^{\beta_q}, \tag{22}$$

which is valid for intermittent systems [32]. The slopes β_q are well fitted by the formula [10]

$$\beta_q = (q-1)^{\nu}, \tag{23}$$

where ν is a scaling exponent. Dependence of ν on squeeze factor r is represented on the Fig.1 at the next values of the parameters of the GL model: $a = -10$, $b = 0.20055$

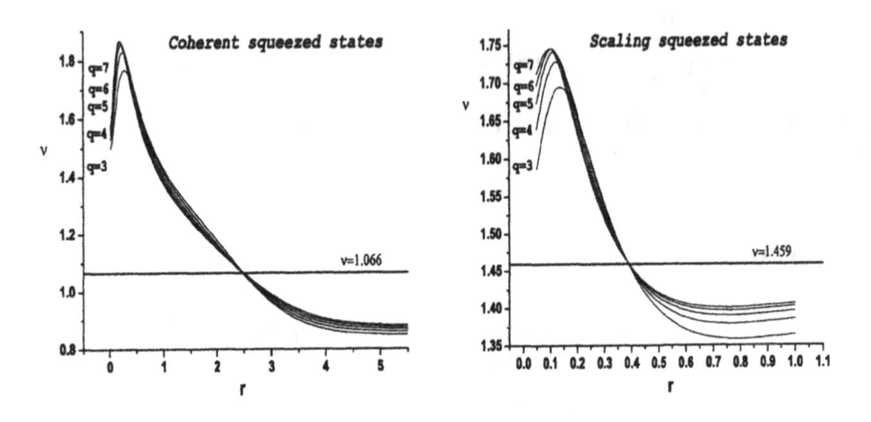

Figure 1. Dependence of the scaling exponent on squeeze factor r.

It is obvious from Fig.1 that we have intermittency in case of the CSS when $\nu = 1.066$ at $r = 2.48$ and $\nu = 1.459$ at $r = 0.3876$ for SSS. Thus scaling behaviours of the normalized factorial moments for the scaling squeezed states are characterized by obtained scaling exponent ν that agrees with experimental NA22 data on particle production in heavy-ion experiments [16]. Parameters at which the scaling exponent values agree with various experimental data [10, 15] are represented in the Tab.1.

In case of the CSS the scaling exponent values are not agree with various experimental data [10, 15] at any values of the parameters a, b, r.

4. Conclusion

We study multiplicity fluctuations and intermittency in second order phase transition from QGP to hadrons within of the GL model. Generalizing P-representation to squeezed state one (in particular, for two types: CSS, SSS) we obtain the explicit expressions for the probability of finding n particles and for the normalized factorial moments F_q which include additional parameters r, θ inherent to the squeezing effect.

Changing new parameters we can more successfully apply GL model for description of the phase transitions. Indeed, at $a = -10, b = 0.20055$ and at $r = 0.3876$ in case of the scaling squeezed states of the hadrons we have intermittency when the value of the scaling exponent is equal to 1.459. Obtained value of the scaling exponent agrees with experimental data [10, 15, 16].

TABLE 1. Parameters at which the scaling exponent values agree with experimental data.

a	$\nu = 1.450$		$\nu = 1.459$		$\nu = 1.550$	
	b	r	b	r	b	r
-1	0.00663	0.32363	0.00592	0.32663	0.00220	0.34375
-2	0.02761	0.31312	0.02455	0.31826	0.00992	0.32266
-3	0.05217	0.32917	0.04988	0.32216	0.02275	0.31387
-4	0.07313	0.35251	0.06956	0.34795	0.03760	0.31758
-5	0.09535	0.36548	0.09052	0.36190	0.05582	0.31572
-6	0.11815	0.37385	0.11213	0.37079	0.07140	0.32305
-7	0.14138	0.37970	0.13398	0.37693	0.08570	0.33040
-8	0.16472	0.38401	0.15609	0.38144	0.09871	0.33740
-9	0.18819	0.38730	0.17834	0.38489	0.11190	0.34261
-10	0.21187	0.38983	0.20055	0.38758	0.12518	0.34664

We hope that squeezed state approach will be available for description of fluctuations in the phase transition from quark-gluon plasma to hadrons in processes where an energy density is very high, for example, in heavy ion collisions at high energy.

References

1. Dokshitzer, Yu. L. , Khoze, V. A., Mueller, A. H., Troyan, S. I. (1991) *Basics of Perturbative QCD,* Frontières, France.
2. Hebbeker, T. (1992) Tests of quantum chromodynamics in hadronic decays of Z^0 bosons produced in e^+e^- annihilation, *Phys. Rep.* **217**, pp. 69–157.
3. Dokshitzer, Yu. L., Khoze, V. A., Troyan, S. I. (1991) Particle spectra in light and heavy quark jets, *J. Phys. G* **17**, pp. 1481–1492.
4. Dremin, I. M., Hwa, R. C. (1994) Quark and gluon jets in QCD: factorial and cumulant moments, *Phys. Rev. D* **49**, pp. 5805–5811.
5. Scalapino, D. J., Sugar, R. L. (1973) A statistical theory of particle production, *Phys. Rev. D* **8**, pp. 2284–2295.
6. Bialas, A., Hwa, R. C. (1991) Intermittency parameters as a possible signal for quark-gluon plasma formation, *Phys. Lett. B* **253**, pp. 436–438.
7. Gausterer, H., Sanielevici, S. (1988) Can the chiral transition in QCD be described by a linear sigma-model in three dimensions, *Phys. Lett. B* **209**, pp. 533–537.
8. Walls, D. F., Milburn, G. J. (1995) *Quantum Optics,* Springer-Verlag, NY. USA.
9. Botke, J. C., Scalapino, D. J. , Sugar, R.L. (1974) Coherent states and particle production, *Phys. Rev. D* **9**, pp. 813–823.
10. Hwa, R. C., Nazirov, M. T. (1992) Intermittency in second-order phase transitions,

Phys. Rev. Lett. **69**, pp. 741–744.
11. Hwa, R. C. (1993) Scaling exponent of multiplicity fluctuation in phase transition, *Phys. Rev. D* **47**, pp. 2773–2781.
12. Babichev, L. F., Klenitsky, D. V., Kuvshinov, V. I. (1995) Intermittency described by generalized Ginzburg-Landau model of first-order phase transitions in HEP, *Proceedings of the 3d International Seminar on Nonlinear Phenomena in Complex Systems*, Minsk, Institute of Physics.
13. Hwa, R. C. (1994) Scaling behaviour in first-order quark-hadron phase transition, *Phys. Rev. C* **50**, pp. 383–387.
14. Babichev, L. F., Klenitsky, D. V., Kuvshinov, V. I. (1995) Intermittency in the Ginzburg-Landau model for first-order phase transitions, *Phys. Lett. B* **345**, pp. 269–271.
15. De Wolf, E. A., Dremin, I. M., Kittel, W. (1996) Scaling laws for density correlations and fluctuations in multiparticle dynamics, *Phys. Rep.* **270**, pp. 1–142.
16. Jain, P. L. et al. (1990) Intermittency in multiparticle production at ultra-relativistic heavy ion collisions, *Phys. Lett. B* **236**, pp. 219-223; Jain, P. L. et al. (1992) Intermittency in relativistic heavy-ion collisions, *Z. Phys. C***53**, pp. 355–360; Jain, P. L. et al.(1993) Factorial moments and multifractal analysis at relativistic energies, *Phys. Rev. C* **48**, pp. 517–521.
17. Bambah, B. A., Satyanarayana, M. V. (1988) Scaling and correlations of squeezed coherent distributions: application to hadronic multiplicities, *Phys. Rev. D* **38**, pp. 2202–2208.
18. Vourdas, A., Weiner, R. M. (1988) Multiplicity distributions and Bose-Einstein correlations in high-energy multiparticle production in the presence of squeezed coherent states, *Phys. Rev. D* **38**, pp. 2209–2217.
19. Dodonov, V. V., Dremin, I. M., Polynkin, P. G., Man'ko, V. I. (1994) Strong oscillations of cumulants of photon distribution function in slightly squeezed states, *Phys. Lett. A* **193**, pp. 209–217.
20. Hirota, O. (1992) *Squeezed light,* Japan, Tokyo.
21. Scully, M. O., Zubairy, M. S. (1997) *Quantum Optics*, Cambridge University Press.
22. Kilin, S. Ya. (1990) *Quantum Optics*, Minsk. (in Russian)
23. Shih, C. C. (1986) Sub-Poissonian distribution in hadronic processes, *Phys. Rev. D* **34**, pp. 2720–2726.
24. Dremin, I. M., Hwa, R. C. (1996) Multiplicity distributions of squeezed isospin states, *Phys. Rev. D* **53**, pp. 1216–1223.
25. Kilin, S. Ya., Kuvshinov, V. I., Firago, S. A. (1993) Squeezed colour states in gluon jet, *Proceeding of the Workshop on Squeezed states and Uncertainty relations*, NASA.
26. Kuvshinov, V. I., Shaporov, V. A. (1999) Gluon squeezed states in QCD jet, *Acta Phys. Pol. B* **30**, pp. 59–68.
27. Kuvshinov, V. I., Shaparau, V. A. (2000) Squeezed states of colour gluons in QCD isolated jet, *Nonlinear Phenomena in Complex Systems* **3**, pp. 28–36.
28. Kuvshinov, V.I., Shaparau, V.A. (2002) Fluctuations and correlations of soft gluons at the nonperturbative stage of evolution of QCD jets, *Phys. Atom. Nucl.* **65**, pp. 309–314.
29. Alner, G. J. et al., (UA5 coll.) (1986) Scaling violations in multiplicity distributions at 200 and 900 GeV, *Phys. Lett. B* **167**, pp. 476–480.
30. Abreu, P. et al., (DELPHI coll.) (1991) Charged particle multiplicity mistributions in Z^0 hadronic decays, *Z. Phys. C* **50**, pp. 185–194.
31. Yuen, H. P. (1976) Two-photon coherent states of the radiation field, *Phys. Rev. A* **13**, pp. 2226–2243.
32. Bialas, A., Peschanski, R. (1986) Moments of rapidity distributions as a measure of short-range fluctuations in high-energy collisions, *Nucl. Phys. B* **273**, pp. 703–718; Bialas, A., Peschanski, R. (1988) Intermittency in multiparticle production at high energy, *Nucl. Phys. B* **308**, pp. 857–867.
33. Prudnikov, A. P., Brychkov, Yu. A., Marichev, O. I. (1983) *Integrals and series. Special functions,* Nauka, Moscow. (in Russian)
34. Ochs, W., Wosiek, J. (1988) Intermittency and jets, *Phys. Lett. B* **214**, pp. 617–620.

SOFT COMPONENT OF PHOTON WAVE FUNCTION IN THE TRANSITIONS $\gamma^*(Q_1^2)\gamma^*(Q_2^2) \to$ QUARK–ANTIQUARK MESON AT MODERATE VIRTUALITIES

L.G. DAKHNO
Petersburg Nuclear Physics Institute,
188300 Gatchina, Sanct-Petersburg district, Russia

The processes involving soft photon, like $\rho^0, \omega, \phi \to \gamma^* \to e^+e^-(\mu^+\mu^-)$ and $\gamma^*(Q_1^2)\gamma^*(Q_2^2) \to (q\bar{q})$ – *meson* at small and moderately large virtualities of photons, $0 \le Q_i^2 \le 1$ (GeV/c)^2, are considered within the technique based on the spectral integration representation for constituent quarks in the intermediate state. Special attention is paid to the photon wave function in the soft region, and analytical expression for it is presented. It is discussed why the soft-photon wave function determination is crucial for current meson spectroscopy problems.

Presently the most challeging task in meson spectroscopy is the discovery of non-$q\bar{q}$ particles. During the last decade tremendous efforts have been paid for the search of the glueball, gg, hybrids $q\bar{q}g$ or multiquark states such as, for example, $q\bar{q}q\bar{q}$. One may state that at the moment the existence of the light scalar glueball near 1500 MeV has been established with rather good confidence, and there is a belief that pseudoscalar and tensor glueballs have been seen near 2000 MeV.

The systematics of meson quark–antiquark states on the (n, M^2)- and (J, M^2)-planes (n is radial quantum number, J is the total spin of a meson with mass squared M^2) appeared to be a powerful tool for the discovery of exotic, non-$q\bar{q}$, states. The fact is that $q\bar{q}$ states can be put, with rather good accuracy, on linear trajectories on the (n, M^2)- and (J, M^2)-planes.

R. Fiore et al. (eds.), Diffraction 2002, 165–174.

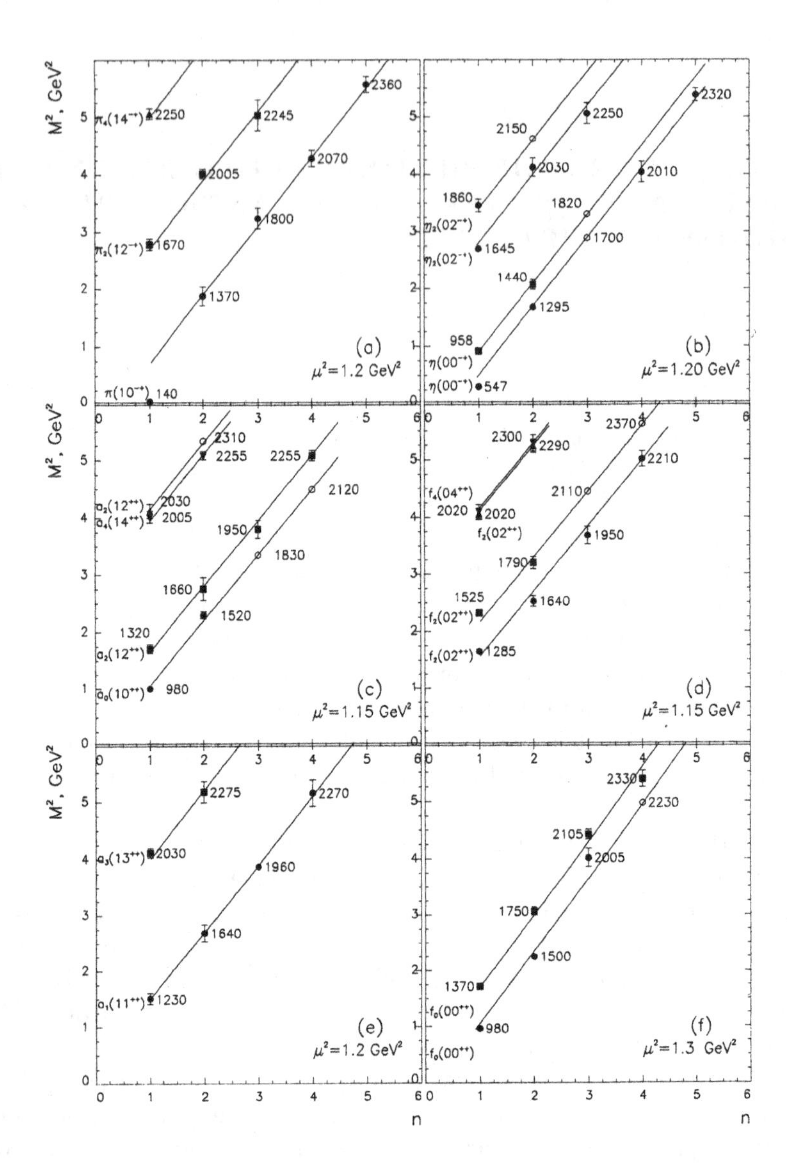

Figure 1. Trajectories on the (n, M^2) planes for the states with $(C = +)$. Open circles stand for the predicted states.

Therefore, the states which do not belong to linear trajectories should be considered as exotic ones.

In terms of $q\bar{q}$-states, the mesons of the nonets $n^{2S+1}L_J$ fill in the following (n, M^2)-trajectories for $M \lesssim 2400$ MeV [1]:

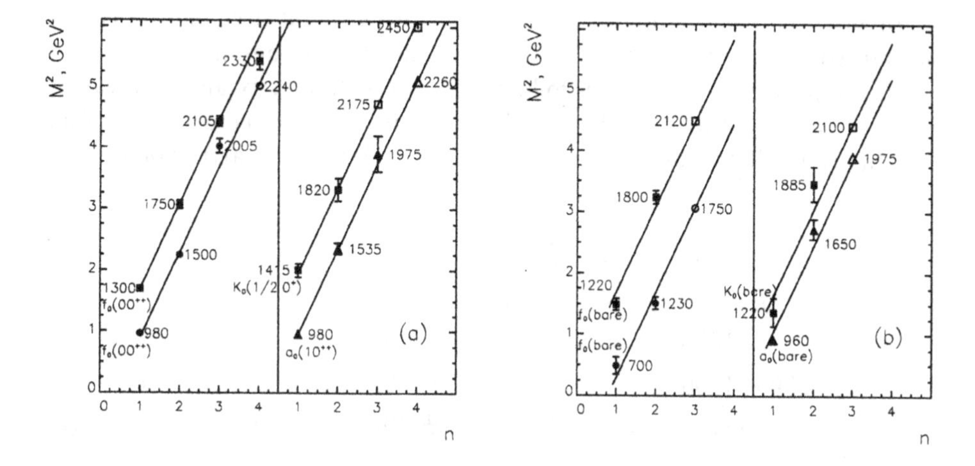

Figure 2. Linear trajectorieson the (n, M^2) plane for scalar resonances (a) and bare scalar states (b). Open labels correspond to the predicted states.

$$
\begin{aligned}
{}^1S_0 &\to \pi(10^{-+}),\ \eta(00^{-+})\ ; \\
{}^3S_1 &\to \rho(11^{--}),\ \omega(01^{--})/\phi(01^{--})\ ; \\
{}^1P_1 &\to b_1(11^{+-}),\ h_1(01^{+-})\ ; \\
{}^3P_J &\to a_J(1J^{++}),\ f_J(0J^{++}),\ J=0,1,2\ ; \\
{}^1D_2 &\to \pi_2(12^{-+}),\ \eta_2(02^{-+})\ ; \\
{}^3D_J &\to \rho_J(1J^{--}),\ \omega_J(0J^{--})/\phi_J(0J^{--}),\ J=1,2,3\ ; \\
{}^1F_3 &\to b_3(13^{+-}),\ h_3(03^{+-})\ ; \\
{}^3F_J &\to a_J(1J^{++}),\ f_J(0J^{++}),\ J=2,3,4\ .
\end{aligned}
\tag{1}
$$

As an example, in Fig. 1 the trajectories on the (n, M^2)-planes are shown for mesons with positive charge parity $C = +$. It should be emphasized that such a classification of mesons became possible after the discovery of resonances at $M \sim 1800 - 2300$ MeV in the analysis of the Crystal Barrel data for the reactions of $p\bar{p}$-annihilation in flight by RAL–PNPI group, see [2] and references therein.

During the last decade the sector of scalar mesons $IJ^{PC} = 00^{++}$ was under intensive investigation, for, due to zero quantum numbers, it was justly considered as the most promising for the discovery of the lightest scalar glueball. The developed K-matrix technique [3] allowed one to carry out the analysis of f_0-mesons in terms of bare state, that is, quark–antiquark state with quantum number of an f_0 resonance without decaying processes accounted for, in other words, without a cloud of real mesons.

An important result which followed from the K-matrix analysis was as follows: bare f_0 states can be also located on linear trajectories (see Fig. 2). One can collate the analysed resonances with the K-matrix poles (bare states):

$$\begin{aligned} f_0(980) &\iff f_0^{bare}(700 \pm 100)\,, \\ f_0(1300) &\iff f_0^{bare}(1220 \pm 40)\,, \\ f_0(1500) &\iff f_0^{bare}(1250 \pm 40)\,, \\ f_0(1200-1600) &\iff f_0^{bare}(1580 \pm 40)\,, \qquad 0^{++} - glueball \\ f_0(1750) &\iff f_0^{bare}(1800 \pm 40)\,. \end{aligned} \tag{2}$$

The broad state $f_0(1200-1600)$ together with its bare-state counterpart, $f_0^{bare}(1580 \pm 40)$, cannot be put on linear (n, M^2) trajectory, so they are not shown in Fig.2. These states are superfluous for linear trajectories, and the K-matrix analysis revealed the gluonium nature of the bare state $f_0^{bare}(1580 \pm 40)$. As to the broad state $f_0(1200-1600)$, this state is a mixture of the gluonium and quarkonium states; by mixing with neighbouring $q\bar{q}$ states the glueball accumulated their widths. This is the reason for the appearance of a broad state $f_0(1200-1600)$. To evaluate the percentage of the gluonium and quarkonium in the broad state $f_0(1200-1600)$ is quite a problem. The matter is that the quarkonium state in the glueball descendant $f_0(1200-1600)$ is nearly the flavour singlet, so the hadronic couplings to the gluonium and quarkonium components are similar to one another.

Therefore, one needs to draw additional information to clarify the content of the broad state, and this information may come from the investigation of radiative meson decays. Also radiative decays allow one to estimate the gluonium component in mesons of the $q\bar{q}$ origin.

During several years the PNPI group has worked upon the study of radiative meson decays such as $\pi^0, \eta, \eta' \to \gamma\gamma^*(Q^2)$ at moderate Q^2. This activity resulted in the elaboration of the method of calculation of form factors and transition form factors like those shown in Fig. 3. In this way the necessity of dealing with photon wave function in the soft region becomes obvious. In the calculation of transition form factors $\pi^0, \eta, \eta' \to \gamma\gamma^*(Q^2)$ a non-trivial structure of the low-energy photon wave function has been suggested. Later on, the use of additional data on the annihilation reaction widths $e^+e^- \to \omega, \rho^0, \phi$ allowed us to obtain more precise, analytical, expression for soft photon wave function, which describes the data on $\pi^0, \eta, \eta' \to \gamma\gamma^*(Q^2)$ and satisfies well the well-known experimental data for the ratio in the soft region: $R(s) = \sigma(e^+e^- \to hadrons)/\sigma(e^+e^- \to \mu^+\mu^-)$ [4].

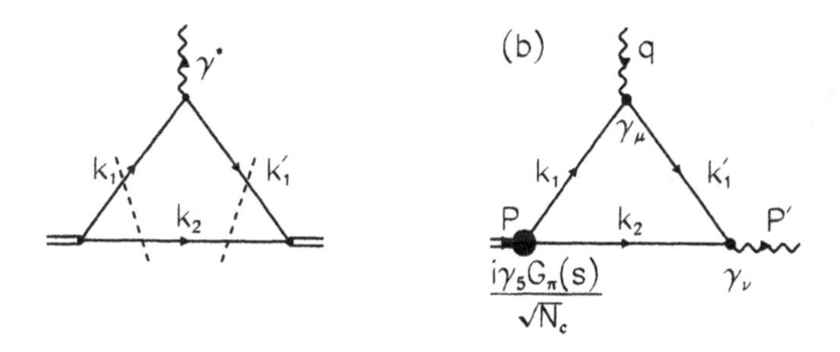

Figure 3. Quark triangle diagrams for meson form factor and meson transition form factor.

Investigations of the processes with real photons $\gamma\gamma \to q\bar{q} - meson$ as well as calculations of pion form factor were performed by using double spectral representation for the triangle quark diagram like shown in Fig. 3. The method initially applied to the deuteron had been then reformulated for constituent quarks.

1. Pion form factor.

In the paper [5] the electromagnetic pion form factor has been studied, and light-cone variables were used for the description of pion wave function. General structure of the amplitude is $A_\mu(q^2) = (p+p')_\mu F_\pi(-q^2)$, where p, p' are pion's incoming/outgoing four-momenta and $F_\pi(-q^2)$ is the commonly used form factor. In this way the pion form factor is represented as a convolution of pion wave functions:

$$F_\pi(-q^2) = Z_\pi \int\limits_{4m^2}^{\infty} \frac{ds}{\pi}\frac{ds'}{\pi} \Psi_\pi(s)\Psi_\pi(s')\Delta_\pi(s,s',q^2) , \tag{3}$$

the wave function Ψ_π normalized so that $F_{\mathrm{meson}}(0) = 1$. The representation of pion wave function, $\Psi_\pi(s) = G_\pi(s)/(s - M_\pi^2)$, is typical for calculations within spectral-representation technique, and the vertex $G_\pi(s)$ is determined by fitting to experimental data. The function $\Delta_{\mathrm{meson}}(s,s',q^2)$ is determined by the quark-loop trace in the intermediate states multiplied by phase-space factor. The invariant energies squared, s, s', for intermediate $q\bar{q}$ states read in the light-cone variables as

$$s = (m^2 + k_\perp^2)/x(1-x) , \qquad s' = (m^2 + (\vec{k}_\perp - x\vec{q})^2)/x(1-x),$$

and $Z_\pi = 1$ is the $\pi^\pm$ charge factor.

Fitting formula (3) to data with the two-exponential parametrization of the wave function Ψ_π,

$$\Psi_\pi(s) = c_\pi \left(\exp(-b_1^\pi s) + \delta_\pi \exp(-b_2^\pi s)\right) , \tag{4}$$

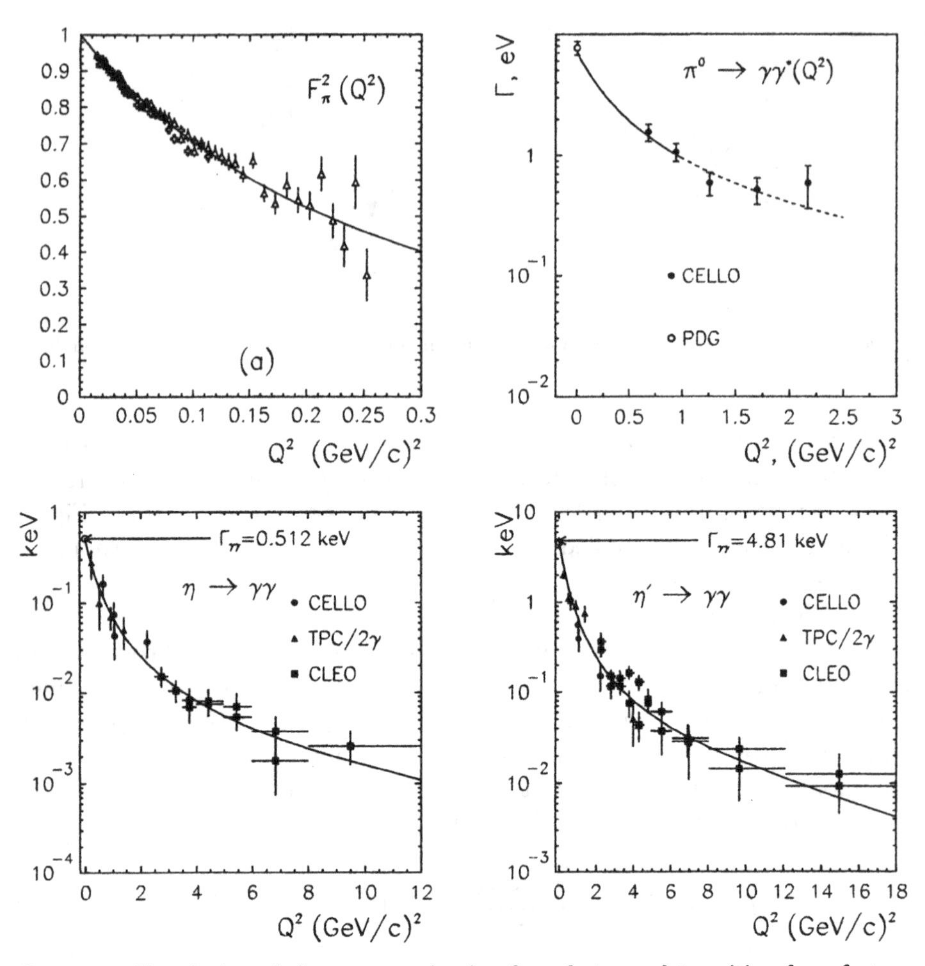

Figure 4. Description of electromagnetic pion form factor and transition form factors $\pi^0, \eta, \eta' \to \gamma\gamma^*(Q^2)$.

one obtains the following parameter values:

$$c_\pi = 209.36,\ \delta_\pi = 0.01381,\ b_1^\pi = 3.57,\ b_2^\pi = 0.4\ . \qquad (5)$$

2. Transition form factors: $\pi, \eta, \eta' \to \gamma^*(Q_1^2)\gamma^*(Q_2^2)$.

Likewise, the formulae for transition form factors of pseudoscalar mesons can be written. In this case general structure of the amplitude is as follows [6]:

$$A_{\mu\nu}(Q_1^2, Q_2^2) = e^2 \epsilon_{\mu\nu\alpha\beta} q^\alpha P^\beta F_{(\pi,\eta,\eta')\to\gamma^*\gamma^*}(Q_1^2, Q_2^2)\ , \qquad (6)$$

and the expression for the transition form factor shown in Fig. 3 is:

$$F_{(\pi,\eta,\eta')\to\gamma^*\gamma^*}(Q_1^2,Q_2^2) = C_\gamma Z_{(\pi,\eta)} \int \frac{ds}{\pi}\frac{ds'}{\pi} \times \qquad (7)$$

$$\times \left(\Psi_{(\pi,\eta)}(s) \frac{G_\gamma(s')}{s'+Q_2^2} \Delta_{\pi,\eta}(s,s',Q_1^2) + \Psi_{(\pi,\eta)}(s) \frac{G_\gamma(s')}{s'+Q_1^2} \Delta_{\pi,\eta}(s,s',Q_2^2) \right) .$$

One should bear in mind that, because of the presence of strange quarkonium component, $s\bar{s}$, in the η, η'-mesons, their form factors are represented through mixing angle θ as

$$F_{\eta\to\gamma^*\gamma^*} = \sin\theta F^{n\bar{n}} - \cos\theta F^{s\bar{s}} \ , \ F_{\eta'\to\gamma^*\gamma^*} = \cos\theta F^{n\bar{n}} + \sin\theta F^{s\bar{s}} \ ,$$

with $n\bar{n} = (u\bar{u} + d\bar{d})/\sqrt{2}$. The structure of the trace factors for π and η, η' mesons is the same though different quark masses enter the functions Δ for non-strange and strange quarks: $\Delta_\eta^{n\bar{n}}(s,s',Q^2,m_q^2)$ and $\Delta_\eta^{s\bar{s}}(s,s',Q^2,m_s^2)$. The charge factors are $Z^{n\bar{n}} = 5/(9\sqrt{2})$, $Z^{s\bar{s}} = 1/9$, $Z^{\pi^0} = 1/(3\sqrt{2})$.

In the calculations of transition form factors of pseudoscalar mesons the vertex function related to non-starnge quarks $n\bar{n}$ has been assumed to be the same as for pion. As to strange component, a different total normalization as compared to formula (5) has been found: $c_{s\bar{s}} = 528.78$.

To determine the photon vertex function it has been chosen in the form:

$$\begin{aligned}
\Psi_\gamma(s,Q_2^2) &= \frac{G_\gamma(s)}{s+Q_2^2}, \\
n\bar{n}: \quad G_\gamma(s) &= c_{n\bar{n}}^\gamma \left(e^{-b_1^\gamma s} + c_2^\gamma e^{-b_2^\gamma s} + \frac{1}{1+e^{-b_\gamma(s-s_0)}} \right) , \\
s\bar{s}: \quad G_\gamma(s) &= c_{s\bar{s}}^\gamma \left(e^{-b_1^\gamma s} + \frac{1}{1+e^{-b_\gamma(s-s_0)}} \right) . \qquad (8)
\end{aligned}$$

The photon wave function used in our calculations has been obtained in a combined study of vector-meson widths in the electron–positron annihilation and in fitting to data on the transition form factors as is seen in Fig. 4. Let us stress that the calculation of widths of vector mesons allowed us to obtain analytical form of the photon vertex function in the soft region, $Q^2 \le 1$ (GeV/c)^2.

3. Electron–positron annihilation into vector meson.

The process which was simultaneously used to find out the photon wave function ($n\bar{n}$ and $s\bar{s}$ components) is the electron-positron annihilation into vector meson (V), the amplitude associated with this process is shown in Fig. 5. Partial width for this reaction reads as follows:

$$m\Gamma_{e^+e^-\to V} = N_c \pi\alpha^2 Z_V^2 A_{e^+e^-\to V}^2 \frac{1}{m_V^4} \left(\frac{4}{3} m_V^2 + \frac{8}{3} m_e^2 \right) \sqrt{\frac{m_V^2 - 4m_e^2}{m_V^2}} \ , \qquad (9)$$

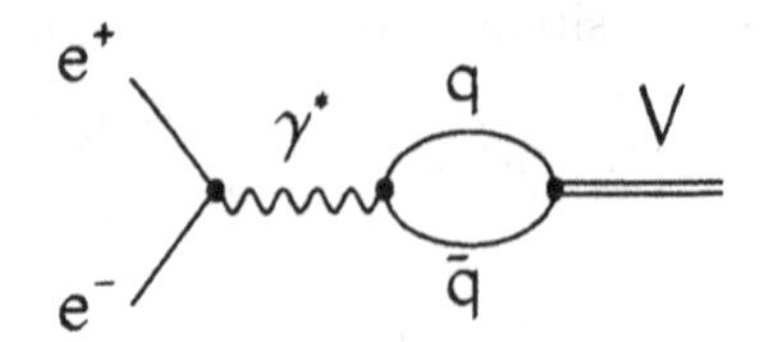

Figure 5. Diagram describing the width of vector mesons.

Here m_V is the vector meson mass, two factors $1/m_V^2$ are associated with the photon propagators, and $\alpha = e^2/(4\pi)$. The Z_V is the quark charge factor for vector mesons: $Z_{\rho^0} = 1/\sqrt{2}$, $Z_\omega = 1/3\sqrt{2}$, $Z_\phi = 1/3$. The integration over electron-positron phase space results in $\sqrt{(m_V^2 - 4m_e^2)/m_V^2}/(16\pi)$, and summing over electron–positron and vector meson spins leads to

$$\frac{1}{3}\,\mathrm{Tr}\left[\gamma_\mu^\perp(\hat{k}_1 + m_e)\gamma_\mu'^\perp(-\hat{k}_2 + m_e)\right] = \frac{4}{3}m_V^2 + \frac{8}{3}m_e^2 \,. \tag{10}$$

The amplitude $A_{e^+e^-\to V}$ is determined in the quark–antiquark loop calculations within spectral-integration technique. In this way we get:

$$A_{e^+e^-\to V} = \sqrt{N_c}\int\limits_{4m^2}^{\infty}\frac{ds'}{\pi}\cdot G_\gamma(s)\Psi_V(s)\frac{1}{16\pi}\sqrt{\frac{s-4m^2}{s}}\left(\frac{8}{3}m^2 + \frac{4}{3}s\right)\,. \tag{11}$$

The process $e^+e^- \to \omega$ was used to define $G_\gamma(s)$ with $n\bar{n}$ part. For the $s\bar{s}$ part we use the process $e^+e^- \to \gamma^* \to s\bar{s} \to \phi$.

4. Photon wave function in the soft region.

Finally, the photon wave function has been defined for moderate virtualities, $Q^2 \le 1$ (GeV/c)^2 [7].

The photon wave function taken in the form (8) satisfies the requirement for the photon–quark interaction: at large s, the vertex G_γ is equal to the unity. The following parameter values, in GeV units, stand for the wave function of Eq. (8):

$$\begin{aligned} n\bar{n}:&\quad c^\gamma_{n\bar{n}} = 32.577,\ c_2^\gamma = -0.0187,\ b_1^\gamma = 4,\ b_2^\gamma = 0.8,\ b^\gamma = 15,\ s_0 = 1.62\,,\\ s\bar{s}:&\quad c^\gamma_{s\bar{s}} = 310.55,\ b_1^\gamma = 4,\ b^\gamma = 15,\ s_0 = 2.15\,. \end{aligned} \tag{12}$$

The use of of Eq. (8) for photon wave function, with the parameters (12) provides a satisfactory description of the experimental data for the electron-positron annihilation into ρ-, ω- and ϕ-mesons as follows:

$$\Gamma^{calc}_{e^+e^-\to\rho^0} = 7.50\ \mathrm{keV} \qquad \Gamma^{exp}_{e^+e^-\to\rho^0} = 6.77 \pm 0.32\ \mathrm{keV}\,,$$

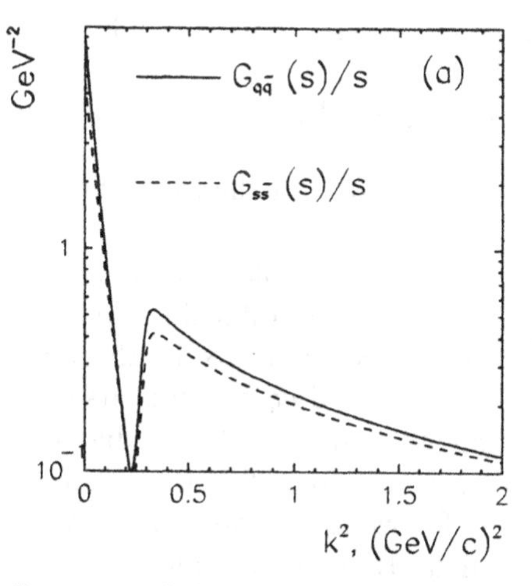

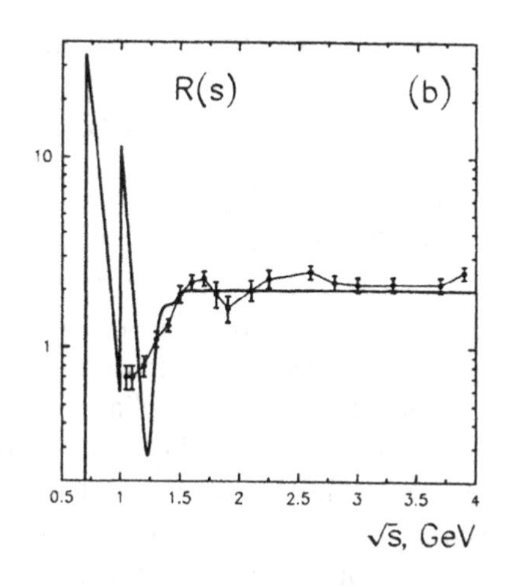

Figure 6. Soft photon wave function components (a) and comparison of the found verticed to the ratio $R(s)$ (b).

$$
\begin{aligned}
&\Gamma^{calc}_{e^+e^-\to\omega} = 0.796\ \text{keV} \qquad && \Gamma^{exp}_{e^+e^-\to\omega} = 0.60 \pm 0.02\ \text{keV}\ , \\
&\Gamma^{calc}_{e^+e^-\to\phi} = 1.33\ \text{keV} \qquad && \Gamma^{exp}_{e^+e^-\to\phi} = 1.32 \pm 0.06\ \text{keV}\ , \\
&\Gamma^{calc}_{\mu^+\mu^-\to\rho} = 7.48\ \text{keV} \qquad && \Gamma^{exp}_{\mu^+\mu^-\to\rho^0} = 6.91 \pm 0.42\ \text{keV}\ , \\
&\Gamma^{calc}_{\mu^+\mu^-\to\phi} = 1.33\ \text{keV} \qquad && \Gamma^{exp}_{\mu^+\mu^-\to\phi} = 1.65 \pm 0.22\ \text{keV}\ .
\end{aligned}
\tag{13}
$$

The agreement with experimental data (see (13) as well as Fig. 4) allows us to conclude that formula (8) is a good representation for photon wave function in the soft region.

5. The ratio $R(s)$ at low energies, $\sqrt{s} \lesssim 3$ GeV: a verification of the structure of the soft photon wave function.

In Fig. 6a, the two components, $q\bar{q}$ and $s\bar{s}$, of the soft photon wave function are depicted. Using analytical representation (8) together with parameter values (12) one can compare our calculations with the experimentally observed ratio (see Fig. 6b):

$$
R(s) = \frac{\sigma(e^+e^- \to hadrons)}{\sigma(e^+e^- \to \mu^+\mu^-)}\ . \tag{14}
$$

The structure of the γ-vertex decaying into two constituent quarks reads:

$$
\gamma^* \to \frac{2}{3}u\bar{u} - \frac{1}{3}d\bar{d} - \frac{1}{3}s\bar{s} = \frac{4}{9}\cdot 3G^2_{u\bar{u}} + \frac{1}{9}\cdot 3G^2_{d\bar{d}}\frac{1}{9}\cdot 3G^2_{s\bar{s}} \tag{15}
$$

the factor 3 accounts for colours. Assuming $G^2_{u\bar{u}} = G^2_{d\bar{d}}$, one can find

$$R(s) = G^2_{\gamma^* \to q\bar{q}} \cdot 5/3 + G^2_{\gamma^* \to s\bar{s}} \cdot 1/3 \ . \qquad (16)$$

A full curve in Fig. 6b reveals reasonable agreement with the experiment.

6. Conclusion.

The search for exotic states, gluonium in the first place, demands a secure method to analyse the content of mesons. The difficulty consists in the fact that quarkonium and gluonium components mix in the real mesons. To perform a reliable separation of the gg and $q\bar{q}$ components one needs to carry out the analysis of experimental data for the production of mesons initiated by $\gamma\gamma^*$ transitions. To this aim form factors and partial widths of transitions are found for the following pseudoscalar, scalar and tensor mesons assuming they are dominantly quark–antiquark states [7]: (i) $\pi, \pi(1300), \eta, \eta', \eta(1290), \eta(1440)$; (ii) the light σ-meson, $f_0(980)$, $f_0(1300)$, $f_0(1500)$, $f_0(1750)$, $a_0(980)$, $a_0(1520)$; (iii) $f_2(1270)$, $f_2(1525)$, $f_2(1580)$, $f_2(1800)$, $a_2(1320)$, $a_2(1660)$, the presentation of the results is not possible due to short length of this article.

References

1. Anisovich, A.V., Anisovich, V.V. and Sarantsev, A.V. (2000) "Systematics of $q\bar{q}$-states in the (n, M^2) and (J, M^2) planes", *Phys. Rev. D* **Vol. no. 62**:051502
2. Anisovich, A.V., Baker, C.A., Batty, C.J., *et al.* (2001) "A partial wave analysis of anti-pp → eta-eta et rest" , *Phys. Lett. B* **Vol. no. 517**, pp. 273–281
3. Anisovich, V.V., Sarantsev, A.V. (2002) "K-matrix analysis of the $(IJ^{PC} = 00^{++})$-wave in the mass region below 1900 MeV", hep-ph/0204328, to be published in *Eur. Physics J. A*
4. Martin, A.D., Outhwaite, J. and Ryskin, M.G. (2001) "Improving $\alpha_{QED}(m_z^2)$ and the charm meson mass by analytical continuation", *Eur. Physics J. C* **Vol. no. 19**, pp. 681–691
5. Anisovich, V.V., Melikhov, D.I. and Nikonov, V.A. (1995) "Quark structure of the pion and pion form factor", *Phys. Rev. D* **Vol. no. 52** pp. 5295–5308
6. Anisovich, V.V., Melikhov, D.I. and Nikonov, V.A. "Photon-meson transition form factors" (1997) *Phys. Rev. D* **Vol. no. 55** pp. 2918–2930
7. Anisovich, A.V., Anisovich, V.V., Dakhno, L.G., Nikonov, V.A. and Sarantsev, A.V. (2002) "Determination of the soft $q\bar{q}$ component of the photon wave function and calculation of the transitions $\gamma^*(Q_1^2)\gamma^*(Q_2^2) \to (q\bar{q}) - meson$ at $0 \le Q_i^2 \le 1$ (GeV/c)^2", in preparation

THE PUZZLE OF HYPERON POLARIZATION

S. M. TROSHIN, N. E. TYURIN
Institute for High Energy Physics, Protvino, Moscow Region, 142281, Russia

In the recent years a number of significant and unexpected spin effects were discovered. They demonstrated that the spin degrees of freedom will be important even at TeV energy scale. Currently spin experiments are conducted at almost all existing accelerators and are planned for those under construction. The experimental data on the spin observables represent a rather small fraction of the data on particle interactions — most of the data are related to spin–averaged ones. Despite that, these data prove to be very important for understanding of particle interactions. They provide information on the spin dependence of the interactions, which in some sense is more fundamental than the dynamics studied using spin–averaged observables. These data *clearly indicate that high energy interaction dynamics depends significantly on the spin degrees of freedom.* With the start of the RHIC spin program, spin studies are moving again to the forefront of high energy physics.

In deep-inelastic scattering data analysis in the framework of perturbative QCD provides information on the longitudinal spin parton densities $\Delta q(x)$. The study of the transverse spin parton densities $\delta q(x)$ has recently become also an equally actual problem. However, in the usual deep-inelastic scattering the transverse spin contribute only as higher–twist contributions. The transverse spin densities can be studied in the Drell-Yan processes or in specific semi-inclusive DIS reactions.

In soft hadronic interactions significant single-spin effects could be expected since the helicity conservation does not work for interactions at large distances, once the chiral $SU(3)_L \times SU(3)_R$ symmetry of the QCD Lagrangian is spontaneously broken. However, the spin asymmetry A_N at low transverse momentum found to be small and decreases with energy. In contrast, the spin asymmetry increases at high transverse momentum in elastic scattering, where we should expect a decreasing behaviour based on the helicity conservation due to the chiral invariance of perturbative QCD. More theoretical work and experimental data will be needed to un-

R. Fiore et al. (eds.), Diffraction 2002, 175–184.

derstand the dynamics of these unexpected single-spin effects observed in elastic scattering and inclusive hyperon and meson production. In the following we consider these issues concentrating on the particular problem of hyperon polarization.

In 1976 it was discovered that in the reaction $pp \to \Lambda X$ highly polarized Λ–hyperons were produced, and their polarization rises with x_F and $p_\perp$. The polarization of Λ is obtained from the angular–distribution analysis of the decay products. Nowadays there is a rather extensive experimental set of data on the hyperon polarization in inclusive reactions. The experimental studies revealed the following regularities:

- In proton–proton interactions, the polarization of inclusively produced Λ–hyperons is negative, its direction is opposite to the vector $\vec{p} \times \vec{p}_\Lambda$, where $\vec{p}$ is the incident particle momentum. The polarization rises with transverse momentum of Λ–particle. For the range of transverse momentum $p_\perp > 0.8$ GeV/c the dependence of P_Λ on the transverse momentum becomes weak. In this region of $p_\perp$, the polarization of Λ grows linearly with x_F. The Λ–hyperon polarization has been measured up to rather high values of $p_\perp \simeq 3.5$ GeV/c. However, even at the maximum values of $p_\perp$ no tendency for a decreasing polarization was observed. It is interesting that spin asymmetry A_N and spin transfer parameter D_{NN} have shown similar $p_\perp$–dependence in Λ–production at FNAL.
- The Λ–hyperon polarization is energy independent in a wide range of beam energy $p = 12 - 2000$ GeV/c.
- In proton–proton interactions, hyperon production reactions reveal the following relations between the hyperon polarizations:

$$P_{\Sigma^0} \simeq P_{\Sigma^-} \simeq P_{\Sigma^+} \simeq -P_\Lambda \simeq -P_{\Xi^0} \simeq -P_{\Xi^-},$$

$$P_{\bar{\Lambda}} \simeq P_p \simeq P_{\Omega^-} \simeq 0, \quad P_\Lambda(p \to \Lambda) \simeq P_{\bar{\Lambda}}(\bar{p} \to \bar{\Lambda}).$$

- For the reactions involving mesons the following relations are observed:

$$P_\Lambda(K^- \to \Lambda) > 0, \quad |P_\Lambda(K^- \to \Lambda)| \gg |P_\Lambda(p \to \Lambda)|$$

and

$$|P_{\bar{\Lambda}}(K^+ \to \bar{\Lambda})| \gg |P_{\bar{\Lambda}}(\bar{p} \to \bar{\Lambda})|.$$

The polarization of Ξ^-–hyperons produced by 800–GeV protons has also been measured at FNAL. Contrary to the regularities observed in the Λ–production, energy dependence of P_{Ξ^-} has been observed for the first time. The quantity P_{Ξ^-} also does not show the x_F–dependence as observed in Λ–polarization data.

To determine the polarization of directly produced hyperons one has to measure the hyperon polarization in exclusive reactions. To this end, the Λ–hyperon polarization has been studied in the exclusive diffractive process

$$p + p \to p + [\Lambda^0 K^+].$$

Results of the measurements demonstrated that Λ–hyperon polarization increases linearly with $p_\perp$ at $\sqrt{s} \simeq 30$ GeV and gets a very large value, $P_\Lambda \simeq 80\%$, at $p_\perp \simeq 1$ GeV/c.

Therefore, since the Λ polarization does not show a decrease at large $p_\perp$ values, the data on hyperon polarization provide a direct evidence for the s–channel helicity nonconservation in hadron collisions in a wide energy and $p_\perp$ range.

Hyperon polarization even the most simple case — the polarization of Λ's — is not understood in pQCD and it seems that in the future it could become an even more serious problem than the nucleon spin problem. Those problems are likely related. One could attempt to connect the spin structure studied in deep–inelastic scattering with the polarization of Λ's observed in hadron production. As is widely known now, only a part (less than one third) of the proton spin is due to spins of the quarks.

Experimental results on the nucleon structure can be interpreted in an effective QCD approach ascribing a substantial part of the hadron spin to orbital angular momentum of the quark matter. It is natural to assume that this orbital angular momentum could be the origin of the asymmetries in hadron production. It is also evident from the recent deep–inelastic scattering data that strange quarks play essential role in the proton structure and in its spin balance. Polarization effects in hyperon production are complimentary to these results and demonstrate also that strange quarks acquire polarization in the course of the hadron interactions. The explicit formula for the asymmetry (similar formula takes place for polarization) in terms of the helicity amplitudes has the following form:

$$A_N = 2\frac{\sum_{X,\lambda_X,\lambda_2} \int d\Gamma_X \mathrm{Im}[F_{\lambda_X;+,\lambda_2} F^*_{\lambda_X;-,\lambda_2}]}{\sum_{X,\lambda_X;\lambda_1\lambda_2} \int d\Gamma_X |F_{\lambda_X;\lambda_1,\lambda_2}|^2},$$

and shows that non-zero single spin asymmetries require the presence of helicity flip and nonflip amplitudes and a phase difference between those amplitudes. A straightforward application of perturbative QCD using collinear factorization scheme, where e. g.

$$A_N d\sigma \sim \sum_{ab\to cd} \int d\xi_A d\xi_B \frac{dz}{z} \delta f_{a/A}(\xi_A) f_{b/B}(\xi_B) \hat{a}_N d\hat{\sigma}_{ab\to cd} D_{C/c}(z) \quad (1)$$

encounters serious difficulties in trying to explain the measured polarization results. This is due to the chiral invariance and vector nature of QCD

Lagrangian which provide important consequences for spin observables and therefore allows for direct and unambiguous experimental test of perturbative QCD. Indeed, the QCD Lagrangian is invariant under transformation of $SU(3)_L \times SU(3)_R$ group ($N_f = 3$) and therefore QCD interactions are the same for left and right quarks:

$$\bar{\psi}\gamma_\mu\psi A^\mu = \bar{\psi}_L\gamma_\mu\psi_L A^\mu + \bar{\psi}_R\gamma_\mu\psi_R A^\mu .$$

Left–handed (right–handed) massless particles will always stay left handed (right–handed). For massless quarks chirality and helicity coincide. If the quarks have a non–zero mass chirality and helicity are only approximately equal at high energies, namely:

$$\psi_{1/2} = \psi_R + O\left(\frac{m}{\sqrt{\hat{s}}}\right)\psi_L, \quad \psi_{-1/2} = \psi_L + O\left(\frac{m}{\sqrt{\hat{s}}}\right)\psi_R$$

where $\pm 1/2$ are the quark helicities. Any quark line entering a Feynman diagram corresponding the QCD Lagrangian emerges from it with unchanged helicity. Quark helicity conservation is the most characteristic feature of this theory. To get quark polarization values $\hat{P}_q \neq 0$ it is necessary that helicity flip amplitude is a non–zero and that phases of helicity flip F_f and non–flip F_{nf} amplitudes are different, since

$$\hat{a}_N, \; \hat{P}_q \propto \mathrm{Im}(F_{nf}F_f^*).$$

In QCD the quark helicity flip amplitude is of order of $m/\sqrt{\hat{s}}$. In the Born approximation the amplitudes are real and therefore one needs to consider the loop diagrams , i.e. the quark helicity flip amplitude will be proportional to

$$F_f^q \propto \frac{\alpha_s m_q}{\sqrt{\hat{s}}} F_{nf}^q,$$

and consequently the quark polarization is vanishingly small in hard interactions:

$$\hat{a}_N, \; \hat{P}_q \propto \frac{\alpha_s m_q}{\sqrt{\hat{s}}}$$

due to large value of the hard scale $\sqrt{\hat{s}} \sim p_\perp$ and small values of α_s and m_q. Here m_q stands for the mass of the current quark. Explicit calculation of s-quark polarization in gluon fusion process yield $\hat{P}_s \leq 4\%$. Thus, a *straightforward collinear factorization leads to very small* P_Λ.

Several modifications of this simple perturbative QCD scheme have been proposed. The factorization formula including higher twist contributions was obtained and it was recently proposed to consider also twist-3 contributions

$$E_{a/A}(\xi_{1A}, \xi_{2A}) \otimes q_{b/B}(\xi_B) \otimes \delta D_{c/\Lambda}(z) \otimes d\hat{\sigma}_{ab\to cd}$$

as a source for the strange quark polarization. Here the function $E_{a/A}(\xi_1, \xi_2)$ is a twist-3 unpolarized distribution, and higher twist partonic correlations could also give additional contributions to fragmentation functions. Since there is no information on higher twist contributions, additional assumptions and models are necessary. Moreover it is not clear what the dynamical origin is of strange quark polarization that arising through the unpolarized quark-gluon correlator $E_{a/A}(\xi_1, \xi_2)$. *The predicted decreasing dependence* $P_\Lambda \sim 1/p_\perp$ *at high* $p_\perp$ *still does not correspond to the experimental trends.*

Note that the situation now appeares to be complicated even for the standard leading twist parton distributions which, are not simply related to the structure functions measured in deep-inelastic scattering as it was commonly accepted until now. The crucial role here belongs to the final-state interactions, which could provide a phase difference and lead to non-zero single-spin asymmetries in deep-inelastic scattering.

Another modification of the collinear factorization takes into account a role of $k_\perp$-effects. By analogy with Collins' suggestion for the fragmentation of a transversely polarized quark it was proposed to consider so called polarizing fragmentation functions. Thus, in this approach, the source of Λ polarization was shifted into the polarizing fragmentation function. Some features of this scheme are:

- a falling $\sim \langle k_\perp \rangle / p_\perp$ dependence of the polarization at large $p_\perp$,
- still need a dynamical model for polarizing fragmentation functions,
- it is not compatible with the e^+e^- - data at LEP, where no significant transverse polarization of the Λ was found.

The overall conclusion is that the dynamics of Λ *polarization as well as of other single spin asymmetries within pQCD (leading twist collinear factorization or its modifications) is far from being settled.* The essential point here is the assumption that at short distances the vacuum is perturbative. The study of the $p_\perp$–dependence of the one–spin asymmetries can be used as a way to reveal the transition from the non–perturbative phase ($P \neq 0$) to the perturbative phase ($P = 0$) of QCD. The very existence of such transition can not be taken for granted since the vacuum, even at short distances, could be filled with the fluctuations of gluon or quark fields. The measurements of the one–spin transverse asymmetries and polarization will be an important probe of the chiral structure of the effective QCD Lagrangian.

At the same time we can note that polarization effects as well as some other recent experimental data demonstrate that hadron interactions have a significant degree of coherence. The persistence of elastic scattering at high energies and significant polarization effects at large angles confirm this. It means that the polarization dynamics has its roots hidden in the genuine non-perturbative sector of QCD. Several models exploiting confinement and chiral symmetry breaking have been proposed. The best known are the model based on the classical string picture (Lund model) where the polarization of a strange quark is the result of angular momentum conservation and the model based on the polarization of strange quark due to the Thomas precession. The main feature of the both models is that these take into account for the consequences of quark confinement.

It is worth noting that the chiral $SU(3)_L \times SU(3)_R$ group of the QCD Lagrangian is not observed in the hadron spectrum. In the real world this group is broken down to $SU(3)_V$ with the appearance of the $N_f^2 - 1 = 8$ Goldstone bosons (π, K and η). Chirality, being a symmetry of QCD Lagrangian, is broken (hidden) by the vacuum state

$$\langle 0|\bar{\psi}\psi|0\rangle = \langle 0|\bar{\psi}_L\psi_R + \bar{\psi}_R\psi_L|0\rangle.$$

The scale of the spontaneous chiral symmetry breaking is $\Lambda_\chi \simeq 4\pi f_\pi \simeq 1$ GeV, where f_π is the pion decay constant. Chiral symmetry breaking generates quark masses:

$$m_U = m_u - 2g_4\langle 0|\bar{u}u|0\rangle - 2g_6\langle 0|\bar{d}d|0\rangle\langle 0|\bar{s}s|0\rangle.$$

Massive quarks are quasiparticles (current quarks surrounded by cloud of quark–antiquark pairs of different flavors). It is worth to stress that in addition to u and d quarks the constituent quarks (U, for example) contain pairs of strange quarks and the ratio of the scalar density matrix elements

$$y = \langle U|\bar{s}s|U\rangle / \langle U|\bar{u}u + \bar{d}d + \bar{s}s|U\rangle \tag{2}$$

is estimated as $y = 0.1 - 0.5$. The spin of the constituent quark J_U is

$$J_U = 1/2 = S_{u_v} + S_{\{\bar{q}q\}} + L_{\{\bar{q}q\}} = 1/2 + S_{\{\bar{q}q\}} + L_{\{\bar{q}q\}}.$$

$$1/2(\Delta\Sigma)_U = 1/2 + S_{\{\bar{q}q\}}, \quad (\Delta\Sigma)_p = (\Delta U + \Delta D)(\Delta\Sigma)_U, \quad (\Delta\Sigma)_p \simeq 0.2,$$

leading to $L_{\{\bar{q}q\}} \simeq 0.4$ Accounting for the axial anomaly in the framework of chiral quark models results in the compensation of the valence quark helicity by helicities of quarks from the cloud in the structure of constituent quark. The specific nonperturbative mechanism of such compensation may be different in different models.

Orbital angular momentum, i.e. orbital motion of quark matter inside constituent quark, could generate the observed asymmetries in inclusive production on a polarized target at moderate and high transverse momenta. The main points of this mechanism are:

- the asymmetry in the pion production reflects the internal spin structure of the constituent quarks, i.e. it arises from the orbital angular momentum of the current quarks inside the constituent quark structure;
- the sign of the asymmetry and its value are proportional to the polarization of the constituent quark inside the polarized nucleon.

The generic behavior of the different asymmetries in inclusive meson and hyperon production

$$P,\ A_N(s,x,p_\perp) = \frac{\sin[\mathcal{P}_{\tilde{Q}}(x)\langle L_{\{\bar{q}q\}}\rangle]W_+^h(s,x,p_\perp)}{W_+^s(s,x,p_\perp)+W_+^h(s,x,p_\perp)}, \tag{3}$$

(where the functions $W_+^{s,h}$ are determined by the interactions at large and small distances) was predicted to have a characteristic $p_\perp$-dependence, in particular: vanishing asymmetry for $p_\perp < \Lambda_\chi$, an increase in the region of $p_\perp \simeq \Lambda_\chi$, and a $p_\perp$-independent (flat) asymmetry for $p_\perp > \Lambda_\chi$. The parameter $\Lambda_\chi \simeq 1$ GeV/c is determined by the scale of spontaneous chiral symmetry breaking. This behavior of the asymmetry follows from the fact that the constituent quarks themselves have a slow (if at all) orbital motion and are in an S-state. Interactions with $p_\perp > \Lambda_\chi$ resolve the internal structure of constituent quark and "feel" the presence of internal orbital momenta inside this constituent quark. The polarization of Λ-hyperons in the model is result of the polarization the constituent quark due to multiple scattering in the effective field. It means that s-quark in the structure of constituent quark will be polarized too, since in the model the spins of quark-antiquark pairs and their angular orbital momenta are exactly compensated:

$$L_{\{\bar{q}q\}} = -S_{\{\bar{q}q\}}. \tag{4}$$

It should be noted that DIS data show that strange quarks are negatively polarized in polarized nucleon, $\Delta s \simeq -0.1$. Also elastic νp-scattering data provide a negative value $\Delta s = -0.15 \pm 0.08$. The presence and polarization of strange quarks inside a hadron should also give an experimental signal in hadronic reactions.

The parent constituent quark and the strange quark have opposite polarization. The size of the constituent quark is determined by the scale of the chiral symmetry breaking, i. e. $R_Q \simeq 1/\Lambda_\chi$. The polarization is significant when the interaction resolves the internal structure of the constituent quark: $r < R_Q$ i. e. for $p_\perp > \Lambda_\chi \simeq 1$ GeV/c. It is useful to note here that current quarks appear in the nonperturbative vacuum and become quasiparticles due to the nonperturbative "dressing up" with a cloud of $\bar{q}q$-pairs. The mechanism of this process could be associated with the strong coupling existing in the pseudoscalar channel.

Thus, this model predicts a similarity of the $p_\perp$-dependencies for the different spin observables: increase up to $p_\perp = \Lambda_\chi$ and flat dependence at $p_\perp > \Lambda_\chi$.

The transition to the partonic picture in this model is described by the introduction of a momentum cutoff $\Lambda = \Lambda_\chi \simeq 1$ GeV/c, which corresponds to the scale of spontaneous chiral symmetry breaking. We shall see that at higher $p_\perp$ the constituent quark to be a cluster of partons, which however should preserve their orbital momenta, i.e. the orbital angular momentum will be retained and the partons in the cluster are correlated. Note, that the short–distance interaction in this approach observes a coherent rotation of correlated $\bar{q}q$–pairs inside the constituent quark and not a gas of the free partons. The nonzero internal orbital momenta of partons in the constituent quark means that there are significant multiparton correlations. Indeed, a high locality of strange sea in the nucleon was found experimentally by the CCFR collaboration at FNAL. This locality serves as a measure of the local proximity of strange quark and antiquark in momentum and coordinate space. It was shown that the CCFR data indicate that the strange quark and antiquark have very similar distributions in momentum and coordinate space.

Experimentally observed persistence and constancy of Λ–hyperon polarization means that chiral symmetry is not restored in the region of energy and values of $p_\perp$ where experimental measurements were performed. Otherwise we would not have any constituent quarks and should expect a vanishing polarization of Λ. It is interesting to perform *Λ–polarization measurements at RHIC and the LHC. It would allow to make a direct check of perturbative QCD and allow to make a cross-check of the QCD background estimations based on perturbative calculations for the LHC.* It is an important problem which still awaits its solution.

Experimentally measurable effects of 10% or higher are expected at c.m.s. energy $\sqrt{s} = 500$ GeV for values of pseudorapidity $|\eta| \geq 2.3$, and for energies 2 TeV and 10 TeV at $|\eta| \geq 3.7$ and $|\eta| \geq 5$ respectively.

The use of a polarized beam will allow to measure the two–spin correlations in hyperon production at TeV energies at the LHC, and to reveal the underlying mechanism. Moreover, measurements of the transverse and longitudinal asymmetries provide information on production mechanism and on hadron spin structure. It seems very promising to study reactions with weakly decaying baryons in the final state such as $p_{\uparrow,\rightarrow} + p \rightarrow \Lambda_{\uparrow,\rightarrow} + X$ and to measure the parameters D_{NN} and D_{LL}. These asymmetries are calculable in the framework of perturbative QCD. The data on D_{NN} and D_{LL} in the fragmentation region at large x_F seem to be interesting also from the point of view of the polarization of the strange sea and strangeness content of the nucleon. The strange sea has a large negative polarization value according to the interpretation of the polarized deep–inelastic scattering results, and it should be revealed in D_{NN} and D_{LL} behavior.

Note, that the corresponding asymmetry in neutral pion production has a value about $5-6\%$ in the same kinematical region.

If two polarized nucleons are available in the initial state then one could measure three–spin parameters such as $(n,n,n,0)$ and $(l,l,l,0)$ in the process: $p_{\uparrow,\rightarrow} + p_{\uparrow,\rightarrow} \rightarrow \Lambda_{\uparrow,\rightarrow} + X$, where the polarization of Λ–hyperon is studied through its decay. Measurements of the three–spin correlation parameters would provide important data for the study of hyperon production dynamics and mechanisms for hyperon polarization. On that basis we expect, as it was already mentioned, similar $p_\perp$–dependence for all spin parameters when initial hadrons have no valence strange quarks.

When one of the colliding hadrons has a valence strange quark, the picture will include two mechanisms: the one described above and another one – polarization of valence strange quarks in the effective field. The latter mechanism will provide Λ's polarized in the opposite direction and can explain the positive large polarization P_Λ in K^-p–interactions.

As it was claimed in on the basis of DIS nonperturbative analysis, that the orbital angular momentum increases with virtuality Q^2. Thus, in the framework of the model studied and according to the relation (4) it would lead to increasing polarization of strange quark. It means, in particular, that the transverse polarization $P_\Lambda(Q^2)$ in the current fragmentation region of the semi–inclusive process $l + p \rightarrow l + \Lambda_\uparrow + X$ will also increase with virtuality, provided the competing mechanism of the polarization of the valence strange quark from the φ–meson gives a small contribution. Regardless of these particular predictions, it seems that the experimental studies of polarization dependence on virtuality could shed new light on this problem.

On the base of the model in one expects zero polarization in the region where QGP has formed, since chiral symmetry is restored and there is no room for quasiparticles such as constituent quarks. The absence or strong diminishing of transverse hyperon polarization can be used therefore as a signal of QGP formation in heavy-ion collisions. This prediction should also be valid for the models based on confinement, e.g. the Lund and Thomas precession model. In particular, the polarization of Λ in heavy–ion collisions in the model based on the Thomas precession was described in where nuclear effects were discussed as well. However, we do not expect a strong diminishing of the Λ–polarization due to the nuclear effects: the available data show a weak A–dependence and are not sensitive to the type of the target. Thus, we could use a vanishing polarization of Λ–hyperons in heavy ion collisions as a sole result of QGP formation provided the corresponding observable is non-zero in proton–proton collisions. The prediction based on this observation would be a decreasing behavior of polarization of Λ with the impact parameter in heavy-ion collisions in the region of energies and densities where QGP was produced: $P_\Lambda(b) \to 0$ at $b \to 0$, since the overlap is maximal at $b = 0$. The value of the impact parameter can be controlled by the centrality in heavy–ion collisions. The experimental program should therefore include measurements of Λ–polarization in pp–interactions first, and then if a significant polarization would be measured, the corresponding measurements could be a useful tool for the QGP detection. Such measurements seem to be experimentally feasible at RHIC and LHC provided it is supplemented with forward detectors.

In conclusion we would like to emphasize the main points of our discussion:

- a universal $p_\perp$–dependence for all spin parameters in Λ–hyperon production which reflects a finite size for constituent quarks is predicted: all spin parameters in the collisions of hadrons without valence strange quarks demonstrate an increase in absolute value up to $p_\perp \simeq 1$ GeV/c and then become flat.
- three–spin correlation parameters – the new observables which can be measured in hyperon production with polarized beams – can provide new insight into the mechanism of hyperon polarization
- hyperon polarization should vanish in heavy–ion collisions when quark–gluon plasma is formed, its decrease with centrality can be considered as a gold–plated signature of QGP formation.

We have not included here figures and lengthy list of references due to space limitations. Interested people can find them in hep-ph/0201267.

One of the authors (S.T.) is very grateful to Laszlo Jenkovszky and Organizing Committee of the Second International "Cetraro" Workshop and NATO Advanced Research Workshop "DIFFRACTION 2002" for the support and warm hospitality in Alushta.

MULTIPARTICLE DYNAMICS OF HADRON DIFFRACTION

A. R. MAŁECKI
KEN Pedagogical University, Kraków, Poland
Instytut Fizyki AP, ul. Podchorążych 2, PL 30-084 Kraków
E-mail: AMALECKI@ULTRA.AP.KRAKOW.PL

Abstract

An approach to inelastic diffraction based on the concept of equivalence of diffractive states is presented. Intermediate transitions inside the equivalence class yield a multi-channel correction which can be factorized producing the diffraction amplitude in the form $N\Delta t$, to be taken in the Bjorken-like 'diffractive limit': $N \to \infty, \Delta t \to 0$ such that $N\Delta t$ is finite. We analyse the contribution of inelastic diffraction to elastic scattering of elementary hadrons and light nuclei, in the range of c.m. energy $\sqrt{s}$= 20 - 1800 GeV. This is done in the framework of a semi-phenomenological model where the diffractive states are built of a two-hadron bulk and some quanta ('diffractons') describing diffractive excitations. The single minimum observed experimentally is explained as an interference effect due to scattering off 'diffractons', present inside the hadronic bulk.

1. Hadron diffraction

The term 'diffraction', originating from classical optics, is frequently used in describing nuclear and hadronic collisions. Its use is justified by striking analogies between light scattering in optics and phenomena which occur in medium-energy nuclear and high-energy hadron collisions. The sufficient energy of collision allows there an opening of a variety of inelastic channels which in turn implies strong absorption since the presence of competing reactions results in a considerable depletion of the particle flux in the elastic channel [1]. Such conditions are analogous to those for diffraction of light by opaque or partially transparent objects in optics.

R. Fiore et al. (eds.), Diffraction 2002, 185–197.

The optical diffraction analogy of high energy hadronic collisions consists in a substantial presence of elastic scattering (and other two-body channels) where very little would naively be expected in violent collisions. More quantitatively, 'diffractive' phenomenon refers to the behaviour of the hadronic differential cross-sections that are strongly peaked in the forward direction and often appears as a series of maxima and minima. Similarily, the diffraction of light on an obstacle leads to a structured penumbra, instead of the darkness expected in geometrical optics, resulting from the alternating constructive and destructive interference of deflected waves.

The great successes of the Glauber-Sitenko model [2,3] in nuclear scattering and of the Chou-Yang model [4] in hadron collisions represent an astonishing evidence that using optical concepts, reflecting merely the fact that nuclei and hadrons have finite sizes, one can achieve a successful description of the basic features of their scattering. Obviously, still many problems, related to the more detailed structure of hadrons, remain. In particular, the hadronic analogy of 'diffractive structure' of the elastic differential cross-section is often obscured since multiple dips and reinforcements may not be present [5]. In fact, numerous dips arising in geometrical models of hadron scattering may be washed out when including the contributions from multi-particle intermediate states [6-8]. Thus the optical resemblances of high-energy hadron diffraction should not be overemphasized. The geometrical picture of diffraction can still be useful for modelling the dominant long-range part of scattering. However, it would be highly desirable to disentangle from the vagueness of geometrical diffraction also phenomena of shorter range related to intrinsic dynamics of colliding particles.

The way in this direction goes through a better understanding of the process of 'inelastic diffraction' [3,9] which involves quasi-elastic transitions with no exchange of intrinsic quantum numbers. The name 'diffraction' refers here merely to the condition of coherence that must (like in elastic diffraction) be satisfied to assure that the interacting particles do not change their character. In elastic scattering where the intrinsic dynamics is hidden inside geometrical shapes, the analogy between optical and hadron diffraction is really very deep. On the contrary, the 'inelastic diffraction' has no classical analogy; it appears as a peculiar quantum phenomenon related to the existence of internal degrees of freedom.

The requirement of quasi-elasticity or 'diffractiveness' of an inelastic transition can be incorporated into a theoretical formalism in either of two ways. In the t-channel approach, inelastic diffraction is described in terms of the exchange of Pomeron, an hypothetical object carrying vacuum quantum numbers. An approach used here, is that of the s-channel. It is based on the presumed relation of equivalence [5-8] between the initial (ground) state and the states involved in diffractive channels. This means

that the Hilbert space of physical states is decomposable into subspaces of diffractive and non-diffractive states. This assumption is based on a phenomenologically confirmed [11] division of inelastic channels:
i) *diffractive* channels are characterized by a slow variation with energy of their cross-sections, i.e. by the energy dependence typical of elastic scattering. Indeed, the states produced in these channels have dominant quantum numbers corresponding to the ground state.
ii) *non-diffractive* channels where cross-sections change rapidly with energy.

2. Diffractive filter

In order to reveal the equivalence of states one must depart from the representation of physical states as eigenstates of the hadronic Hamiltonian in which each state is an equivalence class for itself. One is thus looking for a suitable unitary transformation that allows to expand the physical states in terms of the transposed states.

The natural base for describing 'inelastic diffraction' is obtained through a unitary transformation of physical states such that the transforming operator is reducible in the Hilbert space. In fact, the fundamental point in the description of diffraction is the presumed existence of two orthogonal subspaces of diffractive and non-diffractive states. This requirement can be rephrased by saying that there exist unitary operators U and $U^\dagger$ which are reducible in the Hilbert space of physical states. This implies the existence of a non-trivial subspace $[D]$ such that for any $|j\rangle \in [D]$ also $|Uj\rangle$ and $|U^\dagger j\rangle$ belong to $[D]$. In consequence, for any state $|k\rangle$ belonging to the orthogonal complement $[\sim D]$ also $|Uk\rangle$ and $|U^\dagger k\rangle$ will belong to $[\sim D]$. In terms of the matrix elements this reads:

$$\begin{aligned} \langle k \mid Uj\rangle &= \langle j \mid U^\dagger k\rangle^* = 0, \\ \langle k \mid U^\dagger j\rangle &= \langle j \mid Uk\rangle^* = 0 \end{aligned} \tag{1}$$

for any $|j\rangle \in [D]$ and $|k\rangle \in [\sim D]$.

Alternatively one may say that the operator U can be decomposed into the direct sum of operators which act on orthogonal subspaces (classes of equivalence) of the physical Hilbert space:

$$U = U_D \oplus U_{\sim D} \equiv U_{DF}. \tag{2}$$

Such an operator will be referred to [8] as the unitary diffractive filter U_{DF}.

In the base of physical states the matrix representing U_{DF} has a 'diagonal box' form. Assuming that the initial state $|i\rangle$ belongs to $[D]$ one has then:

$$\begin{aligned} U_{ij} &\neq 0 \rightarrow | j\rangle \in [D], \\ U_{ij} &= 0 \rightarrow | j\rangle \in [\sim D]. \end{aligned} \tag{3}$$

The complementary subspace $[\sim D]$ of states which are non-diffractive with respect to the set $[D]$ may eventually further be decomposed into smaller classes of equivalence. The matrix of U_{DF} would then be made of more than two 'boxes'.

It is convenient to extract from the unitary operator U_{DF} the identity transformation:

$$U_{DF} \equiv 1 - \Lambda. \tag{4}$$

The operator Λ satisfies the relation of normality:

$$\Lambda\Lambda^{\dagger} = \Lambda^{\dagger}\Lambda = \Lambda + \Lambda^{\dagger}. \tag{5}$$

Obviously the reducibility of U_{DF} implies that the corresponding operators Λ and $\Lambda^{\dagger}$ will also be reducible in the space of physical states, i.e. the matrix of Λ is block-diagonal. The property of normality (5) allows their diagonalization.

Writing the operator U_{DF} in a manifestly unitary form:

$$U_{DF} = e^{iM}, \; M = M^{\dagger} \tag{6}$$

we have also at disposal the hermitian operator M. There exist the common eigenstates $|\mu\rangle$ for which

$$M \mid \mu\rangle = 2\,\pi\,\mu \mid \mu\rangle, \tag{7}$$

$$\Lambda \mid \mu\rangle = (1 - e^{2\pi i\lambda}) \mid \mu\rangle \tag{8}$$

where

$$\mu = \lambda + n;\; n = 0, \pm 1, \pm 2, \ldots;\; 0 \leq \lambda < 1. \tag{9}$$

The states $|\mu\rangle = |\lambda + n\rangle$ are infinitely degenerated with respect to the eigenvalue λ. Thus with each value of λ one can associate a subspace $[P_\lambda]$ and the projector operator

$$P_\lambda = \sum_{n=-\infty}^{\infty} \mid \lambda + n\rangle\langle\lambda + n \mid \tag{10}$$

which projects onto this subspace.

This prompts a possible connection between the eigenspaces $[P_\lambda]$ of the operator Λ and the classes of equivalence of physical states. The states will be said to be equivalent *modulo* Λ if they belong to one of the direct sums:

$$[\lambda] = [P_\lambda] \oplus [P_0] \tag{11}$$

for each $\lambda \neq 0$.

Thus for any physical state $|j\rangle \in [\lambda]$ one has:

$$| j\rangle = P_\lambda | j\rangle + P_0 | j\rangle \tag{12}$$

or more explicitly

$$| j\rangle = \sum_{n=-\infty}^{\infty} \varphi_{\lambda j}(n) | \lambda + n\rangle + \sum_{n=-\infty}^{\infty} \varphi_{0j}(n) | n\rangle \tag{13}$$

where $\varphi_{\lambda j}(n) \equiv \langle \lambda + n | j\rangle$.

The form of this expansion implies that the diffractive equivalent states will generally have a complicated structure which may be related to the compositness of the colliding hadrons.

3. Diffractive eigenstates

In this section we comment on standard methods of the s-channel approach to diffraction which are based on the notion of "diffraction eigenstates". The decomposition of the Hilbert space of physical states is there taken for granted, hence the notion of an unitary diffractive filter U_{DF} is not even discussed. Instead, the accent is put on diagonalization of the scattering operator in physical states. Following the classical paper of Good and Walker [12] such a diagonal operator

$$T_0 = U_{GW}^\dagger T U_{GW} \tag{14}$$

exists since the scattering operator is normal. Indeed, the unitarity of the collision operator $S = 1 + iT$ implies:

$$TT^\dagger = T^\dagger T = i(T^\dagger - T).$$

The diagonalization in all physical states

$$T_0 | j\rangle = t_j | j\rangle \tag{15}$$

was replaced with a weaker assumption of diagonalization of T in a particular class of states only [13]. Denoting the chosen subset of states by [D] and its orthogonal complement by [~D] one has for any state $| j\rangle \in [D]$:

$$T_0 \ | j\rangle = t_j | j\rangle + \sum_{|k\rangle \in [\sim D]} t_{kj} \ | k\rangle. \tag{16}$$

Eq. (16) was interpreted as the requirement that the base states of the diffractive sector [D] are subject only to elastic scattering which arises from absorption related to the production of non-diffractive states from [~D].

We claim that the considered in Eq. (2) reducibility of the unitary diffractive filter U_{DF} does not imply that the scattering operator T_0 is also reducible in the space of physical states. In particular, T_0 restricted to the subspace of diffractive states may no longer be normal. Therefore we reject the condition (16) regarding the diagonalization of the scattering operator. First, it may be incompatible with the fundamental assumption regarding the division of states into classes [D] and [~D]. Secondly, the assumption (16) screens a part of dynamics. In fact, considering the most general expression:

$$T_0 \mid j\rangle = \sum_{|k\rangle \in [D]} t_{kj} \mid k\rangle + \sum_{|k\rangle \in [\sim D]} t_{kj} \mid k\rangle \tag{17}$$

one reveals there, besides the absorption of non-diffractive origin, also another source of absorption implied by transitions inside the set of diffractive states. Thus we treat the diagonalization of the scattering operator in Eq.(16) as a redundant assumption which can be incompatible with the fundamental assumption of reducibility of the Hilbert space into the subspaces of diffractive and non-diffractive states.

It should be reminded that the geometrical models of diffraction, with all their merits and limitations, are grounded on the approach of "diffractive eigenstates". In these models there is an implicit assumption that a very fast projectile passing through hadronic medium is outside of the target long before the changes it induces in the medium take place. Thus when the projectile interacts with any of the target constituents, the others are fixed in their positions and can be considered inactive spectators. This means that it is just the states $|\vec{b}_1, \ldots, \vec{b}_n\rangle$, describing the configurations of n hadron constituents with definite impact parameters, which are eigenstates of diffraction $\mid U_{GW} j\rangle$.

The transition amplitude can thus be written in a more familiar form:

$$T_{f\,i}(b) = \sum_{n=1}^{\infty} P_n \int d^2b_1, \ldots, d^2b_n \; \Phi_f^{\star} \; t(\vec{b}, \vec{b}_1, \ldots, \vec{b}_n) \; \Phi_i \tag{18}$$

where Φ_f, Φ_i are the wave functions, e.g. $\Phi_i(\vec{b}_1, \ldots, \vec{b}_n) \equiv \langle \vec{b}_1, \ldots, \vec{b}_n \mid i\rangle$, and P_n is the probability to find the configuration of n constituents.

Assuming the celebrated cluster form of the hadronic profile:

$$t(\vec{b}, \vec{b}_1, \ldots, \vec{b}_n) = i[1 - \prod_{k=1}^{n} [1 - \gamma(\vec{b} - \vec{b}_k)]] \tag{19}$$

one obtains then the nuclear Glauber-Sitenko model [2,3] with $P_n = \delta_{nA}$ and the Chou-Yang model [4] with the Poisson distribution

$$P_n = \frac{\langle n\rangle^n}{n!} exp(-\langle n\rangle) \tag{20}$$

where $\langle n \rangle$ is the mean value of n.

4. Diffractive limit

The states $|U_{DF} j\rangle$ obtained through the unitary transformation which reveals the decomposition of the Hilbert space of states into classes of equivalence constitute a natural base for the description of diffraction. The amplitude of diffractive transitions follows then from

$$T_{fi} \equiv \langle f \mid T \mid i \rangle = \sum_{|j\rangle,|k\rangle \in [D]} U_{fk}\, t_{kj}\, U^{\star}_{ij} \tag{21}$$

where

$$U_{fk} \equiv \langle f \mid U_{DF} \mid k \rangle, \tag{22}$$

$$t_{kj} \equiv \langle U_{DF} k \mid T \mid U_{DF} j \rangle \tag{23}$$

and

$$U_{ij} \equiv \langle i \mid U_{DF} \mid j \rangle. \tag{24}$$

In terms of the normal operator $\Lambda \equiv 1 - U_{DF}$ this reads:

$$T_{fi} = t_{fi}\delta_{fi} - \sum_{|k\rangle \in [D]} \Lambda_{fk}\, t_{ki} - \sum_{|j\rangle \in [D]} t_{fj}\Lambda^{\star}_{ij} + \sum_{|j\rangle,|k\rangle \in [D]} \Lambda_{fk} t_{kj} \Lambda^{\star}_{ij} \tag{25}$$

$$T_{fi} = t_i \delta_{fi} - N_{fi}(T_0)\Lambda_{fi} t_i - \Lambda^{\star}_{if} t_f N^{\star}_{if}(T_0^{\dagger}) + \sum_{|j\rangle \in [D]} N_{fj}(T_0)\Lambda_{fj} t_j \Lambda^{\star}_{ij} \tag{26}$$

where $t_j \equiv t_{jj}$ are the diagonal matrix elements of $T_0 \equiv U_{DF}{}^{\dagger} T U_{DF}$ and the undimensional quantities N_{kj} which serve to reduce the summations [8] are defined as follows:

$$N_{kj}(T_0) \equiv \frac{1}{\Lambda_{kj} t_{jj}} \sum_{|l\rangle \in [D]} \Lambda_{kl} t_{lj}. \tag{27}$$

If the subspace [D] contains a very large number of diffractive states then $N_{kj} \equiv N \to \infty$ for any pair of states $|k\rangle$ and $|j\rangle$. In fact, since Λ is a non-singular operator its matrix elements vary smoothly under the change of diffractive states. This leads to an enormous simplification of Eq.(26) in the limit $N \to \infty$:

$$T_{fi} = t_i \delta_{fi} - N(\Lambda_{fi} t_i + \Lambda^{\star}_{if} t_f - \sum_{|j\rangle \in [D]} \Lambda_{fj} t_j \Lambda^{\star}_{ij}). \tag{28}$$

In general, the effect of non-diagonal transitions inside the diffractive subspace [D] gets factorized. E.g., in the case of elastic scattering one has:

$$T_{ii} = t_i + N \sum_{|j\rangle \in [D]} |\Lambda_{ij}|^2 (t_j - t_i) = t_i + g_i N (t_{av} - t_i) \qquad (29)$$

where $g_i = \sum_{|j\rangle} |\Lambda_{ij}|^2 = 2Re(\Lambda_{ii})$ and

$$t_{av}^{(i)} = \frac{1}{g_i} \sum_{|j\rangle} |\Lambda_{ij}|^2 t_j \qquad (30)$$

is the average value of the diagonal matrix elements t_j.

The expressions of the form $N\Delta t$ where Δt represents diversity of t_j over the subspace of diffractive states [D] are to be considered in the double *diffractive limit* [5-8]:

$$N \to \infty, \quad \Delta t \to 0$$
$$such\ that \quad N\Delta t \quad is\ finite. \qquad (31)$$

The contribution of inelastic diffraction is thus built as an *infinite sum* of the *infinitesimal* contributions from all possible intermediate states belonging to [D].

5. Diffractons

Our numerical analysis of elastic scattering was done in the framework of a model where the diffractive states are built of a two–hadron core (representing the ground state) and some quanta describing diffractive excitations [8]:

$$|j\rangle = |i\rangle + |n; \vec{b}_1 \ldots \vec{b}_n\rangle. \qquad (32)$$

The configurations of these quasi-particles [8] (called *diffractons*) are specified by a number n of constituents and their impact parameters $\vec{b}_1, \ldots, \vec{b}_n$. Thus

$$\frac{1}{g_i} \sum_{|j\rangle \in [D]} |\Lambda_{ij}|^2 \ldots = \sum_{n=1}^{\infty} P_n \int d^2b_1 \ldots d^2b_n \prod_{k=1}^{n} |\Psi(b_k)|^2 \ldots \qquad (33)$$

where $|\Psi(b_k)|^2$ is the density of a spatial distribution of diffractons (with respect to the core) in the impact plane and P_n are probabilities of their number, approximated by Poisson distributions (20).

The diagonal matrix elements of T_0 (in b-space) are specified in terms of the real profile functions. In the diffractive limit (31), i.e. retaining only the single particle terms, we have:

$$N(t_j - t_i) = i(1 - \Gamma_0) \lim_{N\to\infty,\gamma\to 0} N \sum_{k=1}^{n} \gamma(\vec{b} - \vec{b}_k) \tag{34}$$

with $t_i = i\Gamma_0$ representing the hadronic core and γ's corresponding to diffractons. The diffracton model thus explicitly accounts for the *geometrical* diffraction on an absorbing hadronic bulk and the *dynamical* diffraction corresponding to intermediate transitions between diffractive states.

Our analysis, though based on the well-founded theoretical framework, has a semi-phenomenological character. Therefore the shapes of the profiles Γ_0, γ and of the density $|\psi(b)|^2$ are to be assumed. For simplicity, we take them as Gaussians [8,10]. Their strength and size parameters, as well as the coupling constant g_i and the mean number of diffractons $\langle n\rangle$, are to be determined from comparison of the theory with experiments.

6. Elastic scattering of hadrons and nuclei

For collisions of elementary hadrons the elastic scattering profile reads [6,8]

$$\Gamma_{el}(b) = \Gamma_0 + (1 - \Gamma_0)\Gamma_n \tag{35}$$

where the subhadronic profile Γ_n results from Eq.(34):

$$\Gamma_n(b) = Ng_i\langle n\rangle \int d^2s|\psi(s)|^2\gamma(\vec{b} - \vec{s}) \tag{36}$$

In the Gaussian approximation we need for the description of elastic scattering four parameters: the strength σ_0 and size R_0 of the hadronic bulk profile:

$$\Gamma_0(b) = \frac{\sigma_0}{4\pi R_0^2} exp(-\frac{b^2}{2R_0^2}), \tag{37}$$

as well as the corresponding strength σ_n and size R_n of the subhadronic profile.

The value of σ_n results from the diffractive limit:

$$\sigma_n \equiv \lim_{N\to\infty,\sigma_\epsilon\to 0} Ng_i\langle n\rangle\sigma_\epsilon \tag{38}$$

where σ_ϵ represents the strength of the diffracton profiles.

The diffractive elastic scattering amplitude in momentum q-space $T_{ii}(q)$ which is purely imaginary can be enriched with a contribution of the real

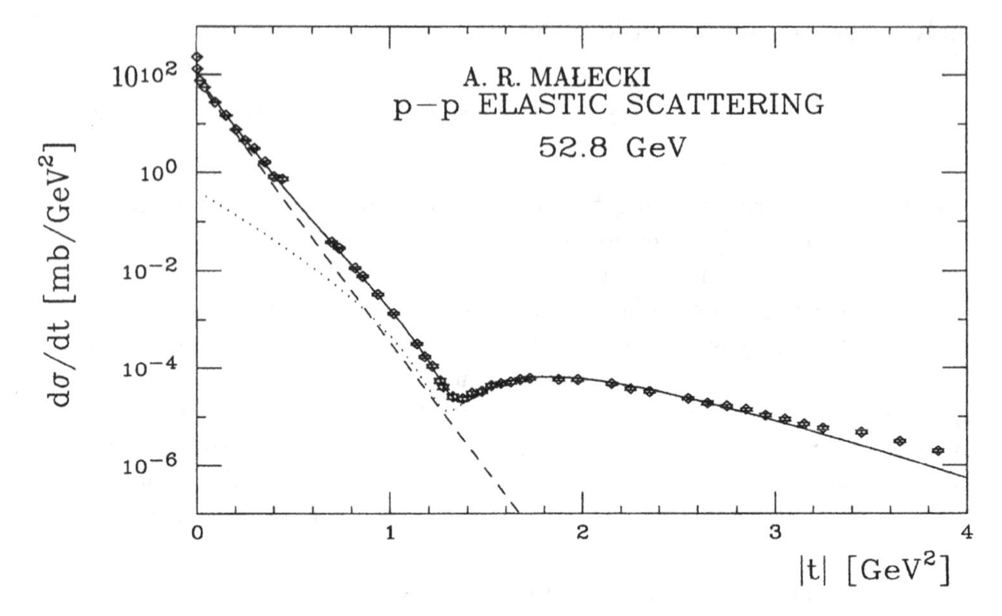

Figure 1. The p-p elastic cross-section at c.m. energy $\sqrt{s} = 52.8$ GeV in the function of the squared momentum transfer $|t|$. The experimental data [14] are compared with the results of our approach (solid curve). The non-diffractive (dashed curve) and diffractive (dotted curve) contributions to the cross-section are shown separately.

part using the prescription of Martin [6]. This introduces the fifth parameter: $\rho \equiv ReT_{ii}(0)/ImT_{ii}(0)$ to be found by comparing the elastic differential cross-section with experimental data.

In Fig.1 we show separately the two contributions to elastic scattering which are present in our approach. The dashed curve represents the term Γ_0 alone while the dotted curve corresponds to the diffractive term, i.e. to the second term in Eq. (35). At small momentum transfers the non-diffractive contribution is dominant. The diffractive term has a single zero which, filled up by the real part of the scattering amplitude, appears as a shallow minimum. Above the dip the non-diffractive contribution is negligible and the diffractive term dominates the elastic cross-section. The solid curve results from the sum of the two contributions in Eq. (35). It was fitted to 44 experimental points from the data on proton-proton elastic scattering at the c.m. energy $\sqrt{s}$=52.8 GeV [14]. The parameters found are $\sigma_0 = 39.4$ mb, $R_0 = 0.70$ fm, $\sigma_n = 5.52$ mb, $R_n = 0.41$ fm, $\rho = 0.066$.

Analogous fits were performed by us to the experimental data on proton-antiproton elastic scattering at the c.m. energies $\sqrt{s}$= 53 GeV, 546-630 GeV

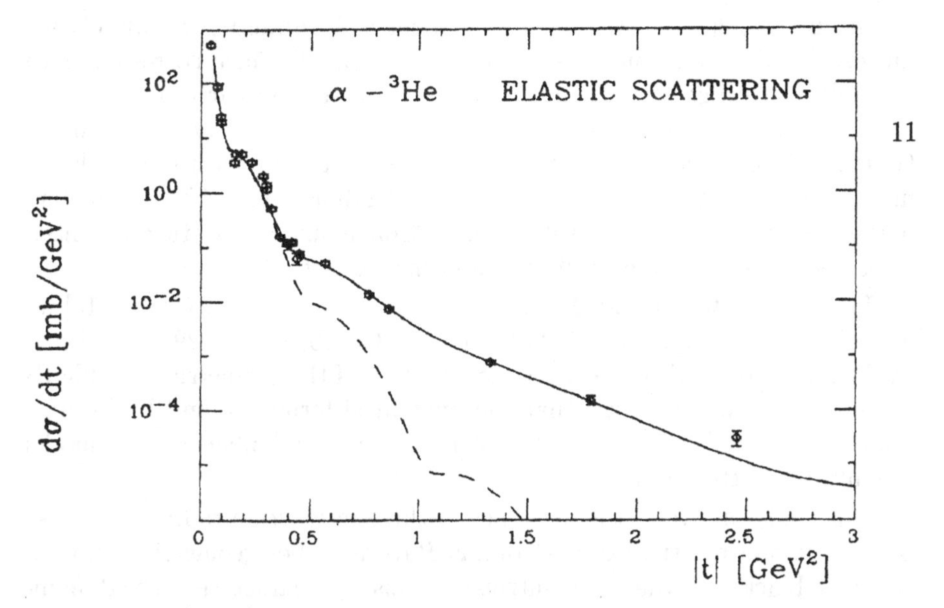

Figure 2. The ^{4}He – ^{3}He elastic cross–section in the function of the squared momentum transfer $|t|$. The experimental data [15] are compared with the results of our approach (solid curve). The non-diffractive contribution to the cross-section is shown separately (dashed curve).

and 1800 GeV. For all these energies the fits are excellent in the whole range of momentum transfer.

For high energies we always had: $\sigma_0 >> \sigma_n$ and $R_0 > R_n$. This is reasonable since the non-diffractive effects dominate a long-range part of scattering and are characterized by large values of the effective coupling strength. The diffractive scattering, on the other hand, is governed by short distance dynamics and small values of the coupling strength.

In the case of nucleus-nucleus scattering there are two sources of dynamical diffraction [10]. At the nuclear level the diffractive excitations correspond to various configurations of nucleons while the partonic compositness of nucleons gives rise to the subnuclear diffraction. Therefore the elastic profile reads:

$$\Gamma_{el}(b) = \Gamma_0 + (1 - \Gamma_0)\Gamma_N + (1 - \Gamma_0)(1 - \Gamma_N)\Gamma_n \qquad (39)$$

where Γ_N and Γ_n refer to nuclear and nucleonic diffractiveness, respectively.

In our calculations we took the geometrical diffraction profile of two nuclei with A and B nucleons as $\Gamma_0 = (1 - \gamma_{AB})^{AB}$ which corresponds to multiple scattering between nucleons frozen in the nuclear ground states. The function γ_{AB} and the dynamical profiles Γ_N and Γ_n were taken in the Gaussian form. The parameters of γ_{AB} are related to the total nucleon–nucleon cross–section and the radii of the colliding nuclei. The parameters of the profiles Γ_N and Γ_n were determined from fitting the elastic differential cross–section based on Eq.(39) to experimental data.

In the presented on Fig. 2 fit to the α–^{3}He data at $p_\alpha = 7$ GeV/c [15] we found $\sigma_N = 77.2$ mb, $R_N = 0.90$ fm, $\sigma_n = 83.9$ mb, $R_n = 0.29$ fm and $\rho =$ - 0.241. The comparison with the contribution of the geometrical profile Γ_0 alone (dashed curve) shows that the dynamical terms Γ_N and Γ_n become important at high momentum transfers where the Glauber-Sitenko model of multiple scattering fails.

Our analysis reveals the existence of three interaction radii in nucleus–nucleus scattering: the external radius R_{AB} describes geometrical diffraction on a black disc, the larger intrinsic radius R_N characterises the dynamics of interacting nucleons while the smaller radius R_n may be interpreted as the size of partonic clusters inside the nucleons. The large value of sub-nucleonic cross-section $\sigma_n > \sigma_N$ could signal an enhanced population of partons and their deconfinement in the whole nuclear interior.

References

[1] Frahn, W. E. (1985) *Diffractive Processes in Nuclear Physics,* Clarendon Press, Oxford.

[2] Glauber, R. J. (1959) High-Energy Collision Theory, in W. E. Brittin and L. G. Dunham (eds.), *Lectures in Theoretical Physics, vol.1*, Interscience, New York, pp. 315 - 414.

[3] Akhiezer, A. I. and Sitenko, A. G. (1957) Diffractional Scattering of Fast Deuterons by Nuclei, *Phys. Rev.* **106**, 1236 - 1246.

[4] Chou, T. T. and Yang, C. N. (1968) Model of Elastic High-Energy Scattering, *Phys. Rev.* **170**, 1591 - 1596.

[5] Etim, E., Małecki, A. and Satta, L. (1987) A New Definition of the Diffractive Limit with Applications to pp and $p\bar{p}$ Scattering, *Phys. Lett.* **B184**, 99 – 104.

[6] Małecki, A. (1989) The Real and Imaginary Part of the Elastic Scattering Amplitude in High Energy Diffraction, *Phys. Lett.* **B221**, 191 – 193.

[7] Małecki, A. (1991) Inelastic Diffraction and Equivalence of States, *Phys. Let* **B267**, 523 – 526.

[8] Małecki, A. R. (1996) Diffractive Limit Approach to Elastic Scattering and Inelastic Diffraction of High Energy Hadrons, *Phys. Rev.* **D54**, 3180 – 3193.
[9] Glauber, R. J. (1955) Deuteron Stripping Processes at High Energies, *Phys. Rev.* **99**, 1515 – 1516.
[10] Małecki, A. (1991) Nuclear and Partonic Dynamics in High Energy Elastic Nucleus-Nucleus Scattering, *Phys. Rev.* **C44**, *R*1273 – 1276.
[11] Goulianos, K. (1983) Diffractive Interactions of Hadrons at High Energies, Phys. *Rep.* **101**, 169 – 219.
[12] Good, M. L. and Walker, W. D. (1960) Diffraction Dissociation of Beam Particles, *Phys. Rev.* **120**, 1857 – 1860.
[13] Białas, A., Czyż, W. and Kotański, A. (1972) A Model of Diffractive Production in Hadron-Hadron Collisions, *Ann. of Phys.* **73**, 439 – 460.
[14] Schubert, K. R. (1979) *Tables of Nucleon - Nucleon Scattering,* Landolt-Börnstein, New Series, Vol. 1/9a, Springer, Berlin.
[15] Satta, L. et al. (1984) Elastic Scattering of α Particles on Light Nuclei at $p_\alpha = 7$ GeV/c, *Phys. Lett.* **B 139**, 263 - 266.

HIGH-ENERGY ELASTIC SCATTERING AND NUCLEON STRUCTURE

VOJTĚCH KUNDRÁT, MILOŠ LOKAJÍČEK
Institute of Physics, AS CR
182 21 Praha 8, Czech Republic

Abstract. The current model of elastic nucleon scattering based on hadronic amplitude with dominant imaginary part is still made use of in analyses of measured data even if it involves assumptions being not in harmony with known experimental facts. These assumptions are critically examined and a more general model is considered. Possible consequences following for elastic and diffractive production channels from such a model approach are discussed. An opaque nucleon structure against a transparent one is shown to be preferred.

1. Introduction

By investigating elastic scattering of α particles on nuclear targets Rutherford discovered one century ago that whole atom mass is concentrated predominately in atomic nucleus. Similarly the high-energy elastic nucleon collisions have been used for probing structure of nucleons. As the contemporary high-energy elastic scattering experiments deal mainly with elastic pp and $\bar{p}p$ collisions we will limit ourselves to the elastic scattering of charged nucleons; for the simplicity, spins of colliding nucleons will not be considered, either.

We shall analyze the series of steps which are standardly used to obtain information about nucleon structure from the measured high-energy elastic scattering data. We shall try to show that the majority of given approaches is based on assumptions that can be hardly theoretically justified or are even in contradiction to experimental facts. They can hardly lead to a true picture of nucleon structure.

The first step of data analysis consists in determinating the momentum transfer t dependence of differential cross section $\frac{d\sigma}{dt}$ from the measured

R. Fiore et al. (eds.), Diffraction 2002, 199–208.

counting rate. It is then possible to write

$$\frac{d\sigma}{dt} = \frac{\pi}{sp^2}|F^{C+N}(s,t)|^2, \tag{1}$$

where $F^{C+N}(s,t)$ is total elastic amplitude (including the influence of Coulomb scattering); s is square of total CMS energy and p is the value of corresponding CMS momentum. In all analyses of interference data the separation of hadronic amplitude $F^N(s,t)$ is currently done with the help of simplified West and Yennie amplitude [1]

$$F^{C+N}(s,t) = \pm\frac{\alpha s}{t} f_1(t) f_2(t) e^{i\alpha\Phi_{WY}} + \frac{\sigma_{tot}}{4\pi} p\sqrt{s}(\rho + i)e^{Bt/2}, \tag{2}$$

with the relative phase

$$\Phi_{WY} = \mp(\ln(-Bt/2) + \gamma); \tag{3}$$

$f_j(t)$, $j = 1,2$ are corresponding dipole form factors, $\alpha = 1/137$ is the fine structure constant and $\gamma = 0.577$ is the Euler constant. The upper (lower) sign corresponds to the pp ($\bar{p}p$) scattering. The hadronic amplitude $F^N(s,t)$ represented by the second term in Eq. (2) is characterized by the total cross section σ_{tot}, diffractive slope B and ratio ρ of real to imaginary parts of hadronic amplitude in forward direction and can be specified by applying Eqs. (2) and (3) to the experimental data.

As shown in Ref. [2] the t dependence of relative phase Φ_{WY} expressed by Eq. (3) is valid provided the t dependence of the modulus $|F^N(s,t)|$ is purely exponential and both the real and imaginary parts of the amplitude $F^N(s,t)$ have the same t dependence at all kinematically allowed t values.

Let us briefly comment justification of these assumptions. The charge independence of strong interactions expressed with the help of isospin conservation in strong interactions [1] relates elastic pp and np hadronic amplitudes to charge-exchange amplitude $np \rightarrow pn$ as

$$F^N_{ch.e.}(np \rightarrow pn) = F^N_{el}(pp) - F^N_{el}(np), \tag{4}$$

which should hold for any s and t. As charge exchange reaction cross section is very small both the elastic hadron scattering amplitudes should be practically identical [2]. The existence of diffractive dip in elastic np differential cross section has shown then that both the hadronic moduli have more complex t dependence then purely exponential one. Similarly, the second assumption assuming the constant ratio $\Re F^N(s,t)/\Im F^N(s,t)$ contradicts

[1]We should like to mention at this occasion that the first paper concerning incorporation of isospin in subnuclear physics was published just 50 years ago; see Ref. [3].

the experimental data exhibiting diffractive minima in all elastic nucleon collisions. It means that the simplified West and Yennie total scattering amplitude (Eqs. (2) and (3)) used for specifying the hadronic amplitude $F^N(s,t)$ may represent a very rough approximation only.

More suitable formula for the total scattering amplitude

$$F^{C+N}(s,t) = \pm\frac{\alpha s}{t} f_1(t) f_2(t) + F^N(s,t)\Big\{1 \mp i\alpha \int_{t_{min}}^{0} dt' \Big[\ln\frac{t'}{t}\Big(f_1(t')f_2(t')\Big)' - \frac{1}{2\pi}\Big(\frac{F^N(s,t')}{F^N(s,t)} - 1\Big) I(t,t')\Big]\Big\}, \quad (5)$$

where

$$I(t,t') = \int_0^{2\pi} d\Phi \frac{f_1(t'')f_2(t'')}{t''} \quad (6)$$

and $t'' = t + t' + 2\sqrt{tt'}\cos\Phi$ has been proposed [4]; $t_{min} = -s + 4m^2$, m is the mass of the nucleon. Eq. (5) holds with the accuracy up to the terms linear in α and may be applied in the whole range of measured momentum transfers. Starting from the impact parameter representation of the elastic scattering amplitude formulated by Adachi and Kotani [5] (see Sec. 3) Eq. (5) may be regarded as mathematically rigorous at any value of s and t.

While the t dependent modulus $|F^N(s,t)|$ is determined to a great extent by the measured differential cross section, the t dependent phase $\zeta^N(s,t)$ remains practically unspecified (limited to some extant by Coulomb - hadronic interference in a small interval of t around zero only). It is then currently assumed that the value of the phase $\zeta^N(s,t)$ slowly increases from its value established in this narrow interval. The corresponding imaginary part is then assumed to vanish at diffractive minimum while the real part is used only to fill in the non-zero value of differential cross section in the minimum. It means that the imaginary part of elastic hadron scattering amplitude is taken as dominant in a very broad interval of t and the t dependence of the phase is assumed to be very week.

2. Dominance of imaginary part

The dominance of the imaginary part of elastic hadron scattering amplitude in a broad interval of momentum transfers seems to be justified from from two reasons. First, it is used in analogy to Fresnel-Kirchoff theory of diffraction in optics; and second, it seems to be supported by the so called asymptotic theorems valid at asymptotic energies.

As to the analogy to optics the existence of many diffractive minima in nuclear diffractive scattering has led Glauber [6] to apply the methods of classical wave mechanics also to the probability waves describing the diffractive collisions in nuclear physics. Frahn [7] has tried then to formulate the general theory of diffractive phenomena in standard quantum mechanics. He has introduced quantum diffraction as a class of quantum phenomena associated with a particular type of quantum measurement intimately connected with the existence of two noncommuting observables having continuous or quasi-continuous spectrum.

If the phase of a given quantum state changes within the diffraction domain very slowly then its Taylor series expansion can be approximated by the linear term only and as a result the Fraunhofer diffraction pattern appears. On the other hand the dominance of the quadratic term in Taylor series expansion of the phase leads to Fresnel diffraction. The so called rainbow scattering in nuclear physics arises when the cubic term of the Taylor series expansion of the phase plays a role; showing that the higher terms describing more complex dependence of the phase within the studied diffraction domain might be very important. However, at difference to optics and nuclear physics only one diffractive minimum (and not a series of diffractive minimima) has been observed in all high-energy elastic hadron collisions. There is not any analogy with optics and consequently, any reason for purely imaginary amplitude.

As to the second reason, the validity of the mentioned theorems derived at asymptotic energies have been also significantly limited [2]. Assuming analycity, crossing symmetry and specific asymptotic energy behavior of the hadronic amplitudes van Hove [8] has shown that the differential cross sections of the processes $\bar{A} + B \rightarrow \bar{A} + B$ and $A + B \rightarrow A + B$ should be practically identical at asymptotic energies for $-t \in (0, t_0)$ without specifying the value of t_0. Adding further assumption about the different asymptotic energy behaviors of the contributions to $A + B \rightarrow A + B$ of exchange of the charge conjugation quantum numbers $C = \pm 1$ in the t channel (A + $\bar{A} \rightarrow B + \bar{B}$) van Hove [9] then argued that both the processes were also described by the same imaginary amplitude in the interval $-t \in (0, t_{0'})$ at asymptotic energies.

The experimental data on pp and $\bar{p}p$ elastic scattering at the ISR energies show, however, that the $\frac{d\sigma}{dt}$ for both the processes may be regarded approximately as the same in forward direction only. Therefore, the theorem of van Hove may be applied at present energies to a very narrow interval around $t = 0$ only and not to such a broad interval of momentum transfers as currently done.

It means, that neither of the mentioned reasons provide sufficient arguments for the dominance of imaginary part of elastic hadron scattering amplitude in a greater interval of t. There is not any argument, either, that imaginary part should go to zero at the diffractive minimum. It is the mere sum of the squares of the real and imaginary parts that should have a minimum at this point.

3. Impact parameter profiles at finite energies

The elastic hadron scattering amplitude $h_{el}(s, b)$ in the impact parameter space is being currently defined by Fourier-Bessel (FB) transformation of the elastic hadron scattering amplitude $F^N(s, t)$:

$$h_{el}(s, b) = \frac{1}{4p\sqrt{s}} \int_{-\infty}^{0} dt\ J_0(b\sqrt{-t})\ F^N(s, t), \tag{7}$$

and vice versa; $J_0(x)$ is the Bessel function of the zeroth order. The elastic impact parameter profile $h_{el}(s, b)$ is then bound with the inelastic impact parameter profile $g_{inel}(s, b)$ - defined as the FB transformation of the inelastic overlap function $G_{inel}(s, t)$ - through the unitarity equation in the impact parameter space

$$\Im h_{el}(s, b) = |h_{el}(s, b)|^2 + g_{inel}(s, b). \tag{8}$$

Formulas (7) and (8) represent the starting basis practically in all contemporary phenomenological model approaches which make use of the impact parameter representation of elastic hadronic amplitudes.

However, as shown by Adachi and Kotani [5] and Islam [10] formula (7) may be regularly applied to at infinite energy only when experimental data from the whole interval $t \in (-\infty, 0)$ may be available. Its use at finite energies evokes some problems as the amplitude $F^N(s, t)$ may be specified in the kinematically allowed interval only: $t \in (t_{min}, 0)$. The shape of $h_{el}(s, b)$ may be significantly influenced by the choice of $F^N(s, t) = \lambda(s, t)$ in the unphysical region; choosing, e.g., $\lambda(s, t) = 0$ the derived function $\Im h_{el}(s, b)$ oscillate around zero at finite energies for higher values of b [5] (for nucleon - nucleon scattering see Ref. [4]). Similarly, also the inelastic overlap function $g_{inel}(s, b)$ oscillates at higher values of b. It means that no direct physical meaning may be attributed to unitarity equation (8) at finite energies.

It has been shown by Islam [11] that the ambiguity in the definition of impact parameter representation of hadronic amplitude at finite energies (depending on the choice of function $\lambda(s, t)$ in the unphysical region of t) may be removed if the scattering amplitudes in both the physical and unphysical domains are linked with the corresponding profiles by means of Sommerfeld-Watson transformation, i.e., by an analytical continuation of the amplitude into the unphysical region. However, the problem has remained practically open when any actual physical theory is not available and one is forced to look for some phenomenological description of the t

dependence of hadronic elastic amplitude. In such a case the solution of the problem may consist in adding an additional real function $c(s,b)$ to the both sides of unitary equation (8); satisfying some conditions (specified in Ref. [12]-[16]) so as the conservation of all physically relevant quantities as cross sections, mean-squares of impact parameters (see later), etc., be conserved. Such a function must be chosen further to guarantee the non-negativity of all profiles to be possible to interpret them as density distributions in the impact parameter space.

The individual profiles (total, elastic and inelastic) in the impact parameter space may be simply characterized by the corresponding mean-squares that can be calculated directly from the t dependent amplitude $F^N(s,t)$ with the help of formulas (see, e.g., [16])

$$\langle b^2(s)\rangle_{el} = 4\,\frac{\int_{t_{min}}^{0} dt\,|t|\,|\frac{d}{dt}F^N(s,t)|^2}{\int_{t_{min}}^{0} dt\,|F^N(s,t)|^2}, \tag{9}$$

$$\langle b^2(s)\rangle_{tot} = 2B(s,0) \tag{10}$$

and

$$\langle b^2(s)\rangle_{inel} = \frac{\sigma_{tot}(s)}{\sigma_{inel}(s)}\langle b^2(s)\rangle_{tot} - \frac{\sigma_{el}(s)}{\sigma_{inel}(s)}\langle b^2(s)\rangle_{el}, \tag{11}$$

where $B(s,0)$ is the diffractive slope at $t = 0$, defined for all values of t as

$$B(s,t) = \frac{2}{|F^N(s,t)|}\,\frac{d}{dt}|F^N(s,t)|. \tag{12}$$

The elastic and inelastic mean-squares should also fulfill the following upper bounds:

$$\langle b^2(s)\rangle_{el} \leq 2B(s,0)\frac{\sigma_{tot}(s)}{\sigma_{el}(s)}; \quad \langle b^2(s)\rangle_{inel} \leq 2B(s,0)\frac{\sigma_{tot}(s)}{\sigma_{inel}(s)}. \tag{13}$$

4. Pumplin's bound

If the hadronic phase $\zeta^N(s,t)$ is constant or exhibits only a weak t dependence it may be shown [17] that elastic hadron scattering in impact parameter space is always central (having its maximum at $b = 0$). E.g., in 'head-on' collisions at the ISR energies two protons should go one through another with 6 % probability without any mutual interaction. The colliding protons have been, therefore, interpreted as transparent objects [18, 19].

Such a property is still being accepted by physical community (see, e.g., Ref. [20]) even if it must be regarded as a puzzle [21]. In fact the existence of this effect follows directly from the assumption of weakly t dependent phase.

The central interpretation of elastic processes differs significantly from the interpretation of diffractive production processes that are regarded as peripheral (see, e.g., [22, 23]) even if both the kinds of diffractive processes exhibit similar dynamical characteristics (diffraction peak, diffractive minimum, etc.) and, consequently, similar dynamical mechanism.

This discrepancy and at the same time the puzzling behavior of elastic processes may be removed if the phase $\zeta^N(s,t)$ is allowed to increase rather strongly with increasing $|t|$ from a small value at $t = 0$; reaching the value of $\pi/2$ at $|t| \sim 0.1$ GeV^2 [4, 17]. Such a t dependence of the hadronic phase $\zeta^N(s,t)$ gives a dominant imaginary part of hadronic amplitude in a very narrow region of t around the forward direction only (in agreement with van Hove theorem [9]).

The fact that both the elastic and diffractive production channels can be considered as peripheral processes seems to contradict Pumplin's bound [24]

$$\sigma_{el}(s,b) + \sigma_{diff}(s,b) \leq \frac{1}{2}\sigma_{tot}(s,b), \qquad (14)$$

assumed to be valid at each s and b; $\sigma_{el}(s,b), \sigma_{diff}(s,b)$ and $\sigma_{tot}(s,b)$ representing the densities of the elastic, diffractive and total scattering. Combining the viewpoint of Good and Walker [25] with the eikonal model and using the naive quark picture Pumplin claimed that in the case of hadron collisions on large nuclei (nucleon correlations being neglected) the elastic collisions are expected to be central. In such a case the whole collision picture requires at least for other diffractive processes to be peripheral. Such a picture was later adopted also in the case of nucleon-nucleon high-energy scattering; for detail see Refs. [26]-[28]. The given picture of high-energy diffractive collisions has been adopted also by Miettinen and Pumplin [29] in their parton model of diffractive scattering (see also Ref. [23]).

However, Pumplin's bound (as an upper bound only) can hardly say anything about actual behavior of collision processes in the impact parameter space. Moreover, it has been derived under two rather strong assumptions: the use of FB transformation formulas valid at asymptotic energies only and the description of all diffractive channels with the help of purely imaginary amplitudes not exceeding (in the Good and Walker scheme) the black disc limit at any value of b and we must ask whether it may be generally valid.

As to FB transformation it is assumed to guarantee the positivity of inelastic overlap function (see, e.g., Eqs. (7) - (9) of Ref. [24]). Due to existence of oscillations it is not fulfilled at finite energies (neither in central nor in peripheral cases). It need not represent any insurmountable problem as the oscillations can be removed with the help of modified unitarity condition (see Sect. 3); however, Pumplin's approach should be significantly modified in a corresponding way.

The other assumption (purely imaginary amplitude) is more crucial for derivation of Pumplin's bound [26]-[28]. The requirement for all diffractive amplitudes to be purely imaginary was weakened by Kroll [30] who has derived Pumplin's bound also for the case when the real parts of all diffractive amplitudes have been taken into account if they are limited by the condition

$$|F_{jj}^{diff}(s,b)|^2 \leq \frac{1}{2}\,\Im F_{jj}^{diff}(s,b), \tag{15}$$

valid at each impact parameter value b. Evidently such an assumption - valid also for the elastic hadronic amplitude - can be hardly fulfilled together with other assumptions.

5. Estimation of the phase from experimental data

It has been mentioned that the complex elastic hadron scattering amplitude $F^N(s,t)$ cannot be quite uniquely established from mere experimental data. While the modulus $|F^N(s,t)|$ may be determined from measured differential cross section the phase might be chosen arbitrarily at least to some extent (cp, e.g., Ref. [12, 16]). Two different possibilities have been demonstrated in Refs. [4, 12, 16] where elastic scattering of pp at 53 GeV and $\bar{p}p$ at 541 GeV have been analysed: giving peripheral or central profiles of elastic hadron scattering in the impact parameter space.

The values of root-mean-squares for total, elastic and inelastic processes determined with the help of formulas (9), (10) and (11) together with the values of upper elastic and upper inelastic bounds (calculated with the help of formulas (13)) are given in Table 1. Inelastic upper bound should be practically saturated in the central case of elastic hadron scattering, while elastic values lie significantly lower. In peripheral case the relations to upper bounds are practically proportional. The actual shapes of all profiles have been presented in Ref. [15]. The value of $\langle b^2(s)\rangle_{inel}$ can be then taken as the measure of hard core of a nucleon.

6. Conclusion

The obtained results have entitled us to prefer the peripheral picture of elastic collisions and, consequently, to the conclusion that nucleons should be regarded in principle as opaque objects, at least inside the diameter of approximately 0.5 fm. Practically, only inelastic collisions occur at such

TABLE 1. Values of root-mean-squares for pp scattering at 53 GeV and $\bar{p}p$ scattering at 541 GeV.

data	profile	$\sqrt{\langle b^2\rangle_{tot}}$ [fm]	$\sqrt{\langle b^2\rangle_{el}}$ [fm]	$\sqrt{\langle b^2\rangle_{inel}}$ [fm]	el. bound [fm]	inel. bound [fm]
pp	per.	1.03	1.80	0.77	2.47	1.13
53 GeV	cent.	1.03	0.68	1.09	2.47	1.13
$\bar{p}p$	per.	1.14	2.21	0.61	2.51	1.28
541 GeV	cent.	1.14	0.76	1.22	2.51	1.28

values of impact parameter. The value seems to be a little bit smaller than the proton charge diameter determined with the help of other methods allowing to establish charge distribution inside protons (cp, e.g., Refs. [31, 32]).

The corresponding region may be regarded as responsible for most inelastic processes while there is a competition between elastic and inelastic processes at higher values of b; the maximum of elastic processes lying at $1.1 \div 1.3$ fm. Elastic collisions prevail then fully at yet higher values of b and must be interpreted as fully peripheral processes.

References

1. West, G.B. and Yennie D.R. (1968) Coulomb interference in high-energy scattering, *Phys. Rev.* **172**, 1413-1422.
2. Kundrát, V. and Lokajíček M. (1996) Description of high-energy elastic hadron scattering in both the Coulomb and hadronic domains, *Mod. Phys. Lett.* **A11**, 2241-2250.
3. Votruba, V. and Lokajíček, M. (1953) On the isotopic spin of elementary particles, *Czech. Journ. Phys.* **2**, 1-12.
4. Kundrát, V. and Lokajíček, M. (1994) High-energy elastic scattering amplitude of unpolarized and charged hadrons, *Z. Phys. C* **63**, 619-629.
5. Adachi, T. and Kotani, T. (1965) An impact parameter formalism, *Progr. Theor. Phys. Suppl., Extra Number*, 316-331; An impact parameter formalism II, *Progr. Theor. Phys.* **463** (1966), 463-484; An impact parameter formalism III, *Progr. Theor. Phys.* **485** (1966), 485-507; Unitarity relation in an impact parameter representation, *Suppl. Theor. Phys.* **37-38** (1966), 297-305.
6. Glauber, R.J.(1958) High-energy collision theory, *Lectures in Theoretical Physics*, Vol. 1 Boulder, Colo., pp. 315-414.
7. Frahn, W.E. (1977), Quantum diffraction, *Riv. Nuovo Cim.* **7**, 499-543.
8. Van Hove, L. (1963) An extension of Pomeranchuk's theorem to diffraction scattering, *Phys. Lett.* **5**, 252-253.

9. Van Hove, L. (1963) Exchange contributions to high energy scattering and imaginary character of the elastic amplitude, *Phys. Lett.* **7**, 76-77.
10. Islam, M.M. (1968) Impact parameter description of high energy scattering, Lectures in theoretical Physics, ed. A. O. Barut and W. E. Brittin, Vol. **10B**, Gordon and Breach, pp.97-156.
11. Islam, M.M. (1976) Impact parameter representation from the Watson-Sommerfeld transform, *Nucl. Phys.* **B104**, 511-532.
12. Kundrát, V., Lokajíček, M. and Krupa, D. (Sept. 7 - 13, 1998) Hadron structure and the ρ value in high-energy elastic hadron collisions, Proc. *Hadron structure '98 Conference*, , Stará Lesná, Slovakia, (eds. D. Bruncko & P. Striženec), pp. 389-395.
13. Kundrát, V., Lokajíček, M. and Krupa, D. (June 28 - July 2 1999) High-energy elastic hadron scattering and space structure of hadrons, Proc. *VIIIth Blois Workshop, International Conference on Elastic and Diffractive Scattering*, Protvino, Russia, (eds. V. A. Petrov and A. V. Prokudin), World Scientific, pp. 333-338.
14. Kundrát, V., Lokajíček, M. and Krupa, D. (2000) High-energy elastic hadron collisions and space structure of hadrons hep-ph/0001047, pp. 1-22.
15. Kundrát, V., Lokajíček, M. and Krupa, D. (June 9 - 15, 2001) Nucleon high-energy profiles, Proc. *IXth Blois Workshop on Elastic and Diffractive Scattering*, Průhonice near Prague, Czech Republic, (eds. V. Kundrát and P. Závada), Institute of Physics AS CR Prague, ISBN 80-238-8243-0, pp. 247-256.
16. Kundrát, V., Lokajíček, M. and Krupa, D. (2002) Impact parameter structure derived from elastic collisions, *Phys. Lett.* **B544**, 132-138.
17. Kundrát, V., Lokajíček, M. Jr. and Lokajíček, M. (1981) Are elastic collisions central or peripheral? *Czech. J. Phys.* **B31**, 1334-1340.
18. Miettinen, H.I. (1974) Impact parameter structure of diffraction scattering, in *Proceedings of the IX th Rencontre de Moriond*, Meribel les Allues, Vol. **1** (ed. J. Tran Thanh Van), Orsay (1974), pp.363-402.
19. Amaldi, U. and Schubert, K.R. (1980) Impact parameter interpretation of proton-proton scattering from a critical review of all ISR data, *Nucl. Phys.* **B166**, 301-320.
20. Shoshi, A.I., Stephen, F.D. and Pirner, H.J. (2002) S-matrix unitarity, impact parameter profiles, gluon saturation and high-energy scattering, *Nucl. Phys.* **A709**, 131-183.
21. Giacomelli, G. and Jacob, M. (1979) Physics at the CERN - ISR, *Phys. Rep.* **55**, 1-132.
22. Cohen-Tannoudji, G. and Maor, U. (1975) Is the pomeron reabsorbed in diffraction scattering? *Phys. Lett.* **57B**, 253-256.
23. De Wolf, E. A. (2002) Diffractive scattering, *J. Phys.* **G28** 1023-1044.
24. Pumplin, J. (1973) Eikonal models for diffraction dissociation on Nuclei, *Phys. Rev.* **D8**, 2899-2903.
25. Good, M.L. and Walker, W.D. (1960) Diffraction dissociation of beam particles, *Phys. Rev.* **120**, 1857-1860.
26. Caneschi, L., Grassberger , P., Miettinen, H. I. and F. Henyey (1975) Unitarity bounds for inelastic diffraction, *Phys. Lett.* **56B**, 359-363.
27. Fialkowski, K. and Miettinen, H.I. (1976) Semitransparent hadrons from multichannel absorption effects, *Nucl. Phys.* **B103**, 247-257.
28. Sukhatme, U.P. and Henyey, F.S. (1976) Unitarity bounds on diffraction dissociation, *Nucl. Phys.* **B108**, 317-326.
29. Miettinen, H.I. and Pumplin, J. (1978) Diffraction scattering and parton structure of hadrons, *Phys. Rev.* **D18**, 1696-1708.
30. Kroll, P. (1977) A comment on unitarity bounds for inelastic diffraction, *Lett. Nuov. Cim.* **19**, 628-632.
31. Karshenboim, S.G. (1999) What do we actually know about the proton radius? *Can. J. Phys.* **77**, 241-266.
32. Rosenfelder, R. (2000) Coulomb corrections to elastic electron proton scattering and the proton charge radius, *Phys. Lett.* **B479**, 381-386.

HIGH DENSITY QCD, SATURATION AND DIFFRACTIVE DIS

I.P. IVANOV
Institute of Mathematics, Novosibirsk, Russia

N.N. NIKOLAEV[A,B)], W. SCHÄFER[A)], B.G. ZAKHAROV[B)]
[A)] *Institut f. Kernphysik, Forschungszentrum Jülich, D-52425 Jülich, Germany*
[B)] *L.D.Landau Institute for Theoretical Physics, Chernogolovka, Russia*

AND

V.R. ZOLLER
Institute for Theoretical and Experimental Physics, Moscow, Russia

Abstract

We review a consistent description of the fusion and saturation of partons in the Lorentz-contracted ultrarelativistic nuclei in terms of a nuclear attenuation of color dipole states of the photon and collective Weizsäcker-Williams (WW) gluon structure function of a nucleus. Diffractive DIS provides a basis for the definition of the WW nuclear glue. The point that all observables for DIS off nuclei are uniquely calculable in terms of the nuclear WW glue amounts to a new form of factorization in the saturation regime.

1. Introduction

Within the QCD parton model the virtual photoabsorption cross section is proportional to the density of partons in the target and vice versa. When DIS is viewed in the laboratory frame, the hadronic properties of photons suggest [1] a nuclear shadowing and depletion of the density of partons, when DIS is viewed in the Breit frame, the Lorentz contraction of an ultrarelativistic nucleus entails a spatial overlap and fusion of partons at $x \lesssim x_A = 1/R_A m_N \sim 0.1 \cdot A^{-1/3}$. This interpretation of nuclear opacity in

R. Fiore et al. (eds.), Diffraction 2002, 209–220.

terms of a fusion and saturation of nuclear partons has been introduced in 1975 [1] way before the QCD parton model. The pQCD link between nuclear opacity and saturation has been considered in ref. [2] and by Mueller [3], the pQCD discussion of fusion of nuclear gluons has been revived by McLerran et al. [4].

Amplitudes of diffractive DIS are intimately related to the unintegrated glue of the target [5, 6]. Because coherent diffractive DIS in which the target nucleus does not break makes precisely 50 per cent of the total DIS events for heavy nuclei at small x [7], diffractive DIS off nuclei offers a unique definition of the collective WW glue of Lorentz-contracted ultarelativistic nuclei [8]. The recent work has shown that all the observables for nuclear DIS are uniquely calculable in terms of the NSS-defined WW nuclear glue [9, 10]. In this overview presented at Diffraction'2002 by one of the authors (N.N.N) we summarize this new development.

2. Quark and antiquark jets in DIS off free nucleons: single particle spectrum and jet-jet decorrelation

In the color dipole approach to DIS [2, 5, 7, 11, 12, 13] the fundamental quantity is the cross section for interaction of the $q\bar{q}$ dipole on a nucleon,

$$\sigma(r) = \alpha_S(r)\sigma_0 \int d^2\boldsymbol{\kappa} f(\boldsymbol{\kappa})\left[1 - \exp(i\boldsymbol{\kappa}\mathbf{r})\right] \tag{1}$$

where $f(\boldsymbol{\kappa})$ is related to the unintegrated glue of the target nucleon by

$$f(\boldsymbol{\kappa}) = \frac{4\pi}{N_c\sigma_0} \cdot \frac{1}{\kappa^4} \cdot \frac{\partial G}{\partial \log \kappa^2}\,. \tag{2}$$

For DIS off a free nucleon, see figs. 1a-1d, $\sigma_N = \int d^2\mathbf{r}dz|\Psi(z,\mathbf{r})|^2\sigma(\mathbf{r})$, and the jet-jet inclusive cross section equals

$$\frac{d\sigma_N}{dzd^2\mathbf{p}_+d^2\boldsymbol{\Delta}} = \frac{\sigma_0}{2} \cdot \frac{\alpha_S(\mathbf{p}^2)}{(2\pi)^2} f(\boldsymbol{\Delta})\left|\langle\gamma^*|\mathbf{p}_+\rangle - \langle\gamma^*|\mathbf{p}_+ - \boldsymbol{\Delta}\rangle\right|^2 \tag{3}$$

where $\mathbf{p}_+$ is the transverse momentum of the quark, $\boldsymbol{\Delta} = \mathbf{p}_+ + \mathbf{p}_-$ is the jet-jet decorrelation momentum, $z_+ = z$ and $z_- = 1 - z$ are the Feynman variables for the quark and antiquark, respectively, and the photon wave functions $\Psi(\mathbf{r})$ and $\langle\mathbf{p}|\gamma^*\rangle$ are found in [2, 5]. The crucial point is that the jet-jet decorrelation is controlled [14] by the unintegrated gluon SF $f(\boldsymbol{\Delta})$.

3. Non-Abelian propagation of color dipoles in nuclear medium

DIS at $x \lesssim x_A$ is dominated by interactions of $q\bar{q}$ states of the photon. The unitarity cuts of the free-nucleon forward Compton diagrams of figs.

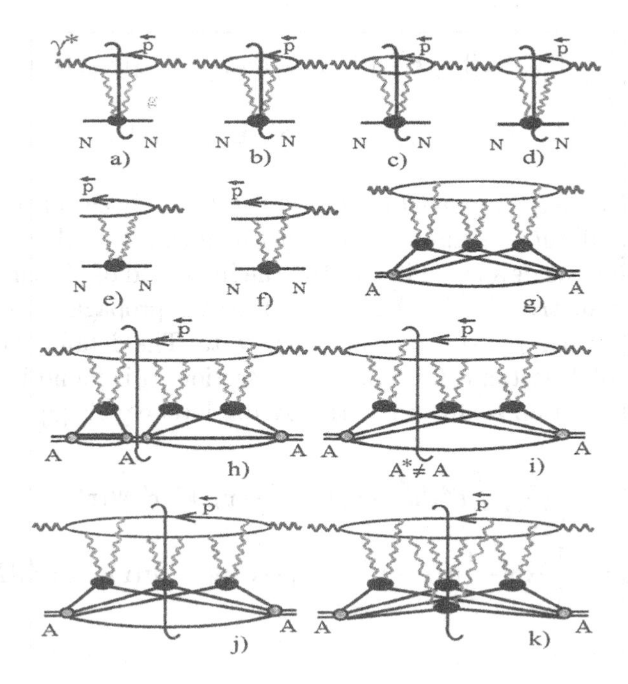

Figure 1. The pQCD diagrams for inclusive (a-d) and diffractive (e,f) DIS off protons and nuclei (g-k). Diagrams (a-d) show the unitarity cuts with color excitation of the target nucleon, (g) - a generic multiple scattering diagram for Compton scattering off nucleus, (h) - the unitarity cut for a coherent diffractive DIS, (i) - the unitarity cut for quasielastic diffractive DIS with excitation of the nucleus A^*, (j,k) - the unitarity cuts for truly inelastic DIS with single and multiple color excitation of nucleons of the nucleus.

1a-1d describe excitation from the color-neutral to color-octet $q\bar{q}$ pair. The unitarity cuts of the nuclear Compton scattering amplitude of fig. 1g which correspond to diffractive and genuine inelastic DIS with color excitation of the nucleus are shown in fig. 1. The elegant multichannel description of the non-Abelian intranuclear evolution of color dipoles is found in [10], here we cite only the principal results.

Let $\mathbf{b}_+$ and $\mathbf{b}_-$ be the impact parameters of the quark and antiquark, respectively, and $S_A(\mathbf{b}_+, \mathbf{b}_-)$ be the S-matrix for interaction of the $q\bar{q}$ pair with the nucleus. We are interested in the 2-body (jet-jet) inclusive inelastic cross section when we sum over all color excitations of the target nucleus and all color states c_{km} of the $q_k\bar{q}_m$ pair:

$$\frac{d\sigma_{in}}{dzd^2\mathbf{p}_+d^2\mathbf{p}_-} = \frac{1}{(2\pi)^4}\int d^2\mathbf{b}'_+d^2\mathbf{b}'_-d^2\mathbf{b}_+d^2\mathbf{b}_-$$
$$\times\exp[-i\mathbf{p}_+(\mathbf{b}_+-\mathbf{b}'_+)-i\mathbf{p}_-(\mathbf{b}_--\mathbf{b}'_-)]\Psi^*(\mathbf{b}'_+-\mathbf{b}'_-)\Psi(\mathbf{b}_+-\mathbf{b}_-)$$

$$\times \left\{ \sum_{A^*} \sum_{km} \langle 1; A | S_A^*(\mathbf{b}'_+, \mathbf{b}'_-) | A^*; c_{km} \rangle \langle c_{km}; A^* | S_A(\mathbf{b}_+, \mathbf{b}_-) | A; 1 \rangle \right.$$
$$\left. - \langle 1; A | S_A^*(\mathbf{b}'_+, \mathbf{b}'_-) | A; 1 \rangle \langle 1; A | S_A(\mathbf{b}_+, \mathbf{b}_-) | A; 1 \rangle \right\} , \qquad (4)$$

where the diffractive component of the final state has been subtracted. According to [9, 10], upon summing over nuclear final states A^* and making use of the technique developed in [15, 16], the integrand of (4) can be represented as an S-matrix $S_{4A}(\mathbf{b}_+, \mathbf{b}_-, \mathbf{b}'_+, \mathbf{b}'_-)$ for the propagation of the two quark-antiquark pairs in the overall singlet state. The detailed description of the matrix of 4-parton color dipole cross section σ_4 is found in [10]. The single-jet cross section evaluated at the parton level equals [9]

$$\begin{aligned}\frac{d\sigma_{in}}{d^2\mathbf{b} d^2\mathbf{p} dz} &= \frac{1}{(2\pi)^2} \int d^2\mathbf{r}' d^2\mathbf{r} \exp[i\mathbf{p}(\mathbf{r}' - \mathbf{r})] \Psi^*(\mathbf{r}') \Psi(\mathbf{r}) \\ &\times \left\{ \exp[-\frac{1}{2}\sigma(\mathbf{r} - \mathbf{r}') T(\mathbf{b})] - \exp[-\frac{1}{2}[\sigma(\mathbf{r}) + \sigma(\mathbf{r}')] T(\mathbf{b})] \right\} \quad (5)\end{aligned}$$

4. The Pomeron-Splitting Mechanism for Diffractive Hard Dijets and Weizsäcker-Williams glue of nuclei

The two distinct diffractive dijet production QCD subprocesses are the classic Landau-Pomeranchuk-Feinberg-Glauber beam-splitting [17] (fig. 1e, fig. 2a) and the Nikolaev-Zakharov pomeron-splitting [5, 6] (fig. 1f, fig. 2b):

$$\begin{aligned}\Phi_0(z, \mathbf{p}) &= \int d^2\mathbf{r}\, e^{-i\mathbf{p}\mathbf{r}}\, \sigma(\mathbf{r})\, \Psi(z, \mathbf{r}) \\ &= \alpha_S(\mathbf{p}^2) \sigma_0 \left[\langle \mathbf{p} | \gamma^* \rangle \int d^2\boldsymbol{\kappa} f(\boldsymbol{\kappa}) - \int d^2\boldsymbol{\kappa} \langle \boldsymbol{\kappa} | \gamma^* \rangle f(\mathbf{p} - \boldsymbol{\kappa}) \right] , \quad (6)\end{aligned}$$

The amplitude for the former mechanism is $\propto \langle \mathbf{p} | \gamma^* \rangle$ and the transverse momentum $\mathbf{p}$ of jets comes from the intrinsic transverse momentum of $q, \bar{q}$ in the beam particle, in the latter jets receive a transverse momentum from gluons in the Pomeron. If the beam particle were a pion, then $\psi(z, \mathbf{p})$ would be much steeper than $f(\mathbf{p})$, and the asymptotics of the convolution integral will be [8]

$$\int d^2\boldsymbol{\kappa} \psi_\pi(z, \boldsymbol{\kappa}) f(\mathbf{p} - \boldsymbol{\kappa}) \approx f(\mathbf{p}) \int d^2\boldsymbol{\kappa} \psi_\pi(z, \boldsymbol{\kappa}) = f(\mathbf{p})\, \phi_\pi(z)\, F_\pi , \qquad (7)$$

and, furthermore, will probe the pion distribution amplitude $\phi_\pi(z)$.

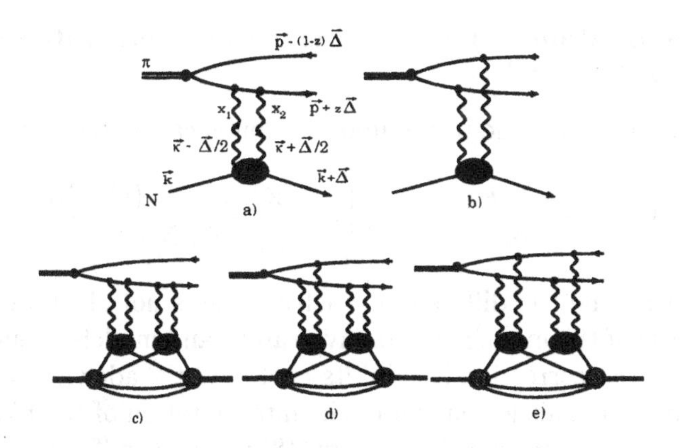

Figure 2. Sample Feynman diagrams for diffractive dijet excitation in πN collisions [diagrams 2a),2b)] and typical rescattering corrections to the nuclear coherent amplitude [diagrams 2c),2d),2e)].

Now we notice that in the color dipole representation the nuclear amplitude is readily obtained [2] by substituting in eq.(6)

$$\sigma(\mathbf{r}) \quad \rightarrow \quad \sigma_A(\mathbf{r}) = 2\int d^2\mathbf{b}\{1-\exp[-\nu_A(\mathbf{b})]\} \tag{8}$$

$$\nu_A(\mathbf{b}) \quad = \quad \frac{1}{2}\alpha_S(r)\sigma_0 T(\mathbf{b}) = \frac{1}{2}\alpha_S(r)\sigma_0\int dz n_A(\mathbf{b},z) \tag{9}$$

Typical nuclear double scattering diagrams of figs. 2c-2e can be classified as shadowing of the pion splitting (fig. 2c), shadowing of single Pomeron splitting (fig. 2d) and double Pomeron splitting (fig. 2e) contributions. In the j Pomeron splitting j exchanged Pomerons couple with one gluon to the quark and with one gluon to the antiquark of the dipole. That involves the j-fold convolution $f^{(j)}(\boldsymbol{\kappa}) = \int \prod_{i=1}^{j} d^2\boldsymbol{\kappa}_i f(\boldsymbol{\kappa}_i)\delta(\boldsymbol{\kappa} - \sum_{i=1}^{j}\boldsymbol{\kappa}_i)$. Now we can invoke the NSS representation [8, 9]

$$\exp\left[-\nu(\mathbf{b})\right] = \int d^2\boldsymbol{\kappa}\Phi(\nu_A(\mathbf{b}),\boldsymbol{\kappa})\exp(i\boldsymbol{\kappa}\mathbf{s}) \tag{10}$$

where $f^{(0)}(\boldsymbol{\kappa}) = \delta(\boldsymbol{\kappa})$, $\Phi(\nu_A(\mathbf{b}),\boldsymbol{\kappa}) = \exp(-\nu_A(\mathbf{b}))f^{(0)}(\boldsymbol{\kappa}) + \phi_{WW}(\mathbf{b},\boldsymbol{\kappa})$ and

$$\phi_{WW}(\mathbf{b},\boldsymbol{\kappa}) = \exp\left[-\nu_A(\mathbf{b})\right]\sum_{j=1}^{\infty}\frac{1}{j!}\nu_A^j(\mathbf{b})f^{(j)}(\boldsymbol{\kappa}) \tag{11}$$

can be identified with the unintegrated nuclear Weizsäcker-Williams glue per unit area in the impact parameter plane [9].

5. Nuclear dilution and broadening of the unintegrated Weizsäcker-Williams glue of nuclei

The hard tail of WW glue per bound nucleon is calculable parameter free:

$$f_{WW}(\mathbf{b},\boldsymbol{\kappa}) = \frac{\phi_{WW}(\mathbf{b},\boldsymbol{\kappa})}{\nu_A(\mathbf{b})} = f(\boldsymbol{\kappa})\left[1 + \frac{2C_A\pi^2\gamma^2\alpha_S(r)T(\mathbf{b})}{C_F N_c \boldsymbol{\kappa}^2}G(\boldsymbol{\kappa}^2)\right] \quad (12)$$

In the hard regime the differential nuclear glue is not shadowed, furthermore, because of the manifestly positive-valued and model-independent nuclear higher twist correction it exhibits nuclear antishadowing property [8]. The application of this formalism to the interpretation of the E791 data on coherent diffraction of pions into dijets [18] is found in [8, 19].

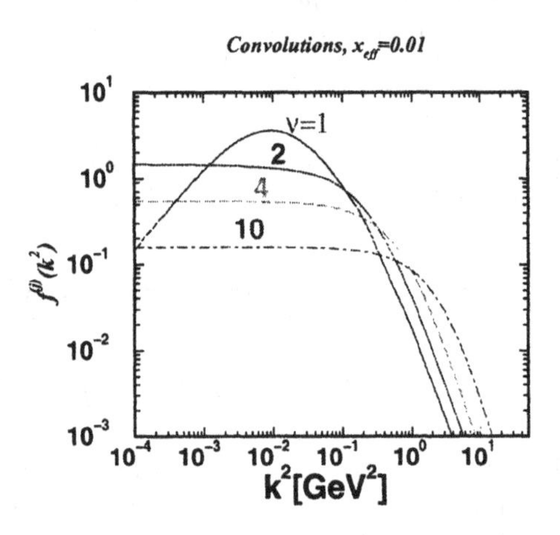

Figure 3. *The dilution for soft momenta and broadening for hard momenta of the multiple convolutions $f^{(j)}(k)$.*

In the soft region one can argue that

$$\phi_{WW}(\boldsymbol{\kappa}) \approx \frac{1}{\pi}\frac{Q_A^2}{(\boldsymbol{\kappa}^2 + Q_A^2)^2}, \quad (13)$$

where the saturation scale $Q_A^2 = \nu_A(\mathbf{b})Q_0^2 \propto A^{1/3}$. Notice a strong nuclear dilution of soft WW glue, $\phi_{WW}(\boldsymbol{\kappa}) \propto 1/Q_A^2 \propto A^{-1/3}$, which must be contrasted to the A-dependence of hard WW glue (12),

$$\phi_{WW}(\boldsymbol{\kappa}) = \nu_A(\mathbf{b})f_{WW}(\boldsymbol{\kappa}) \propto A^{1/3} \times (1 + A^{1/3} \times (HT) + ...). \quad (14)$$

Take the platinum target, $A = 192$. The numerical estimates based on the parameterization [20] for $f(\boldsymbol{\kappa})$ show that for $(q\bar{q})$ color dipoles in the average DIS on this target $\nu_A \approx 4$ and $Q^2_{3A} \approx 0.8\text{GeV}^2$. For the $q\bar{q}g$ Fock states of the photon, which behave predominantly like the dipole made of the two octet color charges, Q^2_{8A} is larger by the factor $C_A/C_F = 9/4$, and for the the average DIS on the Pt target $Q^2_{8A} \approx 2.2$ GeV^2.

6. Nuclear partons in the saturation regime

On the one hand, making use of the NSS representation, the total nuclear photoabsorption cross section (9) can be cast in the form

$$\sigma_A = \int d^2\mathbf{b} \int dz \int \frac{d^2\mathbf{p}}{(2\pi)^2} \int d^2\boldsymbol{\kappa}\phi_{WW}(\boldsymbol{\kappa}) \left|\langle\gamma^*|\mathbf{p}\rangle - \langle\gamma^*|\mathbf{p}-\boldsymbol{\kappa}\rangle\right|^2 \quad (15)$$

and its differential form defines the IS sea in a nucleus,

$$\frac{d\bar{q}_{IS}}{d^2\mathbf{b}d^2\mathbf{p}} = \frac{1}{2}\cdot\frac{Q^2}{4\pi^2\alpha_{em}}\cdot\frac{d\sigma_A}{d^2\mathbf{b}d^2\mathbf{p}}\,. \quad (16)$$

Remarkably, in terms of the WW nuclear glue, all intranuclear multiple-scattering diagrams of fig. 3g sum up to precisely the same four diagrams fig. 3a-3d as in DIS off free nucleons. On the other hand, making use of the NSS representation, after some algebra one finds [9]

$$\begin{aligned}\frac{d\sigma_{in}}{d^2\mathbf{b}d^2\mathbf{p}dz} &= \frac{1}{(2\pi)^2}\left\{\int d^2\boldsymbol{\kappa}\phi_{WW}(\boldsymbol{\kappa})\left|\langle\gamma^*|\mathbf{p}\rangle - \langle\gamma^*|\mathbf{p}-\boldsymbol{\kappa}\rangle\right|^2\right.\\ &\left.-\left|\int d^2\boldsymbol{\kappa}\phi_{WW}(\boldsymbol{\kappa})(\langle\gamma^*|\mathbf{p}\rangle - \langle\gamma^*|\mathbf{p}-\boldsymbol{\kappa}\rangle)\right|^2\right\}, \quad (17)\\ \frac{d\sigma_D}{d^2\mathbf{b}d^2\mathbf{p}dz} &= \frac{1}{(2\pi)^2}\left|\int d^2\boldsymbol{\kappa}\phi_{WW}(\boldsymbol{\kappa})(\langle\gamma^*|\mathbf{p}\rangle - \langle\gamma^*|\mathbf{p}-\boldsymbol{\kappa}\rangle)\right|^2. \quad (18)\end{aligned}$$

Putting the inelastic and diffractive components of the FS quark spectrum together, one finds exactly coincides the IS parton density (16) !

Consider first $\mathbf{p}^2 \lesssim Q^2 \lesssim Q^2_A$ when the nucleus is opaque for all color dipoles in the photon. In this regime the nuclear counterparts of diagrams of figs. 1b,1d,1f can be neglected and diffraction will be dominated by the the Landau-Pomeranchuk mechanism of fig. 1e,2a:

$$\left.\frac{d\bar{q}_{FS}}{d^2\mathbf{b}d^2\mathbf{p}}\right|_D \approx \frac{Q^2}{8\pi^2\alpha_{em}}\int dz\left|\int d^2\boldsymbol{\kappa}\phi_{WW}(\boldsymbol{\kappa})\right|^2|\langle\gamma^*|\mathbf{p}\rangle|^2 \approx \frac{N_c}{4\pi^4}\,. \quad (19)$$

Remarkably, diffractive DIS measures the momentum distribution in the $q\bar{q}$ Fock state of the photon. In contrast to diffraction off free nucleons [5, 6, 21], diffraction off opaque nuclei is dominated by the anti-collinear splitting of hard gluons into soft sea quarks, $\boldsymbol{\kappa}^2 \gg \mathbf{p}^2$. Precisely for this reason one finds the saturated FS quark density, because the nuclear dilution of the WW glue is compensated for by the expanding plateau.

The related analysis of the FS quark density for truly inelastic DIS in the same domain of $\mathbf{p}^2 \lesssim Q^2 \lesssim Q_A^2$ gives

$$\left.\frac{d\bar{q}_{FS}}{d^2\mathbf{b}d^2\mathbf{p}}\right|_{in} = \frac{1}{2}\cdot\frac{Q^2}{4\pi^2\alpha_{em}}\cdot\int dz\int d^2\boldsymbol{\kappa}\phi_{WW}(\boldsymbol{\kappa})\,|\langle\gamma^*|\mathbf{p}-\boldsymbol{\kappa}\rangle|^2$$
$$\approx \frac{Q^2}{8\pi^2\alpha_{em}}\phi_{WW}(0)\int^{Q^2} d^2\boldsymbol{\kappa}\int dz\,|\langle\gamma^*|\boldsymbol{\kappa}\rangle|^2$$
$$= \frac{N_c}{4\pi^4}\cdot\frac{Q^2}{Q_A^2}\cdot\theta(Q_A^2-\mathbf{p}^2) \tag{20}$$

which, as a functional of the photon wave function and nuclear WW gluon distribution, is completely different from the free-nucleon version of (15).

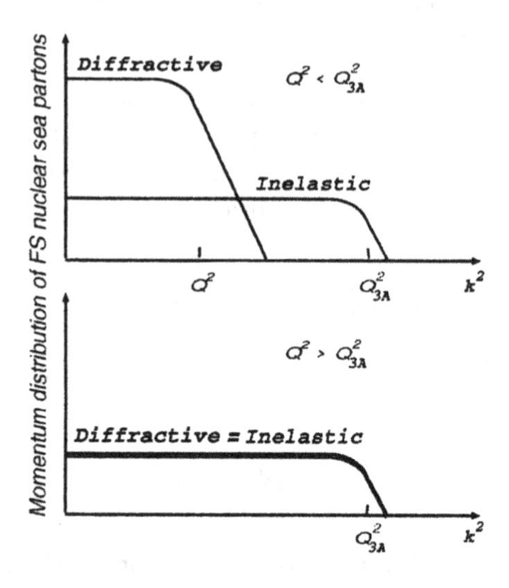

Figure 4. The two-plateau structure of the momentum distribution of FS quarks: (i) $Q^2 \lesssim Q^2_{3A}$: The inelastic plateau is much broader than the diffractive one. (ii) $Q^2 \gtrsim Q^2_{3A}$: inelastic plateau is identical to diffractive one.

Now notice, that in the opacity regime the diffractive FS parton density coincides with the contribution $\propto |\langle\gamma^*|\mathbf{p}\rangle|^2$ to the IS sea parton density from the spectator diagram 1a, whereas the FS parton density for truly inelastic DIS coincides with the contribution to IS sea partons from the diagram of fig. 1c. The contribution from the crossing diagrams 1b,d is negligibly small. Nuclear broadening and unusually strong Q^2 dependence of the FS/IS parton density from truly inelastic DIS, demonstrate clearly a distinction between diffractive and inelastic DIS, see fig. 4.

In the case of soft quarks, $\mathbf{p}^2 \lesssim Q_A^2$, in hard photons, $Q^2 \gtrsim Q_A^2$, the result (19) for diffractive DIS is retained, whereas in the numerator of the result (20) for truly inelastic DIS one must substitute $Q^2 \to Q_A^2$, so that in this case $d\bar{q}_{FS}|_D \approx d\bar{q}_{FS}|_{in}$ and $d\bar{q}_{IS} \approx 2d\bar{q}_{FS}|_D$, see fig. 4. The evolution of soft nuclear sea, $\mathbf{p}^2 \lesssim Q_A^2$, is entirely driven by an anti-collinear splitting of the NSS-defined WW nuclear glue into the sea partons.

The early discussion of the FS quark density in the saturation regime is due to Mueller [22]. Mueller focused on $Q^2 \gg Q_A^2$ and discussed neither a distinction between diffractive and truly inelastic DIS nor a Q^2 dependence and broadening (20) for truly inelastic DIS at $Q^2 \lesssim Q_A^2$.

7. Signals of saturation in exclusive diffractive DIS

The flat $\mathbf{p}^2$ distribution of forward $q, \bar{q}$ jets in truly inelastic DIS in the saturation regime must be contrasted to the $\propto G(\mathbf{p}^2)/\mathbf{p}^2$ spectrum for the free nucleon target. In the diffractive DIS the saturation gain is much more dramatic: flat $\mathbf{p}^2$ distribution of forward $q, \bar{q}$ jets in diffractive DIS in the saturation regime must be contrasted to the $\propto 1/(\mathbf{p}^2)^2$ spectrum for the free nucleon target [5, 6, 21]. In the general case one must compare the saturation scale Q_A to the relevant hard scale $\bar{Q}^2$ for the specific diffractive process. For instance, in exclusive diffractive DIS, i.e., the vector meson production, $\bar{Q}^2 \approx (Q^2 + m_V^2)/4$ and the transverse cross section has been predicted to behave as [23]

$$\sigma_T \propto G^2(x, \bar{Q}^2)(\bar{Q}^2)^{-4} \tag{21}$$

At $\bar{Q}^2 > Q_A^2$ the same would hold for nuclei too, but in the opposite case of $\bar{Q}^2 < Q_A^2$ the $\bar{Q}^2$-dependence is predicted to change to

$$\sigma_T \propto G^2(x, \bar{Q}^2)(Q_A^2)^{-2}(\bar{Q}^2)^{-2} \tag{22}$$

8. Jet-jet decorrelation in DIS off nuclear targets

The derivation of the jet-jet inclusive cross section requires a full fledged non-Abelian multichannel calculation of $S_{4A}(\mathbf{b}_+, \mathbf{b}_-, \mathbf{b}'_+, \mathbf{b}'_-)$. The principal result for the hard, $\mathbf{p}_\pm^2 \gg Q_A^2$, jet-jet cross section is [10]

$$\frac{d\sigma_{in}}{d^2\mathbf{b}dzd^2\mathbf{p}_+d^2\mathbf{\Delta}} = T(\mathbf{b}) \int_0^1 d\beta \int d^2\boldsymbol{\kappa}\Phi(2\beta\lambda_c\nu_A(\mathbf{b}), \mathbf{\Delta} - \boldsymbol{\kappa}) \frac{d\sigma_N}{dzd^2\mathbf{p}_+d^2\boldsymbol{\kappa}}\,. \tag{23}$$

It has a probabilistic form of a convolution of the differential cross section on a free nucleon target with the unintegrated nuclear WW glue $\Phi(2\beta\lambda_c\nu_A(\mathbf{b}), \boldsymbol{\kappa})$, β has a meaning of the fraction of the nuclear thickness which the $(q\bar{q})$ pair propagates in the color-octet state and the nontrivial color factor $\lambda_c = (N_c^2 - 2)/(N_c^2 - 1)$ is a manifestation of the non-Abelian intranuclear propagation of color dipoles.

The nuclear azimuthal decorrelation of hard jets is quantified by

$$\langle \mathbf{\Delta}_\perp^2 \rangle_A \approx \frac{Q_A^2}{2} \left(\frac{Q_A^2 + \mathbf{p}_+^2}{\mathbf{p}_+^2} \log \frac{Q_A^2 + \mathbf{p}_+^2}{Q_A^2} - 1 \right) \tag{24}$$

and the differential out-of-plane momentum distribution

$$\frac{d\sigma}{d\Delta_\perp} \approx \frac{1}{2} \frac{Q_A^2}{(Q_A^2 + \Delta_{\perp 2})^{3/2}} \tag{25}$$

which shows that the probability to observe the away jet at $\Delta_\perp \sim 0$ disappears $\propto 1/Q_A \propto 1/\sqrt{\nu_A}$. For instance, from peripheral DIS to central DIS on a heavy nucleus like Pt, ν_{8A} rises form $\nu_{8A} = 1$ to $\nu_{8A} \sim 13$, so that according to (25) a probability to find the away jet decreases by the large factor ~ 3.5 from peripheral to central DIS off Pt target.

In the generic case the closed form for the jet-jet inclusive cross section is found [10] only for large N_c:

$$\begin{aligned} \frac{d\sigma_{in}}{d^2\mathbf{b}dzd\mathbf{p}_-d\mathbf{\Delta}} &= \frac{1}{2(2\pi)^2}\alpha_S\sigma_0 T(\mathbf{b}) \int_0^1 d\beta \int d^2\boldsymbol{\kappa}_1 d^2\boldsymbol{\kappa}_2 d^2\boldsymbol{\kappa}_3 d^2\boldsymbol{\kappa} f(\boldsymbol{\kappa}) \\ &\times \Phi((1-\beta)\nu_A(\mathbf{b}), \boldsymbol{\kappa}_1)\Phi((1-\beta)\nu_A(\mathbf{b}), \boldsymbol{\kappa}_2) \\ &\times \Phi(\beta\nu_A(\mathbf{b}), \boldsymbol{\kappa}_3)\Phi(\beta\nu_A(\mathbf{b}), \mathbf{\Delta} - \boldsymbol{\kappa}_3 - \boldsymbol{\kappa}) \\ &\times \{\Psi(-\mathbf{p}_- + \boldsymbol{\kappa}_2 + \boldsymbol{\kappa}_3) - \Psi(-\mathbf{p}_- + \boldsymbol{\kappa}_2 + \boldsymbol{\kappa}_3 + \boldsymbol{\kappa})\}^* \\ &\times \{\Psi(-\mathbf{p}_- + \boldsymbol{\kappa}_1 + \boldsymbol{\kappa}_3) - \Psi(-\mathbf{p}_- + \boldsymbol{\kappa}_1 + \boldsymbol{\kappa}_3 + \boldsymbol{\kappa})\} \end{aligned} \tag{26}$$

We emphasize that it is uniquely calculable in terms of the NSS-defined WW glue of the nucleus.

One important implication of (26) is that for minijets with the jet momentum $|\mathbf{p}_-|, |\mathbf{\Delta}| \lesssim Q_A$, the the minijet-minijet inclusive cross section depends on neither the minijet nor decorrelation momentum, i.e., we predict a complete disappearance of the azimuthal decorrelation of jets with the transverse momentum below the saturation scale.

Our experience with application of color dipole formalism to hard hadron-nucleus interactions [15, 16] suggests that our analysis can be readily generalized to mid-rapidity jets. For this reason, we expect similar strong decorrelation of mid-rapidity jets in hadron-nucleus and nucleus-nucleus collisions. To this end, recently the STAR collaboration reported a disappearance of back-to-back high $p_\perp$ hadron correlation in central gold-gold collisions at RHIC [24]. Nuclear enhancement of the azimuthal decorrelation of the trigger and away jets may contribute substantially to the STAR effect.

9. Summary and conclusions

We reviewed a theory of DIS off nuclear targets based on the consistent treatment of propagation of color dipoles in nuclear medium. What is viewed as attenuation in the laboratory frame can be interpreted as a fusion of partons from different nucleons of the ultrarelativistic nucleus. Diffractive attenuation of color single $q\bar{q}$ states gives a consistent definition of the WW unintegrated gluon structure function of the nucleus [8, 9], all other nuclear DIS observables - sea quark structure function and its decomposition into equally important genuine inelastic and diffractive components, exclusive diffraction off nuclei, the jet-jet inclusive cross section, - are uniquely calculable in terms of the NSS-defined nuclear WW glue. This property can be considered as a new factorization which connects DIS in the regimes of low and high density of partons. The anti-collinear splitting of WW nuclear glue is a clearcut evidence for inapplicability of the DGLAP evolution to nuclear structure functions unless $Q^2 \gg Q^2_{8A}$.

N.N.N. is grateful to L. Jenkovszky for the invitation to Diffraction'2002. This work has been partly supported by the INTAS grants 97-30494 & 00-00366 and the DFG grant 436RUS17/89/02.

References

1. Nikolaev N.N. and Zakharov V.I. (1975) Parton model and inelastic scattering of leptons and hadrons off nuclei, *Sov. J. Nucl. Phys.* **21**, p. 227; [*Yad. Fiz.* **21**, p. 434]; Parton model and deep inelastic scattering on nuclei, *Phys. Lett.* **B55**, p. 397 (1975).
2. Nikolaev N.N. and Zakharov B.G. (1991) Color transparency and scaling properties of nuclear shadowing in deep inelastic scattering, *Z. Phys.* **C49**, p. 607
3. Mueller A.H. (1990) Small-x behaviour and parton saturation: a QCD model, *Nucl. Phys.* **B335** p. 115.
4. McLerran L. and Venugopalan R. (1994) Gluon distribution functions for very large nuclei at small transverse momentum, *Phys. Rev.* **D49**, p. 2233; J. Jalilian-Marian et al. (1997) The intrinsic glue distribution at very small x and high densities, *Phys. Rev.* **D55** p. 5414; E. Iancu, A. Leonidov and L. McLerran, Lectures at the Cargèse Summer School, August 6-18, 2001, `arXiv:hep-ph/0202270`.
5. Nikolaev N.N. and B.G. Zakharov B.G. (1992) Pomeron structure function and diffraction dissociation of virtual photons in perturbative QCD, *Z. Phys.* **C53**, p. 331.
6. Nikolaev N.N. and Zakharov B.G. (1994) Splitting the pomeron into two jets: a novel process at HERA, *Phys. Lett.* **B332**, 177.
7. Nikolaev N.N., Zakharov B.G. and Zoller V.R. (1995) Unusual effects of diffraction dissociation for multiproduction in deep inelastic scattering on nuclei, *Z. Phys.* **A351**, p. 435.
8. Nikolaev N.N., Schäfer W. and Schwiete G. (2000) Multiple-pomeron spliiting in QCD - a novel antishadowing effect in coherent dijet production on nuclei, *JETP Lett.* **72** (2000) 583; *Pisma Zh. Eksp. Teor. Fiz.* **72**, p. 583; Coherent production of hard dijets on nuclei in QCD, *Phys. Rev.* **D63**, p. 014020.
9. Nikolaev N.N., Schafer W., Zakharov B.G. and Zoller V.R. (2002) Saturation of nuclear partons: the Fermi statistics or nuclear opacity? *JETP Lett.* **76** p. 195.

10. Nikolaev N.N., Schafer W., Zakharov B.G. and Zoller V.R. (2002) QCD Theory of Decorrelation of Jets in DIS off Nuclei, paper in preparation; Ivanov I.P., Nikolaev N.N., Schafer W., Zakharov B.G. and Zoller V.R. Lectures on Diffraction and Saturation of Nuclear Partons in DIS off Heavy Nuclei, XXXVI St.Petersburg Nuclear Physics Institute Winter School on Nuclear and Particle Physics & VIII St.Petersburg School on Theoretical Physics, St.Petersburg, Repino, February 25 - March 3, 2002, `arXiv: hep-ph/0212161`
11. Nikolaev N.N. and Zakharov B.G. (1994) The pomeron in diffractive deep inelastic scattering, *J. Exp. Theor. Phys.* **78** p. 806; *Zh. Eksp. Teor. Fiz.* **105** (1994) 1498; The triple-pomeron regime and the structure function of the pomeron in the diffractive deep inelastic scattering ar very small x, *Z. Phys.* **C64** (1994) 631.
12. Nikolaev N.N., Zakharov B.G. and Zoller V.R. (1994) The s-channel approach to Lipatov's Pomeron and hadronic cross sections, *JETP Lett.***59**, p. 6
13. Nikolaev N.N. and Zakharov B.G. (1994) On determination of the large $1/x$ gluon distribution at HERA, *Phys. Lett.* **B332**, p. 184
14. Szczurek A., Nikolaev N.N, Schafer W. and Speth J. (2001) Mapping the proton unintegrated gluon distribution in dijets correlations in real and virtual photoproduction at HERA, *Phys. Lett.* **B500**, p. 254
15. Nikolaev N.N., Piller G. and Zakharov B.G. (1995) Quantum coherence in heavy flavor production on nuclei, *J. Exp. Theor. Phys.* **81**, p. 851; Inclusive heavy flavor production from nuclei, *Z. Phys.* **A354**, p. 99 (1996).
16. Zakharov B.G. (1996) Fully quantum treatment of the Landau-Pomeranchuk-Migdal efefct in QED and QCD, *JETP Lett.* **63**, p. 952; Light cone path integral approach to the Landau-Pomeranchuk-Migdal effect, *Phys. Atom. Nucl.* **61**, p. 838 (1998).
17. Landau L.D. and Pomeranchuk I.Ya. (1953) Radiation of γ-quanta in collisions of fast pions with nucleons, *J. Exp. Theor. Phys.* **24**, p. 505; Pomeranchuk I.Ya. and Feinberg E.L. (1953) On external (diffractive) generation of particles in nuclear collisions, *Doklady Akademii Nauk SSSR* **93**, p. 439; Feinberg E.L. and Pomeranchuk I.Ya. (1956) Inelastic diffraction processes at high energies, *Nuovo Cim. (Suppl.)* **4**, 652.
18. Aitala E.M. et al. (2001) Observation of color transparency in diffractive dissociation of pions, *Phys. Rev. Lett.* **86**, p. 4773.
19. Ivanov I.P., Nikolaev N.N., Schafer W., Zakharov B.G. and Zoller V.R. (2002) Diffractive hard dijets and nuclear parton distributions, Proceedings of the Workshop on Exclusive Processes at High Momentum Transfer, Newport News, Virginia, 15-18 May 2002. e-Print Archive: hep-ph/0207045
20. Ivanov I.P. and Nikolaev N.N. (2001) Deep inelastic scattering in k-factorization and the anatomy of the differential gluon structure function of the proton, *Phys. Atom. Nucl.* **64**, p. 753; *Yad. Fiz.* **64**, p. 813; Ivanov I.P. and Nikolaev N.N. (2002) Anatomy of the differential gluon structure function of the proton from the experimental data on $F_{2p}(x,Q^2)$, *Phys. Rev.* **D65**, p:054004.
21. Genovese M., Nikolaev N.N. and Zakharov B.G. (1996) Excitation of open charm and factorization breaking in rapidity gap events at HERA, *Phys. Lett.* **B378** p. 347
22. Mueller A.H. (1999) Parton saturation at small x and in large nuclei, *Nucl. Phys.* **B558**, p. 285; Parton saturation: and overview. Lectures given at Cargese Summer School on QCD Perspectives on Hot and Dense Matter, Cargese, France, 6-18 August 2001. `arXiv:hep-ph/0111244`.
23. Nemchik J., Nikolaev N.N. and Zakharov B.G. (1994) Scanning the BFKL pomeron in elastic production of vector mesons at HERA, *Phys. Lett.* **B341**, p. 228
24. Adler C. et al. (2002), Disappearance of back-to-back high p_T hadron correlations in central $Au + Au$ collisions at $s_{NN}^{1/2} = 200$ GEV, `arXiv: nucl-ex/0210033`.

DGLAP AND BFKL EQUATIONS IN SUPERSYMMETRIC GAUGE THEORIES

A.V. KOTIKOV
Bogoliubov Laboratory of Theoretical Physics,
Joint Institute for Nuclear Research,
141980 Dubna, Russia

AND

L.N. LIPATOV, V.N. VELIZHANIN
Theoretical Physics Department,
Petersburg Nuclear Physics Institute,
Orlova Roscha, Gatchina, 188300, St. Petersburg, Russia

Abstract. We derive DGLAP and BFKL evolution equations in the N=4 supersymmetric gauge theory in the next-to-leading approximation. The eigenvalue of the BFKL kernel in this model turns out to be an analytic function of the conformal spin $|n|$. The corresponding kernel for the Bethe-Salpeter equation has the property of the hermitian separability. The anomalous dimension matrix can be transformed to a triangle form with the use of the similarity transformation for the diagonalization of the anomalous dimension matrix in the leading order. The eigenvalues of these matrices can be expressed in terms of a universal function by an integer shift of its argument. We also investigate in this approximation possible relations between the DGLAP and BFKL equations.

1. Introduction

The Balitsky-Fadin-Kuraev-Lipatov (BFKL) equation [1] and the Dokshitzer-Gribov-Lipatov-Altarelli-Parisi (DGLAP) equation [2] are used now for a theoretical description of structure functions of the deep-inelastic *ep* scattering at small values of the Bjorken variable x. The higher order QCD corrections to the splitting kernels of the DGLAP equation are well known [3]. But the calculation of the next-to-leading order (NLO) corrections to the BFKL kernel was completed comparatively recently [4, 5, 6].

R. Fiore et al. (eds.), Diffraction 2002, 221–233.

In supersymmetric gauge theories the structure of the BFKL and DGLAP equations is simplified significantly. In the case of an extended N=4 SUSY the NLO corrections to the BFKL equation were calculated in ref. [6] for arbitrary values of the conformal spin n. Below these results are presented in the dimensional reduction ($\overline{\text{DR}}$) scheme [7] which does not violate the supersymmetry. The analyticity of the eigenvalue of the BFKL kernel as a function of the conformal spin $|n|$ gives a possibility to relate in the leading logarithmic approximation (LLA) the DGLAP and BFKL equations in this model (see [6]). It was shown [8], that the eigenvalue of the BFKL kernel has the property of hermitian separability similar to the holomorphic separability.

Let us introduce the unintegrated parton distributions (UnPD) $\varphi_a(x, k_\perp^2)$ (hereafter $a = q, g, \varphi$ for the spinor, vector and scalar particles, respectively) and the (integrated) parton distributions (PD) $n_a(x, Q^2)$. In DIS $Q^2 = -q^2$ and $x = Q^2/(2pq)$ are the Bjorken variables, $k_\perp$ is the transverse component of the parton momentum and q and p are the photon and hadron momenta, respectively.

The DGLAP equation after the Mellin transformation of partonic distributions

$$n_a(j, Q^2) = \int_0^1 dx\ x^{j-1} n_a(x, Q^2)$$

can be written as follows [2]

$$\frac{d}{d\ln Q^2} n_a(j, Q^2) = \sum_b \gamma_{ab}(j) n_b(j, Q^2)\,.$$

The Mellin moment of the splitting kernel $\gamma_{ab}(j)$ coincides with the anomalous dimension matrix for the twist-2 operators. [1] These operators are constructed as bilinear combinations of the fields which describe corresponding partons

$$\begin{aligned}
O^g_{\mu_1,\ldots,\mu_j} &= \hat{S} G_{\rho\mu_1} D_{\mu_2} D_{\mu_3} \ldots D_{\mu_{j-1}} G_{\rho\mu_j}\,, \\
\tilde{O}^g_{\mu_1,\ldots,\mu_j} &= \hat{S} G_{\rho\mu_1} D_{\mu_2} D_{\mu_3} \ldots D_{\mu_{j-1}} \tilde{G}_{\rho\mu_j}\,, \\
O^q_{\mu_1,\ldots,\mu_j} &= \hat{S} \overline{\Psi} \gamma_{\mu_1} D_{\mu_2} \ldots D_{\mu_j} \Psi\,, \\
\tilde{O}^q_{\mu_1,\ldots,\mu_j} &= \hat{S} \overline{\Psi} \gamma_5 \gamma_{\mu_1} D_{\mu_2} \ldots D_{\mu_j} \Psi\,, \\
O^\varphi_{\mu_1,\ldots,\mu_j} &= \hat{S} \bar{\Phi} D_{\mu_1} D_{\mu_2} \ldots D_{\mu_j} \Phi\,,
\end{aligned} \tag{1}$$

where the spinor Ψ and field tensor $G_{\rho\mu}$ describe gluinos and gluons, respectively. The last expression is constructed from the covariant derivatives D_μ

[1] As in Ref. [4, 6], the anomalous dimensions differ from those used usually in the description of DIS by a factor (-2), i.e. $\gamma_{ab}(j) = (-1/2)\gamma_{ab}^{DIS}(j)$.

of the scalar field Φ appearing in extended supersymmetric models. The symbol $\hat{S}$ implies a symmetrization of the tensor in the Lorenz indices $\mu_1, ..., \mu_j$ and a subtraction of its traces.

On the other hand, the BFKL equation relates the unintegrated gluon distributions with small values of the Bjorken variable x:

$$\frac{d}{d\ln(1/x)}\varphi_g(x,k_\perp^2) = 2\omega(-k_\perp^2)\,\varphi_g(x,k_\perp^2) + \int d^2k'_\perp K(k_\perp,k'_\perp)\varphi_g(x,k_\perp^2),$$

where $\omega(-k_\perp^2) < 0$ is the gluon Regge trajectory [1].

The matrix elements of $O^a_{\mu_1,...,\mu_j}$ and $\tilde{O}^a_{\mu_1,...,\mu_j}$ are related to the moments of the distributions $n_a(x,Q^2)$ and $\Delta n_a(x,Q^2)$ for unpolarized and polarized partons in a hadron h in the following way

$$\int_0^1 dx\, x^{j-1} n_a(x,Q^2) = \langle h|\tilde{n}^{\mu_1}...\,\tilde{n}^{\mu_j}\, O^a_{\mu_1,...,\mu_j}|h\rangle, \quad a=(q,g,\varphi),$$

$$\int_0^1 dx\, x^{j-1} \Delta n_a(x,Q^2) = \langle h|\tilde{n}^{\mu_1}...\,\tilde{n}^{\mu_j}\, \tilde{O}^a_{\mu_1,...,\mu_j}|h\rangle, \quad a=(q,g). \quad (2)$$

Here the vector $\tilde{n}^\mu$ is light-like: $\tilde{n}^2 = 0$. Note, that in the deep-inelastic ep scattering we have $\tilde{n}^\mu \sim q^\mu + xp^\mu$.

The quantum numbers appeared in the solution of the BFKL equation being the integer conformal spin $|n|$ and the quantity $1+\omega$ (ω is an eigenvalue of the BFKL kernel) coincide respectively with the total numbers of transversal and longitudinal Lorentz indices of the tensor $O^a_{\mu_1,...,\mu_J}$ with the rank $J = 1+\omega+|n|$. The corresponding matrix elements can be expressed through the solution of the BFKL equation

$$\tilde{n}^{\mu_1}...\,\tilde{n}^{\mu_{1+\omega}}\, l_\perp^{\mu_{2+\omega}}...l_\perp^{\mu_j} \langle P|O^g_{\mu_1,...,\mu_j}|P\rangle \sim \int_0^1 dx x^\omega \int d^2k_\perp \left(\frac{k_\perp}{|k_\perp|}\right)^n \varphi_g(x,k_\perp^2),$$

It is important, that the anomalous dimension matrices $\gamma_{ab}(j)$ and $\tilde{\gamma}_{ab}(j)$ for the twist-2 operators $O^a_{\mu_1,...,\mu_j}$ and $\tilde{O}^a_{\mu_1,...,\mu_j}$ do not depend on various projections of indices due to the Lorentz invariance. But generally the Lorentz spin j is less then J. Thus, it looks reasonable to extract some additional information concerning the parton x-distributions satisfying the DGLAP equation from the analogous $k_\perp$-distributions satisfying the BFKL equation.

In LLA the integral kernel of the BFKL equation is the same in all supersymmetric gauge theories. Due to the Möbius invariance in the impact parameter representation the solution of the homogeneous BFKL equation has the form (see [9])

$$E_{\nu,n}(\overrightarrow{\rho_{10}},\overrightarrow{\rho_{20}}) \equiv \langle \phi(\overrightarrow{\rho_1})\, O_{m,\tilde{m}}(\overrightarrow{\rho_0})\, \phi(\overrightarrow{\rho_2})\rangle = \left(\frac{\rho_{12}}{\rho_{10}\rho_{20}}\right)^m \left(\frac{\rho^*_{12}}{\rho^*_{10}\rho^*_{20}}\right)^{\widetilde{m}},$$

where $m = 1/2 + i\nu + n/2$ and $\widetilde{m} = 1/2 + i\nu - n/2$ are conformal weights related to the eigenvalues of the Casimir operators of the Möbius group. We introduced also the complex variables $\rho_k = x_k + iy_k$ in the transverse subspace and used the notation $\rho_{kl} = \rho_k - \rho_l$.

For the principal series of the unitary representations the quantities ν and n are respectively real and integer numbers. The projection n of the conformal spin $|n|$ can be positive or negative, but the eigenvalue of the BFKL equation in LLA [1]

$$\omega = \omega^0(n,\nu) \quad = \quad \frac{g^2 N_c}{2\pi^2}\left(\Psi(1) - Re\,\Psi\left(\frac{1}{2} + i\nu + \frac{|n|}{2}\right)\right) \tag{3}$$

depends only on $|n|$. The Möbius invariance takes place also for the Schrödinger equation describing the composite states of several reggeized gluons [10]. As a consequence of the relation

$$\omega^0(n,\nu) = \omega^0(m) + \omega^0(\widetilde{m})\,,\; \omega^0(m) = \frac{g^2 N_c}{8\pi^2}\Big(2\Psi(1) - \Psi(m) - \Psi(1-m)\Big)\,.$$

one obtains the property of the holomorphic separability $H = h + h^*$, $[h, h^*] = 0$ for the Hamiltonian of an arbitrary number of reggeized gluons in the multi-colour QCD [11]. In the same limit the BFKL dynamics is completely integrable [12],[13] and the holomorphic Hamiltonian h coincides with the local Hamiltonian for an integrable Heisenberg spin model [14]. Moreover the theory turns out to be invariant under the duality transformation [15]. Presumably the remarkable mathematical properties of the reggeon dynamics in LLA are consequences of the extended N=4 supersymmetry [16]. It was argued also in Ref. [16] that generalized DGLAP equations for the matrix elements of quasi-partonic operators [17] for N=4 SUSY are integrable. For the case of the Odderon a non-trivial integral of motion [12] was used to construct the energy spectrum and wave functions of the three-gluon state [18]. A new Odderon solution with the intercept equal to 1 was found in [19]. Recently the energy spectrum for reggeon composite states in the multi-color QCD was obtained with the use of the Baxter equation (see refs. [20] and [21]). The effective action for reggeon interactions is known [22]. It gives a possibility to calculate next-to-leading corrections to the trajectory and couplings of the reggeons [4]-[6].

The solution of the inhomogeneous BFKL equation in the LLA approximation can be constructed in the impact parameter representation and for $\overrightarrow{\rho_{1'}} \to \overrightarrow{\rho_{2'}}$ we obtain [9]

$$< \phi(\overrightarrow{\rho_1})\,\phi(\overrightarrow{\rho_2})\,\phi(\overrightarrow{\rho_{1'}})\,\phi(\overrightarrow{\rho_{2'}}) >$$

$$\sim \sum_n C(\nu_\omega,|n|)\,\frac{E_{\nu_\omega,|n|}(\overrightarrow{\rho_{11'}},\overrightarrow{\rho_{21'}})}{\omega'(|n|,\nu_\omega)}\,|\rho_{1'2'}|^{1+2i\nu_\omega}\left(\frac{\rho_{1'2'}}{\rho^*_{1'2'}}\right)^{|n|/2}$$

where ν_ω is a solution of the equation $\omega = \omega^0(|n|,\nu)$ with $Im\,\nu_\omega < 0$.

The above asymptotics has a simple interpretation in terms of the Wilson operator-product expansion of two reggeon fields produced the local operator $O_{\nu_\omega,|n|}(\overrightarrow{\rho_{1'}})$ having the transverse dimension $\Gamma_\omega = 1 + i\nu_\omega$ calculated in units of a squared mass. The corresponding tensor has a mixed projections of a gauge-invariant tensor O with $J = 1+\omega+|n|$ indices. Note, that because Γ_ω is real in the deep-inelastic regime $\rho_{12} \to 0$, the operator $O_{\nu_\omega,|n|}(\overrightarrow{\rho_{1'}})$ belongs to an exceptional series of unitary representations of the Möbius group (see [23]).

The anomalous dimension $\gamma(j)$ obtained from the BFKL equation in LLA (see (3)) has the poles

$$\Gamma_\omega = 1 + \frac{|n|}{2} - \gamma(j), \quad \gamma(j)|_{\omega\to 0} = \frac{g^2 N_c}{4\pi^2\omega}\,. \tag{4}$$

The operator $O_{\nu_\omega,|n|}$ for $|n| = 1, 2, ...$ has the twist higher than 2 because its anomalous dimension γ is singular at $\omega \to 0$. In the paper [6] the analytic continuation of the BFKL anomalous dimension $\gamma(|n|,\omega)$ to the points $|n| = -r - 1$ $(r = 0, 1, 2, ...)$ was suggested to calculate the singularities of the anomalous dimension of the twist-2 operators in negative integer $J = 1 + \omega + |n| \to -r$. Because for positive $|n|$ and $\omega \to 0$ the quantity $\gamma(|n|,\omega)$ corresponds to higher twist operators, to obtain γ for the twist-2 operators, one should push $\Delta|n| = |n| + r + 1$ to zero at $\omega \to 0$ sufficiently rapidly $\Delta|n| = C(r)\omega^2$. In LLA the results obtained from the BFKL equation

$$\gamma(j)|_{j\to -r} = \frac{g^2 N_c}{4\pi^2}\frac{1}{j+r} \tag{5}$$

for $r = -1, 0, 1, ...$ coincide with the direct calculation of the eigenvalues of the anomalous dimension matrix $\gamma_{a,b}$ for N=4 SUSY in accordance with the fact, that only in this theory the eigenvalue of the BFKL equation is an analytic function of $|n|$ [6]. It is important that the anomalous dimension is the same for all twist-2 operators entering in the N=4 supermultiplet up to a shift of its argument by an integer number, because this property leads to an integrability of the evolution equations for quasi-partonic operators [17] in the multi-color limit $N_c \to \infty$ (see [16]).

In the NLO approximation for the BFKL equation there is a difficulty related to an appearance of the double-logarithmic (DL) terms leading to triple poles at $j = -r$ for even r. The origin of the DL terms can be understood in a simple way using as an example the process of the forward annihilation of the e^+e^- pair in the $\mu^+\mu^-$ pair in QED [24]. For this process the t-channel partial wave f_ω $(\omega = j)$ in the DL approximation can be written as follows

$$f_\omega = \frac{\gamma^{-1}}{\omega} - \frac{\alpha_{em}}{2\pi\left(\frac{1}{\omega-\gamma}+\frac{1}{\gamma}\right)}.$$

By expanding perturbatively the position of the pole in γ we shall generate the triple pole term in the anomalous dimension $\Delta\gamma \sim \alpha_{em}^2/\omega^3$.

It is important to note, that the equation for the pole position of f_ω is similar to the BFKL equation in the modified leading logarithmic approximation

$$\omega \sim 2\Psi(1) - \Psi(1+|n|+\omega-\gamma) - \Psi(\gamma)\,,\ |n| = -1,-2,\ldots,$$

2. NLO corrections to the BFKL kernel in the N=4 SUSY

Let us introduce the new variables $M = \gamma + \frac{|n|}{2}$, $\widetilde{M} = \gamma - \frac{|n|}{2}$, $\gamma = \frac{1}{2} + i\nu$. Then the eigenvalue relation for the BFKL equation in the $\overline{\mathrm{DR}}$-scheme can be written in the hermitially separabel form [8]

$$\begin{aligned} 1 &= \frac{4\hat{a}}{\omega}\left(2\Psi(1) - \Psi(M) - \Psi(1-\widetilde{M}) + \hat{a}\left(\phi(M) + \phi(1-\widetilde{M})\right)\right) \\ &\quad -2\hat{a}\left(\rho(M) + \rho(1-\widetilde{M})\right),\ \hat{a} = \overline{a} + \frac{\overline{a}^2}{3},\ \overline{a} = \frac{g^2 N_c}{16\pi^2}, \qquad (6) \end{aligned}$$

where $\overline{a}$ and $\hat{a}$ are expressed through the Yang-Mills constants g in the $\overline{\mathrm{MS}}$ and $\overline{\mathrm{DR}}$-schemes, respectively, and

$$\rho(M) = \beta'(M) + \frac{1}{2}\zeta(2), \qquad (7)$$

$$\phi(M) = 3\zeta(3) + \Psi''(M) - 2\Phi_2(M) + 2\beta'(M)\left(\Psi(1) - \Psi(M)\right). \qquad (8)$$

Here

$$\Psi(M) = \frac{\Gamma'(M)}{\Gamma(M)},\ \beta'(z) = \sum_{l=0}^{\infty}\frac{(-1)^{l+1}}{(l+z)^2},$$

$$\Phi_2(M) = \sum_{k=0}^{\infty}\frac{\left(\beta'(k+1) + (-1)^k\Psi'(k+1)\right)}{k+M}.$$

In the right-hand side of the eigenvalue equation the first contribution corresponds to the singularity at $l=-1$ generated by the pole of the Legendre function $Q_l(x)$ in the kernel at $\omega = 1+l \to 0$ and the last term appears from the regular part of the Born contribution. Note, that M and $1-\widetilde{M}$ coincide with the anomalous dimensions appearing in the asymptotical expressions for the BFKL kernel in the limits when the gluon virtualities are large:

$q_1^2 \to \infty$ and $q_2^2 \to \infty$, respectively. Because $1 - \widetilde{M} = M^*$, the hermitial separability guarantees the symmetry of ω for the principal series of the unitary representations of the Möbius group to the substitution $\nu \to -\nu$ and the hermicity of the BFKL hamiltonian. It turns out, however, that the holomorphic separability corresponding to the symmetry $m \leftrightarrow \tilde{m}$ is violated in the NLO approximation [8].

Analogously to refs. [4, 6] one can calculate the eigenvalues of the BFKL kernel in the case of the non-symmetric choice for the energy normalization s_0 in eq.(15) related to the interpretation of the NLO corrections in the framework of the renormalization group. For the scale $s_0 = q^2$ natural for the deep-inelastic scattering we obtain the corresponding eigenvalue equation

$$\begin{aligned} 1 &= \frac{4\hat{a}}{\omega}\left(2\Psi(1) - \Psi(\gamma) - \Psi(J-\gamma) + \hat{a}\left(\phi(\gamma) + \phi(J-\gamma)\right)\right) \\ &\quad +2\hat{a}\left(p(\gamma) + p(J-\gamma)\right), \quad J = 1 + |n| + \omega - \gamma . \end{aligned} \tag{9}$$

Here J is the total number of the Lorentz indices of the corresponding operator which does not coincide generally with its Lorentz spin j and

$$p(\gamma) = \Psi'(\gamma) - \rho(\gamma) = \Psi'(\gamma) - \beta'(\gamma) - \frac{1}{2}\zeta(2) = 2\sum_{k=0}^{\infty}\frac{1}{(\gamma+2k)^2} - \frac{1}{2}\zeta(2) .$$

By solving the above equation one can obtain for the anomalous dimension at physical $|n|$

$$\gamma = \frac{4\hat{a}}{\omega}\Big(1 + \omega\left(\Psi(1) - \Psi(|n|+1)\right)\Big) + \frac{(4\hat{a})^2}{\omega^2}\left(\Psi(1) - \Psi(|n|+1) + \frac{\omega}{2}c(n)\right),$$

where

$$c(n) = \Psi'(|n|+1) - \Psi'(1) - \beta'(|n|+1) + \beta'(1).$$

At $n = 0$ the correction $\sim \hat{a}$ to γ is absent, but for other n the result is non-zero and is divergent for $1 + |n| \to -r$, which is related to the appearance of the double-logarithmic terms $\Delta\gamma \sim \hat{a}^2/\omega^3$ in γ near the points $j = -r$ (see Introduction). We remind, that for positive integer $|n|$ we calculate the anomalous dimensions γ of the higher twist operators (with an anti-symmetrization between n transverse and $1 + \omega$ longitudinal indices). The singularities of the anomalous dimensions of the twist-2 operators according to [6] can be obtained only in the limit when $\Delta|n| \sim \omega^2$ for $\omega \to 0$.

3. Anomalous dimension matrix in the N=4 SUSY

In LLA anomalous dimension matrices in the N=4 SUSY have the following form (see [25]): for tensor twist-2 operators

$$
\begin{aligned}
\gamma_{gg}^{(0)}(j) &= 4\left(\Psi(1)-\Psi(j-1)-\frac{2}{j}+\frac{1}{j+1}-\frac{1}{j+2}\right),\\
\gamma_{qg}^{(0)}(j) &= 8\left(\frac{1}{j}-\frac{2}{j+1}+\frac{2}{j+2}\right), \quad \gamma_{\varphi g}^{(0)}(j) = 12\left(\frac{1}{j+1}-\frac{1}{j+2}\right),\\
\gamma_{gq}^{(0)}(j) &= 2\left(\frac{2}{j-1}-\frac{2}{j}+\frac{1}{j+1}\right), \quad \gamma_{q\varphi}^{(0)}(j) = \frac{8}{j},\\
\gamma_{qq}^{(0)}(j) &= 4\left(\Psi(1)-\Psi(j)+\frac{1}{j}-\frac{2}{j+1}\right), \quad \gamma_{\varphi q}^{(0)}(j) = \frac{6}{j+1},\\
\gamma_{\varphi\varphi}^{(0)}(j) &= 4\left(\Psi(1)-\Psi(j+1)\right), \quad \gamma_{g\varphi}^{(0)}(j) = 4\left(\frac{1}{j-1}-\frac{1}{j}\right), \qquad (10)
\end{aligned}
$$

for the pseudo-tensor operators:

$$
\begin{aligned}
\tilde{\gamma}_{gg}^{(0)}(j) &= 4\left(\Psi(1)-\Psi(j+1)-\frac{2}{j+1}+\frac{2}{j}\right),\\
\tilde{\gamma}_{qg}^{a,(0)}(j) &= 8\left(-\frac{1}{j}+\frac{2}{j+1}\right), \quad \tilde{\gamma}_{gq}^{(0)}(j) = 2\left(\frac{2}{j}-\frac{1}{j+1}\right),\\
\tilde{\gamma}_{qq}^{(0)}(j) &= 4\left(\Psi(1)-\Psi(j+1)+\frac{1}{j+1}-\frac{1}{j}\right). \qquad (11)
\end{aligned}
$$

Note, that in the N=4 SUSY multiplet there are twist-2 operators with fermion quantum numbers but their anomalous dimensions are the same as for the bosonic components of the corresponding supermultiplet (cf. ref. [17]). It is possible to construct 5 independent twist-two operators with a multiplicative renormalization. The corresponding parton distribution momenta and their LLA anomalous dimensions have the form $(\Psi(j+1)-\Psi(1)\equiv S_1(j))$ [25]:

$$
\begin{aligned}
& n_I(j) = n_g^j + n_q^j + n_\varphi^j, \quad \gamma_I^{(0)}(j) = -4S_1(j-2) \equiv \gamma_+^{(0)}(j),\\
& n_{II}(j) = -2(j-1)n_g^j + n_q^j + \frac{2}{3}(j+1)n_\varphi^j, \quad \gamma_{II}^{(0)}(j) = -4S_1(j) \equiv \gamma_0^{(0)}(j),\\
& n_{III}(j) = -\frac{j-1}{j+2}n_g^j + n_q^j - \frac{j+1}{j}n_\varphi^j, \quad \gamma_{III}^{(0)}(j) = -4S_1(j+2) \equiv \gamma_-^{(0)}(j),\\
& n_{IV}(j) = 2\Delta n_g^j + \Delta n_q^j, \quad \gamma_{IV}^{(0)}(j) = -4S_1(j-1) \equiv \tilde{\gamma}_+^{(0)}(j),\\
& n_V(j) = -(j-1)\Delta n_g^j + \frac{j+2}{2}\Delta n_q^j, \quad \gamma_V^{(0)}(j) = -4S_1(j+1) \equiv \tilde{\gamma}_-^{(0)}(j),
\end{aligned}
$$

Thus, we have one supermultiplet of operators with the same anomalous dimension $\gamma^{LLA}(j)$ proportional to $\Psi(1)-\Psi(j-1)$. The momenta of the corresponding linear combinations of the parton distributions can be obtained from the above expressions by an appropriate shift of their argument j to obtain this universal anomalous dimension $\gamma^{LLA}(j)$. Moreover, the coefficients in these linear combinations for N=4 SUSY can be found from the super-conformal invariance (cf. Ref [17]). However, in two-loop approximation these coefficients are slightly renormalized [26] due to the breaking of the conformal invariance [27]. In the paper [8] using some plausible arguments an universal anomalous dimension in two-loops for N=4 SUSY in the $\overline{\mathrm{DR}}$-scheme was suggested. Other anomalous dimensions are obtained by an integer shift of its arguments. These results were justified by a direct calculation of the anomalous dimension matrix in ref. [26]. With the use of the basis for the multiplicatively renormalizable operators obtained in LLA in [8] one can transform this matrix to a triangle form. The diagonal elements of the triangle matrix are expressed in terms of the universal anomalous dimension $\gamma(j)$ by an appropriate integer shift of its argument:

$$\gamma(j) = -\frac{\alpha_s N_c}{\pi} S_1(j-2) + \left(\frac{\alpha_s N_c}{4\pi}\right)^2 \hat{Q}(j-2), \tag{12}$$

where

$$\hat{Q}(j) = -\frac{4 S_1(j)}{3} + 16 S_1(j) S_2(j) + 8 S_3(j) - 8 \tilde{S}_3(j) + 16 \tilde{S}_{1,2}(j),$$

$$S_k(n) = \sum_{i=1}^{n} \frac{1}{i^k}, \qquad \tilde{S}_k(n) = \sum_{i=1}^{n} \frac{(-1)^i}{i^k}, \qquad \tilde{S}_{k,l}(n) = \sum_{i=1}^{n} \frac{1}{i^k} \tilde{S}_l(i).$$

The analytical continuation of functions $\gamma_{ab}^{(1)}$ $(a, b = g, q, \varphi)$ and $\tilde{\gamma}_{ab}^{(1)}$ $(a, b = g, q)$ to the complex values of j can be done analogously to refs. [28, 8].

It is possible to redefine the coupling constant $\alpha_s \to \alpha_s(1 - \alpha_s N_c/(12\pi))$ to remove in $\hat{Q}(j)$ the term proportional to $S_1(j)$. For the anomalous dimension matrix one can obtain a number of linear relations [26].

4. Relation between the DGLAP and BFKL equations

As we have discussed already above, in the case of N=4 SUSY the BFKL eigenvalue is analytic in $|n|$ and one can continue the anomalous dimensions to the negative values of $|n|$. It gives a possibility to find the singular contributions of the anomalous dimensions of the twist-2 operators not only at $j = 1$ but also at other integer points $j = 0, -1, -2....$ As it was discussed already in the Introduction, in the Born approximation we obtain $\gamma = 4\,\hat{a}\,(\Psi(1) - \Psi(j-1))$ which coincides with the result of the direct calculations (see [16, 25] and the discussions in subsection 4.2). Thus, in the case of N=4 the BFKL equation presumably contains the information sufficient for restoring the kernel of the DGLAP equation. Below we investigate the relation between these equations in the NLO approximation.

Let us start with an investigation of singularities of the anomalous dimensions obtained from the DGLAP equation. By presenting the Lorentz spin j as $\omega - r$, where $r = -1, 0, 1, ...$ and pushing $\omega \to 0$ we can calculate the singular behavior of the universal anomalous dimension $\gamma(j)$ after the redefinition of the coupling constant $\alpha_s \to \alpha_s(1 - \alpha_s N_c/(12\pi))$

$$\begin{aligned}\gamma(j) &= G\left[\frac{1}{\omega} - S_1(r+1) + O(\omega)\right] \\ &+ G^2 \begin{cases} \frac{1}{\omega^3} - 2S_1(r+1)\frac{1}{\omega^2} - \hat{S}_2(r+1)\frac{1}{\omega} + O(\omega^0) & \text{if } r = 2m \\ S_2(r+1)\frac{1}{\omega} + O(\omega^0) & \text{if } r = 2m+1, \end{cases}\end{aligned} \tag{13}$$

where $G = \alpha_s N_c/\pi$. Let us consider initially the BFKL equation in a modified LLA i.e. $\omega^{MLLA} = G(2\Psi(1) - \Psi(\gamma) - \Psi(J-\gamma))$. In the limit $J = 1 + |n| + \omega \to -r + \omega$ by inverting this equation one can obtain

$$\gamma = \frac{G}{\omega} + G^2\left[\frac{1}{\omega^3} - S_1(r)\frac{1}{\omega^2} - \hat{S}_2(r)\frac{1}{\omega}\right] + O(G^3), \tag{14}$$

i.e. this result coincides after the shift $r \to r+1$ with the singular part of the corresponding DGLAP result for the even values of r with the exception of the coefficient in the front of G^2/ω^2.

In a general case the next-to-leading corrections contain the divergencies at $|n| \to -r-1$. Their appearance is related to the presence of the double-logarithms. Indeed the eigenvalue relation for the Bethe-Salpeter equation can be written near $\gamma = 0$ and $J \simeq j = -r + \omega$ in the form

$$1 = G\frac{1}{\gamma(\omega - \gamma)} + O(G^2).$$

Because the first contribution in the right hand side contains additional singularities in comparison with the pole $1/\omega$ in the physical case of the positive $|n|$, we should subtract from the correction $O(G^2)$ the terms appeared in its first iteration:

$$G^2\frac{1}{\gamma^2(\omega-\gamma)^2} \simeq G^2\left(\frac{1}{\gamma^2\omega^2} + \frac{2}{\gamma\omega^3}\right).$$

This subtraction leads to the final result for even r, which is in an agreement with the fact, that the double-logarithms in the universal anomalous dimension $\gamma(j)$ exist at even negative j. For the odd r the divergency $\alpha^2/(\gamma\omega^3)$ in $O(G^2)$ is absent in an accordance with the absence in $\gamma(j)$ of the DL terms

$\sim \alpha^2/\omega^3$ at odd negative j. For a more accurate comparison of the singularities of the BFKL and DGLAP equations in two-loop approximation one needs to calculate in the BFKL kernel non-singular terms at $j \to -r$.

5. Conclusion

Above we reviewed the LLA and NLL results for the eigenvalue of the kernels of the BFKL and DGLAP equations in the N=4 supersymmetric gauge theory and constructed the operators with a multiplicative renormalization [8]. These anomalous dimensions can be obtained from the universal anomalous dimension $\gamma^{univ}(j)$ by a shift of its argument $j \to j + k$. The NLO corrections to the anomalous dimension matrix were found with the use of the plausible arguments [8] and by the direct methods [26].

Note that recently the LLA anomalous dimensions in this theory for large $\alpha_s N_c$ were constructed in ref. [29] in the limit $j \to \infty$ from the superstring model with the use of the Maldacena correspondence [30, 31, 32]. Also in N=4 SUSY at large $\alpha_s N_c$ the Pomeron coincides with the graviton [33]. It will be interesting to obtain these results directly from the DGLAP and BFKL equation. Already in the perturbation theory as it was demonstrated above, the BFKL dynamics has remarkable properties: analyticity in the conformal spin $|n|$, Möbius invariance, holomorphic (and hermitian) separability and integrability in a generalized LLA. On the other hand, for the DGLAP dynamics the anomalous dimensions for all twist-2 operators are proportional in LLA to the function $\Psi(1) - \Psi(j-1)$ up to an integer shift of its arguments, which corresponds to the eigenvalue of a pair Hamiltonian in the integrable Heisenberg spin model [6, 34]. The investigation of the N=4 supersymmetric model should be continued in the perturbation theory and for large $\alpha_s N_c$ because it is helpful for understanding of QCD.

References

1. Fadin, V.S., Kuraev, E.A., and Lipatov, L.N. (1975) On the Pomeranchuk singularity in asymptotically free theories, *Phys. Lett.* **B60**, 50–52.
   Lipatov, L.N. (1976) Reggeization of the vector meson and the vacuum singularity in nonabelian gauge theories, *Sov. J. Nucl. Phys.* **23**, 338–345.
   Kuraev, E.A., Lipatov, L.N., and Fadin, V.S. (1976) Multi - Reggeon processes in the Yang-Mills theory, *Sov. Phys. JETP* **44**, 443–450.
   Kuraev, E.A., Lipatov, L.N., and Fadin, V.S. (1977) The Pomeranchuk singularity in nonabelian gauge theories, *Sov. Phys. JETP* **45**, 199–204.
   Balitsky, I.I. and Lipatov, L.N. (1978) The Pomeranchuk singularity in quantum chromodynamics, *Sov. J. Nucl. Phys.* **28**, 822–829.
   Balitsky, I.I. and Lipatov, L.N. (1979) Calculation of meson meson interaction crosssection in quantum chromodynamics. (In Russian), *JETP Lett.* **30**, 355.
2. Gribov, V.N. and Lipatov, L.N. (1972) Deep inelastic e p scattering in perturbation theory, *Yad. Fiz.* **15**, 781–807.

Gribov, V.N. and Lipatov, L.N. (1972), e+ e- pair annihilation and deep inelastic e p scattering in perturbation theory, *Yad. Fiz.* **15**, 1218–1237.
Lipatov, L.N. (1975) The parton model and perturbation theory, *Sov. J. Nucl. Phys.* **20**, 94–102.
Altarelli, G. and Parisi, G. (1977) Asymptotic freedom in parton language, *Nucl. Phys.* **B126**, 298.
Dokshitzer, Y.L. (1977) Calculation of the structure functions for deep inelastic scattering and e+ e- annihilation by perturbation theory in quantum chromodynamics. (In Russian), *Sov. Phys. JETP* **46**, 641–653.

3. Curci, G., Furmanski, W., and Petronzio, R. (1980) Evolution of parton densities beyond leading order: the nonsinglet case, *Nucl. Phys.* **B175**, 27.
Furmanski, W. and Petronzio, R. (1980) Singlet parton densities beyond leading order, *Phys. Lett.* **B97**, 437.
Floratos, E.G., Kounnas, C. and Lacaze, R. (1981) Higher order QCD effects in inclusive annihilation and deep inelastic scattering, *Nucl. Phys.* **B192**, 417.
Lopez, C. and Yndurain, F.J. (1980) Behavior of deep inelastic structure functions near physical region endpoints from QCD, *Nucl. Phys.* **B171**, 231.
Lopez, C. and Yndurain, F.J. (1981) Behavior at x = 0, 1, sum rules and parametrizations for structure functions beyond LO, *Nucl. Phys.* **B183**, 157.
Mertig, R. and van Neerven, W.L. (1996) The calculation of the two loop spin splitting functions P(ij)(1)(x), *Z. Phys.* **C70**, 637–654.
Vogelsang, W. (1996) The spin-dependent two-loop splitting functions, *Nucl. Phys.* **B475**, 47–72.
4. Fadin, V.S. and Lipatov, L.N. (1998) BFKL Pomeron in the next-to-leading approximation, *Phys. Lett.* **B429**, 127–134.
5. Ciafaloni, M. and Camici, G. (1998) Energy scale(s) and next-to-leading BFKL equation, *Phys. Lett.* **B430**, 349–354.
6. Kotikov, A.V. and Lipatov, L.N. (2000) NLO corrections to the BFKL Equation in QCD and in supersymmetric gauge theories, *Nucl. Phys.* **B582**, 19–43.
7. Siegel, W. (1979) Supersymmetric dimensional regularization via dimensional reduction, *Phys. Lett.* **B84**, 193.
8. Kotikov, A.V. and Lipatov, L.N. (2001) DGLAP and BFKL evolution equations in the N = 4 supersymmetric gauge theory, hep-ph/0112346
Kotikov, A.V. and Lipatov, L.N. (2002) DGLAP and BFKL equations in the N = 4 supersymmetric gauge theory, hep-ph/0208220
9. Lipatov, L.N. (1986) The bare Pomeron in quantum chromodynamics, *Sov. Phys. JETP* **63**, 904–912.
10. Bartels, J. (1980) High-energy behavior in a nonabelian gauge theory. 2. First corrections to T(n-¿m) beyond the leading lns approximation, *Nucl. Phys.* **B175**, 365.
Kwiecinski, J. and Praszalowicz, M. (1980) Three gluon integral equation and odd c singlet Regge singularities in QCD, *Phys. Lett.* **B94**, 413.
11. Lipatov, L.N. (1990) Pomeron and Odderon in QCD And a two-dimensional conformal field theory, *Phys. Lett.* **B251**, 284–287.
12. Lipatov, L.N. (1993) High-energy asymptotics of multicolor QCD and two- dimensional conformal field theories, *Phys. Lett.* **B309**, 394–396.
13. Lipatov, L.N. (1994) High-energy asymptotics of multicolor QCD and exactly solvable lattice models, *JETP Lett.* **59**, 596–599.
14. Faddeev, L.D. and Korchemsky, G.P. (1995) High-energy QCD as a completely integrable model, *Phys. Lett.* **B342**, 311–322.
15. Lipatov, L.N. (1999) Duality symmetry of Reggeon interactions in multicolour QCD, *Nucl. Phys.* **B548**, 328–362.
16. Lipatov, L.N. (1997) Evolution equations in QCD, Perspectives in Hadronic Physics, in: *Proc. of the ICTP conf.* (World Scientific, Singapore, 1997).
17. Bukhvostov, A.P., Frolov, G.V., Lipatov, L.N. and Kuraev, E.A. (1985) Evolution equations for quasi - partonic operators, *Nucl. Phys.* **B258**, 601–646.

18. Wosiek, J. and Janik, R.A. (1997) Solution of the odderon problem for arbitrary conformal weights, *Phys. Rev. Lett.* **79**, 2935–2938.
Janik, R.A. and Wosiek, J. (1999) Solution of the odderon problem, *Phys. Rev. Lett.* **82**, 1092–1095.
19. Bartels, J., Lipatov, L.N. and Vacca, G.P. (2000) A new odderon solution in perturbative QCD, *Phys. Lett.* **B477**, 178–186.
20. De Vega, H.J. and Lipatov, L.N. (2001) Interaction of Reggeized gluons in the Baxter-Sklyanin representation, *Phys. Rev.* **D64**, 114019.
de Vega, H.J. and Lipatov, L.N. (2002) Exact resolution of the Baxter equation for reggeized gluon interactions, *Phys. Rev.* **D66**, 074013.
21. Derkachov, S.E., Korchemsky, G.P. and Manashov, A.N. (2001) Noncompact Heisenberg spin magnets from high-energy QCD. I: Baxter Q-operator and separation of variables, *Nucl. Phys.* **B617**, 375–440.
Derkachov, S.E., Korchemsky, G.P., Kotanski, J. and Manashov, A.N. (2002) Noncompact Heisenberg spin magnets from high-energy QCD. II: Quantization conditions and energy spectrum, *Nucl. Phys.* **B645**, 237–297.
22. Lipatov, L.N. (1995) Gauge invariant effective action for high-energy processes in QCD, *Nucl. Phys.* **B452**, 369–400.
Lipatov, L.N. (1999) Hamiltonian for Reggeon interactions in QCD, *Phys. Rept.* **320**, 249–260.
23. Klimyk, A. and Schmudgen, K. (1997) Quantum groups and their representations, Berlin, Germany: Springer 552 p.
24. Gorshkov, V.G., Gribov, V.N., Lipatov, L.N. and Frolov, G.V. (1966) Double logarithmic asymptotics of quantum electrodynamics, *Phys. Lett.* **22**, 671–673.
25. Lipatov, L.N. (2001) Next-to-leading corrections to the BFKL equation and the effective action for high energy processes in QCD, *Nucl. Phys. Proc. Suppl.* **99A**, 175–179.
26. Kotikov, A.V., Lipatov, L.N. and Velizhanin, V.N. (2002) Anomalous dimensions of Wilson operators in N=4 supersymmetric gauge theory, *Phys. Rev. Lett.*, in print.
27. Belitsky, A.V. and Muller, D. (2002) Superconformal constraints for QCD conformal anomalies, *Phys. Rev.* **D65**, 054037.
28. Kazakov, D.I. and Kotikov, A.V. (1988) Total alpha-s correction to deep inelastic scattering cross-section ratio, R = sigma-L / sigma-t in QCD. Calculation of longitudinal structure function, *Nucl. Phys.* **B307**, 721.
29. Gubser, S.S., Klebanov, I.R. and Polyakov, A.M. (2002) A semi-classical limit of the gauge/string correspondence, *Nucl. Phys.* **B636**, 99–114.
30. Maldacena, J.M. (1998) The large N limit of superconformal field theories and supergravity, *Adv. Theor. Math. Phys.* **2**, 231–252.
31. Gubser, S.S., Klebanov, I.R. and Polyakov, A.M. (1998) Gauge theory correlators from non-critical string theory, *Phys. Lett.* **B428**, 105–114.
32. Witten, E. (1998) Anti-de Sitter space, thermal phase transition, and confinement in gauge theories, *Adv. Theor. Math. Phys.* **2**, 505–532.
33. Brower, R.C., Mathur, S.D. and Tan, C.-I. (2000) Glueball spectrum for QCD from AdS supergravity duality, *Nucl. Phys.* **B587**, 249–276.
Janik, R.A. and Peschanski, R. (2002) Reggeon exchange from AdS/CFT, *Nucl. Phys.* **B625**, 279–294.
34. Frolov, G.V., Gribov, V.N. and Lipatov, L.N. (1970) On Regge poles in quantum electrodynamics, *Phys. Lett.* **B31**, 34.
Gribov, V.N., Lipatov, L.N. and Frolov, G.V. (1971) The leading singularity in the j plane in quantum electrodynamics, *Sov. J. Nucl. Phys.* **12**, 543.
Cheng, H. and Wu, T.T. (1970) Logarithmic factors in the high-energy behavior of quantum electrodynamics, *Phys. Rev.* **D1**, 2775–2794.
Cheng, H. and Wu, T.T. (1987) Expanding protons: scattering at high-energies, Cambridge, USA: MIT-PR. 285p.

JUSTIFICATION OF THE BFKL APPROACH IN THE NLA

V.S. FADIN
Budker Institute for Nuclear Physics
630090, Lavrenteva st. 11, Novosibirsk, Russia

Abstract. For self-consistency of the BFKL approach to the description of small-x processes a set of the "bootstrap" conditions must be fulfilled. The conditions appear from the requirement of compatibility of the gluon Reggeization with the s-channel unitarity. The set of these conditions in the next-to-leading approximation (NLA) is presented. It is argued that fulfillment of the whole set gives a possibility to prove the gluon Reggeization in the NLA.

1. Introduction

In the BFKL approach [1] the scattering amplitude for the process $A + B \longrightarrow A' + B'$ at large center of mass energy $\sqrt{s}$ and fixed momentum transfer $\sqrt{-t}$, $s \gg |t|$, may be symbolically written as the convolution

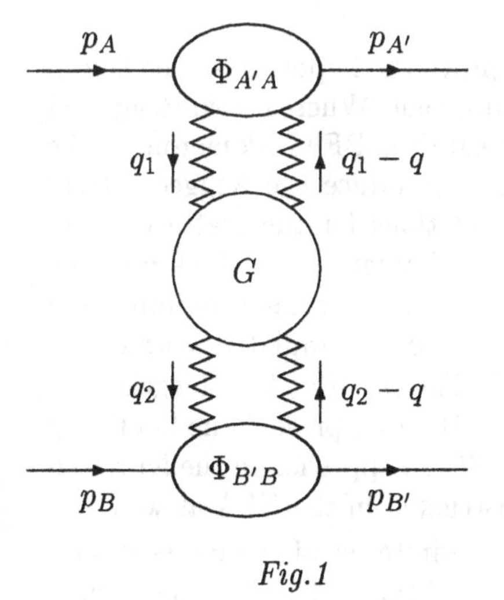

Fig.1

$$\Phi_{A'A} \otimes G \otimes \Phi_{B'B} \tag{1}$$

(see Fig.1), where the impact factors $\Phi_{A'A}$ and $\Phi_{B'B}$ depict the transitions $A \to A'$ and $B \to B'$ caused by the scattering on the Reggeized gluons, while G is the Green's function describing the interaction of two such gluons. All dependence on nature of the particles A, A' (B, B') is contained in the impact factor $\Phi_{A'A}$ ($\Phi_{B'B}$), which are energy independent, while dependence on energy is determined by the universal (process independent) Green's function G,

R. Fiore et al. (eds.), Diffraction 2002, 235–245.

which is determined by the BFKL equation. Hence, it is clear that the BFKL approach is based on the gluon Reggeization. To derive the representation (1) and the BFKL equation the Reggeized form for the set of the amplitudes with the gluon quantum numbers in channels with fixed (not increasing with s) transferred momenta was assumed. In the leading logarithmic approximation (LLA), when only the leading terms $(\alpha_S \ln s)^n$ are resummed [1], the set is given by the scattering amplitudes in the multi-Regge kinematics (MRK), which means large (growing with s) invariant masses of any pair of produced particles and fixed transverse momenta (for brevity, we include the Regge kinematics in the MRK). The Reggeized form of these amplitudes is proved [2], so that in the LLA the BFKL approach has a steady basis.

Now the approach is developed in the NLA, when the terms $\alpha_S(\alpha_S \ln s)^n$ are also resummed. The kernel of the BFKL equation for the forward scattering ($t = 0$ and color singlet in the t-channel) in the next-to-leading order (NLO) is found [3],[4]. The calculation of the NLO kernel for the non-forward scattering [5] is not far from completion (see [6],[7]). The impact factors of gluons [8] and quarks [9] are calculated in the NLO and the impact factors of the physical (color singlet) particles are under investigation [10],[11],[12],[13],[14].

The development is founded on two assumptions. The first is that the Reggeized form of the MRK amplitudes remains valid in the NLA (with the gluon Regge trajectory and the Reggeon vertices taken in the NLO). The second concerns the production amplitudes in the quasi-multi-Regge kinematics (QMRK), where a pair of produced particles has fixed invariant mass. They also are supposed to have the Reggeized form. Both assumptions still remain a hypothesis.

Evidently, it is extremely desirable to prove the hypothesis. The key to the proof is given by the "bootstrap" requirement. Whereas two Reggeized gluons of Fig.1 in the color singlet state create the BFKL Pomeron, in the antisymmetric color octet state they must reproduce the Reggeon itself. This requirement leads to the bootstrap relations for the scattering amplitudes. To justify the BFKL approach the bootstrap relations must be satisfied for all amplitudes involved there. On the other hand, fulfillment of all these relations secures the Reggeized form of the radiative corrections order by order in the perturbation theory. The proof of the Reggeization in the LLA was constructed [2] on this way. Here we present the bootstrap relations for all amplitudes involved in the BFKL approach in the NLA and argue that an analogous proof can be constructed in the NLA as well.

The bootstrap requirement leads to an infinite set of the bootstrap relations for the multi-particle production amplitudes. On the other hand, all the amplitudes are expressed in terms of the gluon trajectory and a finite number of the Reggeon vertices. Therefore it is extremely nontrivial to satisfy all these relations. Nevertheless, it occurs that all of them can be fulfilled if the vertices and trajectory submit to several bootstrap conditions.

2. The hypothesis of the gluon Reggeization

For the elastic scattering process $A+B \to A'+B'$ in the Regge kinematics ($s \simeq -u \to \infty$ and fixed t) the hypothesis means that the amplitude with the colour octet state in the t-channel and negative signature, *i.e.* odd under $s \leftrightarrow u$ exchange (in the following we consider only such amplitudes omitting explicit indications in order not to entangle denotations), has the Regge form (see Fig.2)

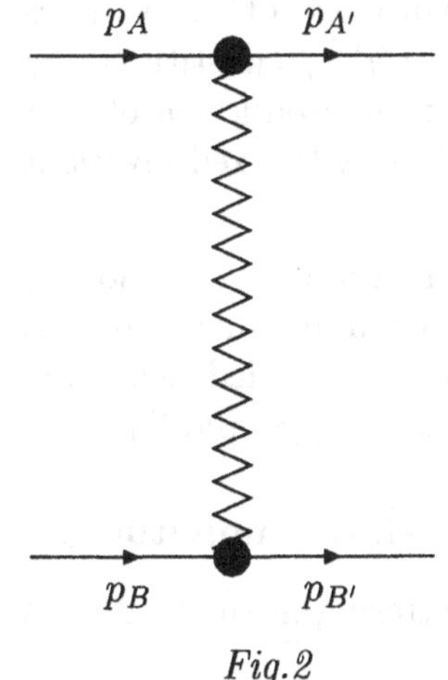

Fig.2

$$\mathcal{A}_{AB}^{A'B'} = \Gamma^c_{A'A}\left[\left(\frac{-s}{-t}\right)^{j(t)} - \left(\frac{s}{-t}\right)^{j(t)}\right]\Gamma^c_{B'B} \; ; \quad (2)$$

$j(t)$ is the Reggeized gluon trajectory, $j(t) = 1 + \omega(t)$, $j(0) = 1$, $\Gamma^c_{A'A}$ - particle-particle-Reggeon (PPR) vertices; $\Gamma^c_{A'A} = g\langle A'|T^c|A\rangle\Gamma_{A'A}$, T^c - colour group generators in the fundamental (for quarks) or the adjoint (for gluons) representation. The vertices and the trajectory in the leading order (LO) are easily calculated with this assumption. In the NLO it is not so easy. It was done in [15]-[18] for the vertices and in [19]-[22] for the trajectory.

Multi-particle amplitudes have a complicated analytical structure. It is not simple even in the MRK (see, for instance, [15],[23]). Fortunately, only real parts of these amplitudes are used in the BFKL approach in the NLA as well as in the LLA. We restrict ourselves also by consideration of the real parts, although it is not explicitly indicated below. For the process $A+B \;\to\; \tilde{A}+\tilde{B}+n$ in the MRK it was assumed that (see Fig.3)

$$\mathcal{A}_{AB}^{\tilde{A}\tilde{B}+n} = 2s\Gamma^{c_1}_{\tilde{A}A}\left[\prod_{i=1}^{n}\frac{1}{t_i}\gamma^{P_i}_{c_ic_{i+1}}(q_i,q_{i+1})\left(\frac{s_i}{\sqrt{\vec{k}^2_{i-1}\vec{k}^2_i}}\right)^{\omega(t_i)}\right]$$

$$\frac{1}{t_{n+1}}\left(\frac{s_{n+1}}{\sqrt{\vec{k}^2_n\vec{k}^2_{n+1}}}\right)^{\omega(t_{n+1})}\Gamma^{c_{n+1}}_{\tilde{B}B} \; ; \;\; t_i = q_i^2 \simeq q_{i\perp}^2 = -\vec{q}_i^{\;2} \; , \quad (3)$$

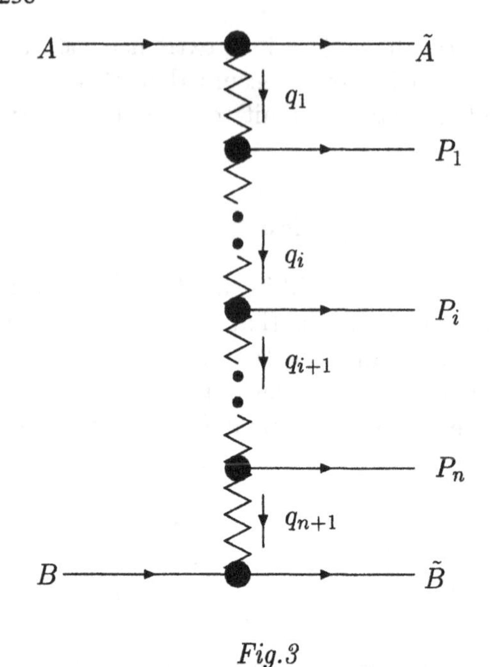

Fig.3

where $s_i = (k_{i-1} + k_i)^2$, k_i are momenta of the produced particles, $k_0 \equiv p_{\tilde{A}}$, $k_{n+1} \equiv p_{\tilde{B}}$, $\gamma^{P_i}_{c_i c_{i+1}}(q_i, q_{i+1})$ are the Reggeon-Reggeon-particle (RRP) vertices. Note that in the LLA only gluons can be produced. In the NLO $\gamma^G_{ab}(q_1, q_2)$ was calculated in [15],[24],[25], [26] and [27]. The production amplitudes in the QMRK have the same form (3) as in the MRK with one of the vertices $\gamma^{P_i}_{c_i c_{i+1}}$ or $\Gamma^c_{\tilde{P}P}$ substituted by a vertex for production of pair of particles with fixed invariant mass.

Note that because the QMRK in the unitarity relations leads to loss of large logarithms, scales of energies $\sqrt{s_i}$ (see (3)) are unimportant in this kinematics; moreover, the trajectory and the vertices are needed there only in the LO. The QMRK vertices were obtained in [28]-[30],[8] and [9].

3. The bootstrap relations for elastic and inelastic amplitudes

To derive the bootstrap relation for the elastic scattering in the NLA it is sufficient to note that with this accuracy

$$\frac{1}{-2\pi i} disc_s \left(\ln^n(-s) + \ln^n s\right) = \frac{1}{2}\frac{\partial}{\partial \ln s} \Re \left[\ln^n(-s) + \ln^n s\right] , \tag{4}$$

where $disc_s$ denotes the s-channel discontinuity and $\Re$ -the real part. Therefore we have

$$\frac{1}{-2\pi i s} disc_s \, \mathcal{A}^{A'B'}_{AB} = \frac{1}{2}\frac{\partial}{\partial \ln s} \Re \left[\frac{1}{s} \mathcal{A}^{A'B'}_{AB}\right] . \tag{5}$$

Using in the R.H.S. the Reggeized form of the elastic amplitude (2), we come to the bootstrap relation:

$$\frac{1}{-2\pi i s} disc_s \mathcal{A}^{A'B'}_{AB} = \frac{1}{2}\omega(t) \Re \left[\frac{1}{s} \mathcal{A}^{A'B'}_{AB}\right] . \tag{6}$$

Note that it is not exact relation; we can not guarantee that it is valid in approximations higher than the NLA.

The important point is that the L.H.S. of (6) can be calculated putting the amplitudes (3) in the unitarity condition. Since the amplitudes are expressed through the vertices of the Reggeon interactions, the relation (6) imposes strong restrictions on these vertices. The restrictions can be formulated as the bootstrap conditions for the color octet impact factors and the BFKL kernel in the NLO [5]. Fulfillment of these conditions was shown in [8] and [9] for the impact factors of gluons and quarks, in [6] and [31] for the quark and gluon contributions to the kernel.

Although these conditions are very strict, their fulfillment cannot be considered as a proof of the Reggeization. Evidently, there are bootstrap relations for the production amplitudes, and one can expect that restrictions which they give are more strong. To derive these relations we need to use analytical properties of the production amplitudes. Unfortunately, we know not much about them. But, fortunately, in the NLA we don't need much.

Let us turn to the discontinuities in s_1, s_2 and s of the one-particle production amplitude $A + B \rightarrow A' + B' + G$ in the MRK. Here we meet two complications in comparison with the elastic case. Firstly, the discontinuities are not pure imaginary (e.g., the discontinuity in s can have, in turn, discontinuities in s_1 and s_2). The second is that due to the relation

$$\frac{s_1 s_2}{s} = \vec{k}_1^{\,2} \tag{7}$$

the amplitude (without the common factor s, see (3)) contains dependence on s_1, s_2 and s not only through powers of (large) logarithms of these variables. We overtake these complications considering only imaginary parts of the discontinuities and taking appropriate combinations of the discontinuities.

Remind that we consider only negative signatures in both t_i- channels, i.e. in the part of the amplitude which is antisymmetric with respect to any of the substitutions $s_1 \rightarrow -s_1$ and $s_2 \rightarrow -s_2$. Due to the relation (7), which is hold in all physical channels, this part is also antisymmetric with respect to $s \rightarrow -s$.

An important observation is that a discontinuity of a product fg is expressed through discontinuities of f and g:

$$f_+ g_+ - f_- g_- = \frac{1}{2}(f_+ - f_-)(g_+ + g_-) + \frac{1}{2}(f_+ + f_-)(g_+ - g_-)\,.$$

The second important observation is that the sum of discontinuities of $F(\vec{k}_1^{\,2})$ on s_1 and s is zero. From these two observations it follows that for such sum instead of analytical functions of $\vec{k}_1^{\,2}$ we can take their real parts. The same is true for the sum of the discontinuities on s_2 and s and for the difference of the discontinuities on s_1 and s_2.

Now, if the variables s_1, s_2 and s don't enter into the combination $s_1 s_2/s = \vec{k}^{\,2}$, they can enter only as $\hat{S}\ln^{n_1}(-s_1)\ln^{n_2}(-s_2)\ln^{n_3}(\pm s)$, with $n_1+n_2+n_3=n$, where n is less or equal the order of perturbation theory and $\hat{S}$ is the operator of symmetrization with respect to exchanges $s_1 \leftrightarrow -s_1\ ,\ \ s \leftrightarrow -s$ and $s_2 \leftrightarrow -s_2\ ,\ \ s \leftrightarrow -s$. Note that the terms containing products of $\ln(-s_i)\ln(s_i)$, where s_i can be s_1, s_2 or s, are forbidden, on the same ground as the terms containing $\ln(-s)\ln(s)$ are forbidden in the elastic amplitudes. Since in the NLO we need to keep only the first two leading total powers of n, calculating imaginary part of discontinuity in anyone of variables s_1, s_2 or s we can take only real parts of logarithms of other variables. It means that with our accuracy

$$\Re\left[\frac{1}{-2\pi i}disc_{s_i}\left(\hat{S}\ln^{n_1}(-s_1)\ln^{n_2}(-s_2)\ln^{n_3}(\pm s)\right)\right]$$
$$=\frac{1}{2}\frac{\partial}{\partial\ln s_i}\Re\left[\hat{S}\ln^{n_1}(-s_1)\ln^{n_2}(-s_2)\ln^{n_3}(\pm s)\right]\ , \tag{8}$$

where s_i can be s_1, s_2 or s and the partial derivative is taken at fixed $s_j \neq s_i$.

Therefore we have, for example

$$\Re\left[\frac{1}{-2\pi i s}\left(disc_{s_2}+disc_s\right)\mathcal{A}_{AB}^{A'GB'}\right]=\frac{1}{2}\left(\frac{\partial}{\partial\ln s_2}+\frac{\partial}{\partial\ln s}\right)\Re\left[\frac{1}{s}\mathcal{A}_{AB}^{A'GB'}\right]\ , \tag{9}$$

where in the right part the first derivative is taken at fixed s_1 and s and the second at fixed s_1 and s_2. Using that

$$\left(\frac{\partial}{\partial\ln s_2}+\frac{\partial}{\partial\ln s}\right)f(s_1,s_2,s)=\frac{\partial}{\partial\ln s_2}f(s_1,s_2,\frac{s_1 s_2}{\vec{k}_1^{\,2}})\ , \tag{10}$$

we come to

$$\Re\left[\frac{1}{-2\pi i s}\left(disc_{s_2}+disc_s\right)\mathcal{A}_{AB}^{A'GB'}\right]=\frac{1}{2}\frac{\partial}{\partial\ln s_2}\Re\left[\frac{1}{s}\mathcal{A}_{AB}^{A'GB'}\right]\ , \tag{11}$$

where in the right part the amplitude is considered as function of s_1, s_2 and $\vec{k}_1^{\,2}$.

The requirement of the Reggeized form (3) of the amplitude in the R.H.S. gives us the bootstrap relation:

$$\Re\left[\frac{1}{-2\pi i}\left(disc_{s_2}+disc_s\right)\mathcal{A}_{AB}^{A'GB'}\right]=\frac{1}{2}\omega(t_2)\Re\,\mathcal{A}_{AB}^{A'GB'}\ . \tag{12}$$

The bootstrap relations for the amplitudes of multi-particle production in the MRK can be derived in the same way. We obtain for them

$$\Re\left[\frac{1}{-2\pi i s}\left(\sum_{l=k+1}^{n+1}disc_{s_{kl}}-\sum_{l=0}^{k-1}disc_{s_{lk}}\right)\mathcal{A}_{AB}^{A'B'+n}\right]$$

$$= \frac{1}{2}\left(\frac{\partial}{\partial \log s_{k,k+1}} - \frac{\partial}{\partial \log s_{k-1,k}}\right) \Re \left[\frac{1}{s}\mathcal{A}_{AB}^{A'B'+n}(s_{i-1,i})\right] , \qquad (13)$$

where $s_{ij} = (k_i + k_j)^2$ and in the right part the amplitudes are expressed in terms of $s_{i-1,i}$, which are considered as independent variables.

If the amplitudes have the Reggeized form (3) we obtain the bootstrap relations:

$$\Re \left[\frac{1}{-2\pi i}\left(\sum_{l=k+1}^{n+1} disc_{s_{kl}} - \sum_{l=0}^{k-1} disc_{s_{lk}}\right) \mathcal{A}_{AB}^{A'B'+n}\right]$$

$$= \frac{1}{2}\left(\omega(t_{k+1}) - \omega(t_k)\right) \Re\, \mathcal{A}_{AB}^{A'B'+n} . \qquad (14)$$

These relations impose much more strong restrictions on the Reggeon vertices than (6). Their fulfilment means the proof of the Reggeization hypothesis in the NLA, since the energy dependence of the amplitudes can be calculated order by order in perturbation theory using equations (13). Indeed, the discontinuities entering in the L.H.S. of these equations in some order in the coupling constant g can be expressed with the help of the unitarity relations through the multi-particle amplitudes in lower orders in g. Fulfillment of the bootstrap relations means that the energy dependence has the Regge form.

4. The bootstrap conditions for the Reggeon vertices

The bootstrap relations impose restrictions on the gluon trajectory and the Reggeon vertices (bootstrap conditions). To formulate these conditions it is convenient to introduce operators in transverse momentum representation. From t-channel point of view we have to consider two interacting Reggeized gluons with "coordinates" in the transverse momentum space $\vec{r}$ and $\vec{q} - \vec{r}$, where $\vec{q}$ is the total transverse momentum in the t-channel. Let us introduce $\hat{\vec{r}}$ as the operator of "coordinate" of one of the Reggeized gluons in the transverse momentum space: $\hat{\vec{r}}\,|\vec{q}_i\rangle = \vec{q}_i|\vec{q}_i\rangle$. The total transverse momentum $\vec{q}$ is considered as the c-number. With the normalization $\langle\vec{q}_1|\vec{q}_2\rangle = \vec{q}_1^{\,2}(\vec{q}_1 - \vec{q})^{\,2}\delta^{(D-2)}(\vec{q}_1 - \vec{q}_2)$ we have

$$\langle A|B\rangle = \langle A|\vec{k}\rangle\langle\vec{k}|B\rangle = \int \frac{d^{D-2}k}{\vec{k}^2(\vec{k} - \vec{q})^2} A(\vec{k})B(\vec{k}) . \qquad (15)$$

The impact factors $\Phi_{A'A}(\vec{q}_1, \vec{q})$ appear in this formalism as the wave functions of the t-channel states $\langle\vec{q}_1|A'A\rangle$, and the BFKL kernel $\mathcal{K}(\vec{q}_2, \vec{q}_1, \vec{q})$ as the matrix element $\langle\vec{q}_2|\hat{\mathcal{K}}|\vec{q}_1\rangle$. Calculating the L.H.S. of (6) from the unitarity condition, as it is done in the BFKL approach, we obtain in these denotations

$$\frac{-2s}{(2\pi)^{D-1}}\langle A'A|\left(\frac{s}{-t}\right)^{\hat{\mathcal{K}}}|B'B\rangle = \frac{\omega(t)}{2}\,2s\,\Gamma_{A'A}\left(\frac{s}{-t}\right)^{\omega(t)}\Gamma_{B'B}\,. \tag{16}$$

If this equation were exact, then it should be [32]

$$\left(\hat{\mathcal{K}} - \omega(t)\right)|A'A\rangle = 0\,, \quad |A'A\rangle = \Gamma_{A'A}|R_\omega\rangle\,, \tag{17}$$

with an universal (process independent) eigenstate $|R_\omega\rangle$ of the kernel with the eigenvalue $\omega(t)$. These equations are known as the strong bootstrap conditions [33],[34]. Fulfilment of the condition for the impact factors was verified in [32], for the quark contribution to the kernel in [6],[34],[35]. For the gluon contribution verification requires much more efforts, but recently it was also done [36].

In fact, since our accuracy is restricted, the conditions (17) can not be derived from (6). But it occurs, that they must be satisfied for fulfillment of the bootstrap relations for inelastic amplitudes. Moreover, these relations lead to new bootstrap conditions. Take the gluon production amplitude $A + B \rightarrow A' + B' + G$ in the MRK. The unitarity condition in the s_2-channel gives

$$\frac{disc_{s_2}\mathcal{A}_{AB}^{A'GB'}}{-2\pi i} = \frac{-2s}{(2\pi)^{D-1}}\,\Gamma_{A'A}\frac{1}{t_1}\left(\frac{s_1}{-t_1}\right)^{\omega(t_1)}\langle GR|\left(\frac{s_2}{-t_2}\right)^{\hat{\mathcal{K}}}|B'B\rangle\,, \tag{18}$$

where $\langle GR|$ is the t_2-channel state of the produced gluon G and the t_1-channel Reggeized gluon R. The wave function of this state (it is the generalization of the impact factor for the case when instead of on mass-shell particle we have the Reggeized gluon) is expressed in terms of the Reggeon vertices. Within the NLO

$$\langle GR|r_\perp\rangle_{ij} = \frac{f^{jcc'}}{\sqrt{N_c}}\sum_{\{f\}}\int\frac{ds_f}{2\pi}\theta\left(s_\Lambda - s_f\right)\gamma_{ic}^{\{f\}}(q_1,r)\Gamma_{G\{f\}}^{c'} + \frac{1}{2}\langle GR|r_\perp\rangle_{ij}^{(B)}$$
$$\times\left[\left(\omega(q_1^2)+\omega(r_\perp^2)\right)\ln\left(\frac{k_\perp^2}{(q_1-r)_\perp^2}\right) + \omega((q_2-r)_\perp^2)\ln\left(\frac{k_\perp^2}{(q_2-r)_\perp^2}\right)\right]$$
$$-\frac{1}{2}\int\frac{d^{D-2}r_{1\perp}}{r_{1\perp}^2(q_2-r_1)_\perp^2}\mathcal{K}_r^{(B)}(r_\perp,r_{1\perp};q_{2\perp})\langle GR|r_{1\perp}\rangle_{ij}^{(B)}\ln\left(\frac{s_\Lambda^2}{k_\perp^2(r-r_1)_\perp^2}\right)\,, \tag{19}$$

where i and j are the colour indices of the Reggeized gluons in the t_1 and t_2 channels, the sum over $\{f\}$ is performed over all the contributing states

$\{f\}$, s_f are their squared invariant masses, the superscript (B) denotes the leading order, $\mathcal{K}_r$ means the part of the BFKL kernel related to the real particle production. The intermediate parameter s_Λ must be taken tending to infinity, so that the dependence on s_Λ disappears.

The s-channel discontinuity can be presented as

$$\frac{disc_s \mathcal{A}_{AB}^{A'GB'}}{-2\pi i} = \frac{-2s}{(2\pi)^{D-1}} \langle A'A| \left(\frac{s_1}{-t_2}\right)^{\hat{\mathcal{K}}} \hat{\mathcal{G}} \left(\frac{s_2}{-t_2}\right)^{\hat{\mathcal{K}}} |B'B\rangle \, , \tag{20}$$

where $\hat{\mathcal{G}}$ is the operator of the gluon production with change of total two-Reggeon state momentum from q_1 to q_2;

$$\begin{aligned}
\langle r_{1\perp}|\hat{\mathcal{G}}(q_1,q_2)|r_{2\perp}\rangle_{ij} = \frac{f^{iaa'}f^{jbb'}}{N_c} \Big[&\gamma^G_{a'b'}(q_1 - r_{1\perp}, q_2 - r_{2\perp})\delta^{D-2}(r_{1\perp} - r_{2\perp}) \\
&\times r_{1\perp}^2 \delta_{ab} + (q_1 - r_1)_\perp^2 \delta^{D-2}((q_1 - r_1)_\perp - (q_2 - r_2)_\perp)\delta_{a'b'}\gamma^G_{ab}(r_1, r_2) \\
&+ \int_{\frac{\vec{k}'^2}{s_\Lambda}}^{1-\frac{\vec{k}^2}{s_\Lambda}} \frac{dx'}{2x'(1-x')} \sum_{G'} \gamma_{ab}^{\{G(k)G'(k')\}}(r_1, r_2)\, \gamma_{a'b'}^{G'(-k')}(q_1 - r_1, q_2 - r_2) \\
&+ \int_{\frac{\vec{k}'^2}{s_\Lambda}}^{1-\frac{\vec{k}^2}{s_\Lambda}} \frac{dx'}{2x'(1-x')} \sum_{G'} \gamma_{ab}^{G'(k')}(r_1, r_2)\gamma_{a'b'}^{\{G(k)G'(-k')\}}(q_1 - r_1, q_2 - r_2)\Big] \\
&- \int \frac{d^{D-2}r_\perp \langle r_\perp|\hat{\mathcal{G}}(q_1,q_2)|r_{2\perp}\rangle^{(B)}}{2r_\perp^2\,(r-q_1)_\perp^2} \mathcal{K}_r^{(B)}(r_{1\perp}, r_\perp; q_{1\perp}) \ln\left(\frac{s_\Lambda^2}{(\vec{r}-\vec{r}_1)^2\vec{k}^{\,2}}\right) \\
&- \int \frac{d^{D-2}r_\perp \langle r_{1\perp}|\hat{\mathcal{G}}(q_1,q_2)|r_\perp\rangle^{(B)}}{2r_\perp^2\,(r-q_2)_\perp^2} \mathcal{K}_r^{(B)}(r_\perp, r_{2\perp}; q_{2\perp}) \ln\left(\frac{s_\Lambda^2}{(\vec{r}-\vec{r}_2)^2\vec{k}^{\,2}}\right) \, ,
\end{aligned} \tag{21}$$

where x' is the part of longitudinal momentum of $\{GG'\}$ carried by G'. The bootstrap relation (12) can be satisfied only if the strong bootstrap conditions (17) are fulfilled. Besides this, the relation requires fulfillment of new bootstrap condition:

$$\frac{-2t_2}{(2\pi)^{D-1}} \left[t_1 \langle R_\omega|\hat{\mathcal{G}}|R_\omega\rangle + \langle RG|R_\omega\rangle\right] = \omega(t_2)\gamma_{ij}^G(q_1, q_2) \, . \tag{22}$$

Fulfillment of this condition is not checked yet.

It occurs that fulfillment of these conditions secures fulfillment of the bootstrap relations for all the MRK amplitudes, that is quite nontrivial and crucial for the proof of the gluon Reggeization.

The bootstrap relations for the QMRK amplitudes lead to the bootstrap conditions for the Reggeon vertices describing productions of pairs of particles $\{P_1P_2\}$ with rapidities of the same order [37]. If the pair is produced in the region of fragmentation of the particle A, the production is described by the vertices $\Gamma^c_{\{P_1P_2\}A}(q)$, q is momentum of the Reggeized gluon and c is its colour index. The bootstrap condition for them

$$\frac{igf^{cab}}{(2\pi)^{D-1}}\int\frac{d^{D-2}r_\perp}{r_\perp^2(q-r)_\perp^2}\sum_{\{i\}}\Gamma^a_{\{i\}A}(r)\Gamma^b_{\{P_1P_2\}\{i\}}(q-r)=\frac{\omega(t)}{t}\Gamma^c_{\{P_1P_2\}A}(q)\,. \tag{23}$$

If the pair is far away in the rapidity space from the colliding particles, then it is created by two Reggeized gluons, and its production is described by the vertices $\gamma^{\{P_1P_2\}}_{c_1c_2}(q_1,q_2)$ (P_1P_2 can be gg or $q\bar{q}$). The bootstrap condition for these vertices is:

$$\frac{igf^{jbb'}}{(2\pi)^{D-1}}\int\frac{d^{D-2}r_\perp}{r_\perp^2(q_2-r)_\perp^2}\left[\frac{igf^{ia'b}q_{1\perp}^2}{(q_1-r)_\perp^2}\gamma^{\{P_1P_2\}}_{a'b'}(q_1-r,q_2-r)+\sum_G\gamma^G_{ib}(q_1,r)\right.$$
$$\times\Gamma^{b'}_{\{P_1P_2\}G}+\sum_{P_1'}\Gamma^{b'}_{P_1P_1'}\gamma^{\{P_1'P_2\}}_{ib}(q_1,r)+\sum_{P_2'}\Gamma^{b'}_{P_2P_2'}\gamma^{\{P_1P_2'\}}_{ib}(q_1,r)$$
$$\left.+\frac{igf^{ia'a}q_{1\perp}^2}{(q_1-k_1')_\perp^2k_{1\perp}'^2}\gamma^{P_2}_{ab}(q_1-k_1',r)\gamma^{P_1}_{a'b'}(k_1',q_2-r)\right]=\frac{\omega(t_2)}{t_2}\gamma^{\{P_1P_2\}}_{ij}(q_1,q_2)\,, \tag{24}$$

where k_i (k_i') are momenta of the particles P_i (P_i'). Note that the last term in the L.H.S. does contribute only in the case of the two-gluon production.

In [37] it is shown that the conditions (23),(24) are satisfied.

5. Summary

The gluon Reggeization, which is the basis of the BFKL approach, was proved in the LLA, but still remains a hypothesis in the NLA. The hypothesis can be proved using the bootstrap requirement, i.e. the demand of compatibility of the Reggeized form of the amplitudes with the s-channel unitarity. The requirement leads to an infinite set of the bootstrap relations for the scattering amplitudes constructed from the Reggeon vertices and the gluon Regge trajectory. Their fulfillment guarantees the Reggeized form of the radiative corrections order by order in the perturbation theory. It occurs that all these relations can be satisfied if the vertices and the trajectory submit to several bootstrap conditions. Fulfilment of all these conditions opens a way for the proof of the gluon Reggeization in the NLA. Only one of these conditions is not yet checked.

Acknowledgments: Work was supported in part by INTAS and in part by the Russian Fund of Basic Researches. I thank the Alexander von Humboldt foundation for the research award, the Universität Hamburg and DESY for their warm hospitality while a part of this work was done.

References

1. V.S. Fadin, E.A. Kuraev and L.N. Lipatov, Phys. Lett. **B 60** (1975) 50; E.A. Kuraev, L.N. Lipatov and V.S. Fadin, Zh. Eksp. Teor. Fiz. **71** (1976) 840 [Sov. Phys. JETP **44** (1976) 443]; **72** (1977) 377 [**45** (1977) 199]; Ya.Ya. Balitskii and L.N. Lipatov, Sov. J. Nucl. Phys. **28** (1978) 822.
2. Ya.Ya. Balitskii, L.N. Lipatov and V.S. Fadin, in Proceedings of Leningrad Winter School on Physics of Elementary Particles, Leningrad, 1979, 109 (in Russian).
3. V.S. Fadin and L.N. Lipatov, Phys. Lett. **B429** (1998) 127.
4. M. Ciafaloni and G. Camici, Phys. Lett. **B430** (1998) 349.
5. V.S. Fadin and R. Fiore, Phys. Lett. **B440** (1998) 359.
6. V.S. Fadin, R. Fiore and A. Papa, Phys. Rev. **D60** (1999) 074025.
7. V.S. Fadin and D.A. Gorbachev, Pis'ma v Zh. Eksp. Teor. Fiz. **71** (2000) 322 [JETP Letters **71** (2000) 222]; Phys. Atom. Nucl. **63** (2000) 2157 [Yad. Fiz. **63** (2000) 2253].
8. V.S. Fadin, R. Fiore, M.I. Kotsky and A. Papa, Phys. Rev. **D61** (2000) 094005.
9. V.S. Fadin, R. Fiore, M.I. Kotsky and A. Papa, Phys. Rev. **D61** (2000) 094006.
10. V.S. Fadin and A.D. Martin, Phys. Rev. **D60** (1999) 114008.
11. V. Fadin, D. Ivanov and M. Kotsky, In: New Trends in High-Energy Physics, Ed. L.L. Jenkovszky, Kiev, 2000, pp. 190-194, hep-ph/0007119.
12. J. Bartels, S. Gieseke and C.F. Qiao, Phys. Rev. **D63** (2001) 056014.
13. V. Fadin, D. Ivanov, M. Kotsky, hep-ph/0106099.
14. J. Bartels, S. Gieseke, A. Kyrieleis, Phys. Rev. **D65** (2000) 014006, hep-ph/0107152.
15. V. S. Fadin and L. N. Lipatov, Nucl. Phys. B **406** (1993) 259.
16. V. S. Fadin and R. Fiore, Phys. Lett. B **294** (1992) 286.
17. V. S. Fadin, R. Fiore and A. Quartarolo, Phys. Rev. D **50** (1994) 2265.
18. V. S. Fadin, M. I. Kotsky and R. Fiore, Phys. Lett. B **359** (1995) 181.
19. V.S. Fadin, Zh. Eksp. Teor. Fiz. Pis'ma **61** (1995) 342; V.S. Fadin, R. Fiore, A. Quartarolo, Phys. Rev. **D53** (1996) 2729; M.I. Kotsky and V.S. Fadin, Yad. Fiz. **59** (6) (1996) 1; V.S. Fadin, R. Fiore and M.I. Kotsky, Phys. Lett. **B359** (1995) 181.
20. V.S. Fadin, R. Fiore and M.I. Kotsky, Phys. Lett. **B387** (1996) 593.
21. J. Blumlein, V. Ravindran and W.L. van Neerven, Phys. Rev. **D58** (1998) 091502.
22. V. Del Duca and E.W.N. Glover, JHEP **0110** (2001) 035, hep-ph/0109028.
23. J. Bartels, Nucl. Phys. **B 175** (1980) 365.
24. V. S. Fadin, R. Fiore and A. Quartarolo, Phys. Rev. D **50** (1994) 5893.
25. V.S. Fadin, R. Fiore and M.I. Kotsky, Phys. Lett. **B389** (1996) 737.
26. V. Del Duca and C. R. Schmidt, Phys. Rev. D **59** (1999) 074004.
27. V.S. Fadin, R. Fiore and A. Papa, Phys. Rev. **D63** (2001) 034001, hep-ph/0008006.
28. L.N. Lipatov, V.S. Fadin, Zh. Eksp. Teor. Fiz. Pis'ma **49** (1989) 311 [Sov. Phys. JETP Lett. **49** (1989) 352]; Yad. Fiz. **50**, (1989) 1141 [Sov. J. Nucl. Phys. **50** (1989) 712].
29. V.S. Fadin and L.N. Lipatov, Nucl. Phys. **B477** (1996) 767.
30. V. S. Fadin, R. Fiore, A. Flachi and M. I. Kotsky, Phys. Lett. B **422** (1998) 287 [arXiv:hep-ph/9711427]; V. S. Fadin, M. I. Kotsky, R. Fiore and A. Flachi, Phys. Atom. Nucl. **62** (1999) 999 [Yad. Fiz. **62** (1999) 1066].
31. V.S. Fadin, R. Fiore and M.I. Kotsky, Phys. Lett. **B494** (2000) 100.
32. V.S. Fadin, R. Fiore, M.I. Kotsky and A. Papa, Phys. Lett. **B495** (2000) 329.
33. M. Braun, hep-ph/9901447.
34. M. Braun and G.P. Vacca, Phys. Lett. **B447** (2000) 156.
35. M. Braun and G.P. Vacca, Phys. Lett. **B454** (1999) 319.
36. V. S. Fadin and A. Papa, Nucl. Phys. B **640** (2002) 309.
37. V.S Fadin, M.G. Kozlov and A.V. Reznichenko, to be published.

FULFILLMENT OF THE STRONG BOOTSTRAP CONDITION

A. PAPA
Dipartimento di Fisica, Università della Calabria,
and INFN, Gruppo collegato di Cosenza
87036 Arcavacata di Rende, Cosenza, Italy

Abstract. The self-consistency of the assumption of Reggeized form of the production amplitudes in multi-Regge kinematics, which are used in the derivation of the BFKL equation, leads to strong bootstrap conditions. The fulfillment of these conditions opens the way to a rigorous proof of the BFKL equation in the next-to-leading approximation. The strong bootstrap condition for the kernel of the BFKL equation for the octet color state of two Reggeized gluons is one of these conditions. We show that it is satisfied in the next-to-leading approximation.

1. Gluon Reggeization and bootstrap conditions

The property of gluon Reggeization plays an essential role in the derivation of the Balitsky-Fadin-Kuraev-Lipatov (BFKL) equation [1] for the cross sections at high energy $\sqrt{s}$ in perturbative QCD. The simplest realization of the gluon Reggeization is in the elastic process $A + B \longrightarrow A' + B'$ with exchange of gluon quantum numbers in the t-channel, whose amplitude in the Regge limit (i.e. for $s \to \infty$ and $|t|$ not growing with s) takes the form

$$\left(\mathcal{A}_8^-\right)_{AB}^{A'B'} = \Gamma^c_{A'A} \left[\left(\frac{-s}{-t}\right)^{j(t)} - \left(\frac{s}{-t}\right)^{j(t)} \right] \Gamma^c_{B'B} \, . \tag{1}$$

Here c is a color index, $\Gamma^c_{P'P}$ are the particle-particle-Reggeon (PPR) vertices, not depending on s, and $j(t) = 1 + \omega(t)$ is the Reggeized gluon trajectory.

In the leading logarithmic approximation (LLA), which means resummation of the terms of the form $\alpha_s^n (\ln s)^n$, this form of the amplitude has been proved [1]. In the next-to-leading approximation (NLA), which means

R. Fiore et al. (eds.), Diffraction 2002, 247–256.

resummation of the terms $\alpha_s^{n+1}(\ln s)^n$, the form (1) has been checked in the first three orders of perturbation theory and is only assumed to be valid to all orders.

On the other hand, the amplitude for the elastic scattering process $A + B \longrightarrow A' + B'$ (for any color representation in the t-channel resulting from the composition of two color octet representations) can be determined from s-channel unitarity, by expressing its imaginary part in terms of the inelastic amplitudes $A+B \longrightarrow \tilde{A}+\tilde{B}+\{n\}$ and $A'+B' \longrightarrow \tilde{A}+\tilde{B}+\{n\}$, and then by reconstructing the full amplitude by use of the dispersion relations. It turns out (see for instance [1, 2]) that this amplitude can be written as

$$(\mathcal{A}_{\mathcal{R}})_{AB}^{A'B'} = \frac{i\,s}{(2\pi)^{D-1}} \int \frac{d^{D-2}q_1}{\vec{q}_1^{\,2}\vec{q}_1^{\,\prime\,2}} \int \frac{d^{D-2}q_2}{\vec{q}_2^{\,2}\vec{q}_2^{\,\prime\,2}} \int_{\delta-i\infty}^{\delta+i\infty} \frac{d\omega}{\sin(\pi\omega)} \left[\left(\frac{-s}{s_0}\right)^{\omega} - \tau\left(\frac{s}{s_0}\right)^{\omega}\right] \sum_{\mathcal{R},\nu} \Phi_{A'A}^{(\mathcal{R},\nu)}(\vec{q}_1,\vec{q};s_0)\, G_{\omega}^{(\mathcal{R})}(\vec{q}_1,\vec{q}_2,\vec{q})\, \Phi_{B'B}^{(\mathcal{R},\nu)}(-\vec{q}_2,-\vec{q};s_0)\,, \quad (2)$$

where $D = 4+2\epsilon$ is the space-time dimension, $\mathcal{A}_{\mathcal{R}}$ stands for the scattering amplitude with the representation $\mathcal{R}$ of the color group in the t-channel, the index ν enumerates the states in the irreducible representation $\mathcal{R}$ and $G_{\omega}^{(\mathcal{R})}$ is the Mellin transform of the Green function for the Reggeon-Reggeon scattering. The signature τ is positive (negative) for symmetric (antisymmetric) representation $\mathcal{R}$. The parameter s_0 is an arbitrary energy scale introduced in order to define the partial wave expansion of the scattering amplitudes through the Mellin transform. The dependence on this parameter disappears in the full expressions for the amplitudes. $\Phi_{P'P}^{(\mathcal{R},\nu)}$ are the so-called impact factors.

The Green function obeys the generalized BFKL equation

$$\begin{aligned} \omega G_{\omega}^{(\mathcal{R})}(\vec{q}_1,\vec{q}_2,\vec{q}) &= \vec{q}_1^{\,2}\vec{q}_1^{\,\prime\,2}\delta^{(D-2)}(\vec{q}_1-\vec{q}_2) \\ &+ \int \frac{d^{D-2}q_r}{\vec{q}_r^{\,\prime\,2}\vec{q}_r^{\,2}} \mathcal{K}^{(\mathcal{R})}(\vec{q}_1,\vec{q}_r;\vec{q})\, G_{\omega}^{(\mathcal{R})}(\vec{q}_r,\vec{q}_2;\vec{q})\,, \end{aligned} \quad (3)$$

where we have introduced the notation $q_i' \equiv q - q_i$. The kernel $\mathcal{K}^{(\mathcal{R})}$ consists of two parts, a "virtual" part, related with the Reggeized gluon trajectory, and a "real" part, related to particle production:

$$\begin{aligned} \mathcal{K}^{(\mathcal{R})}(\vec{q}_1,\vec{q}_2;\vec{q}) &= \left[\omega\left(-\vec{q}_1^{\,2}\right) + \omega\left(-\vec{q}_1^{\,\prime\,2}\right)\right] \vec{q}_1^{\,2}\,\vec{q}_1^{\,\prime\,2}\,\delta^{(D-2)}(\vec{q}_1-\vec{q}_2) \\ &+ \mathcal{K}_r^{(\mathcal{R})}(\vec{q}_1,\vec{q}_2;\vec{q})\,. \end{aligned} \quad (4)$$

In the LLA, the Reggeized gluon trajectory is needed at 1-loop accuracy and the only contribution to the "real" part of the kernel is from the production of one gluon at Born level in the collision of two Reggeons. In the

NLA, the Reggeized gluon trajectory is needed at 2-loop accuracy and the "real" part of the kernel takes contributions from one-gluon production at 1-loop level and from two-gluon and $q\bar{q}$-pair production at Born level. The representation (2) for the elastic amplitude must reproduce with NLA accuracy the representation (1), in the case of exchange of gluon quantum numbers (i.e. color octet representation and negative signature) in the t-channel. This leads to two so-called "soft" bootstrap conditions [2]: the first of them involves the kernel of the generalized non-forward BFKL equation in the octet color representation; the second one involves the impact factors in the octet color representation. Besides providing a stringent check of the gluon Reggeization in the NLA, the soft bootstrap conditions are important since they test, at least in part, the correctness of the year-long calculations which lead to the NLA BFKL equation. The first bootstrap condition has been verified at arbitrary space-time dimension, for the part concerning the quark contribution to the kernel (in massless QCD) [3] and, in the $D \to 4$ limit, for the part concerning the gluon contribution to the kernel [4]. The second bootstrap condition is process-dependent and should therefore be checked for every new impact factor which is calculated[1]. It has been verified explicitly at arbitrary space-time dimension for quark and gluon impact factors in QCD with massive quarks [6, 7].

The derivation of the BFKL equation in the NLA involves also the assumption of Reggeized form for production amplitudes in multi-Regge and quasi-multi-Regge kinematics. The compatibility of these amplitudes with s-channel unitarity leads to the so-called strong bootstrap conditions [1]. The fulfillment of these conditions opens the way to a proof of the gluon Reggeization in the NLA [1]. Among these conditions, there appear the two ones suggested by Braun and Vacca [8] and derived from the assumption of gluon Reggeization in the (unphysical) particle-Reggeon scattering amplitude with gluon quantum numbers in the t-channel [9]. They can be presented in the form (octet color representation understood)

$$\int \frac{d^{D-2}q_2}{\vec{q}_2^{\,2}\vec{q}_2^{\,\prime\,2}} \mathcal{K}(\vec{q}_1,\vec{q}_2;\vec{q}) R(\vec{q}_2,\vec{q}) = \omega(t) R(\vec{q}_1,\vec{q}) \,, \tag{5}$$

$$\Phi^a_{A'A}(\vec{q}_1,\vec{q}) = \frac{-ig\sqrt{N}}{2} \Gamma^a_{A'A}(q) R(\vec{q}_1,\vec{q}) \,. \tag{6}$$

The last condition fixes the process dependence of the impact factor: it is proportional to the corresponding effective vertex, with a universal coefficient function R [9]. The II strong bootstrap condition has been verified in the case of gluon and quark impact factors in the NLA [9]. Let us consider

[1]See, however, Ref. [5] where a general proof for arbitrary process has been sketched.

the I strong bootstrap condition. Using Eq. (4), it can be written in the form

$$\int \frac{d^{D-2}q_2}{\vec{q}_2^{\,2}\vec{q}_2^{\,\prime\,2}} \mathcal{K}_r(\vec{q}_1,\vec{q}_2;\vec{q}) R(\vec{q}_2,\vec{q}) = \left(\omega(t) - \omega(t_1) - \omega(t_1')\right) R(\vec{q}_1,\vec{q}) \;, \tag{7}$$

where $t = q^2 = -\vec{q}^{\,2}$, $t_i = q_i^2 = -\vec{q}_i^{\,2}$ and $t_i' = q_i'^{\,2} = -\vec{q}_i^{\,\prime\,2}$, with $i = 1, 2$. At the leading order, we have:

$$\int \frac{d^{D-2}q_2}{\vec{q}_2^{\,2}\vec{q}_2^{\,\prime\,2}} \mathcal{K}_r^{(0)}(\vec{q}_1,\vec{q}_2;\vec{q}) = \omega^{(1)}(t) - \omega^{(1)}(t_1) - \omega^{(1)}(t_1') \;, \tag{8}$$

with

$$\mathcal{K}_r^{(0)}(\vec{q}_1,\vec{q}_2;\vec{q}) = \frac{g^2 N_c}{2(2\pi)^{D-1}} f_B(\vec{q}_1,\vec{q}_2;\vec{q}) \;,$$

$$f_B(\vec{q}_1,\vec{q}_2;\vec{q}) = \frac{\vec{q}_1^{\,2}\vec{q}_2^{\,\prime\,2} + \vec{q}_2^{\,2}\vec{q}_1^{\,\prime\,2}}{\vec{k}^{\,2}} - \vec{q}^{\,2} \;, \quad \vec{k} = \vec{q}_1 - \vec{q}_2$$

and

$$\omega^{(1)}(t) = \frac{g^2 N_c t}{2(2\pi)^{D-1}} \int \frac{d^{D-2}q_1}{\vec{q}_1^{\,2}\vec{q}_1^{\,\prime\,2}} = -g^2 \frac{N_c \Gamma(1-\epsilon)}{(4\pi)^{D/2}} \frac{\Gamma^2(\epsilon)}{\Gamma(2\epsilon)} (\vec{q}^{\,2})^{\epsilon} \,.$$

It is trivial to see that the leading order bootstrap condition (8) is satisfied. In the NLA (taking into account Eq. (8)), we have instead

$$\int \frac{d^{D-2}q_2}{\vec{q}_2^{\,2}\vec{q}_2^{\,\prime\,2}} \left[\mathcal{K}_r^{(1)}(\vec{q}_1,\vec{q}_2,\vec{q}) + \mathcal{K}_r^{(0)}(\vec{q}_1,\vec{q}_2,\vec{q}) \left(R^{(1)}(\vec{q}_2,\vec{q}) - R^{(1)}(\vec{q}_1,\vec{q})\right)\right.$$
$$= \omega^{(2)}(t) - \omega^{(2)}(t_1) - \omega^{(2)}(t_1') \,. \tag{9}$$

The fulfillment of the above relation for the quark part has been already verified [8, 9], while the fulfillment for the gluon part is much more complicated and is the subject of this paper.

2. Proof of the I strong bootstrap condition for the gluon part of the kernel

The ingredients for the calculation are the following (gluon part is understood everywhere):

$$\begin{aligned} \omega^{(2)}(t) &= \left[\frac{g^2 N_c \Gamma(1-\epsilon)(\vec{q}^{\,2})^{\epsilon}}{(4\pi)^{D/2}\epsilon}\right]^2 \left[\frac{11}{3} + \left(2\psi'(1) - \frac{67}{9}\right)\epsilon \right. \\ &+ \left.\left(\frac{404}{27} + \psi''(1) - \frac{22}{3}\psi'(1)\right)\epsilon^2\right] + O(\epsilon)] \;, \end{aligned} \tag{10}$$

$$R^{(1)}(\vec{q}_1,\vec{q}) = \frac{\omega^{(1)}(t)}{2}\left[\frac{\epsilon\Gamma(1+2\epsilon)(\vec{q}^{\,2})^{1-\epsilon}}{2\Gamma^2(1+\epsilon)}\int\frac{d^{D-2}k}{\Gamma(1-\epsilon)\pi^{1+\epsilon}}\frac{\ln(\vec{q}^{\,2}/\vec{k}^{\,2})}{(\vec{k}-\vec{q}_1)^2(\vec{k}+\vec{q}_1^{\,\prime})^2}\right.$$

$$+\left(\left(\frac{\vec{q}_1^{\,2}}{\vec{q}^{\,2}}\right)^{\epsilon}+\left(\frac{\vec{q}_1^{\,\prime 2}}{\vec{q}^{\,2}}\right)^{\epsilon}-1\right)\left(\frac{1}{2\epsilon}+\psi(1+2\epsilon)-\psi(1+\epsilon)+\frac{11+7\epsilon}{2(1+2\epsilon)(3+2\epsilon)}\right)$$

$$\left.-\frac{1}{2\epsilon}+\psi(1)+\psi(1+\epsilon)-\psi(1-\epsilon)-\psi(1+2\epsilon)\right]\,, \qquad (11)$$

$$\mathcal{K}_r^{(1)} = \frac{\bar{g}^4}{\pi^{1+\epsilon}\Gamma(1-\epsilon)}\Big(\mathcal{K}_1+\mathcal{K}_2+\mathcal{K}_3\Big)\,, \qquad \bar{g}^2 \equiv \frac{g^2N_c\Gamma(1-\epsilon)}{(4\pi)^{D/2}}\,,$$

with

$$\mathcal{K}_1 = -f_B(\vec{q}_1,\vec{q}_2;\vec{q})\frac{(\vec{k}^{\,2})^{\epsilon}}{\epsilon}\left[\frac{11}{3}+\left(2\psi'(1)-\frac{67}{9}\right)\epsilon\right.$$

$$\left.+\left(\frac{404}{27}+7\psi''(1)-\frac{11}{3}\psi'(1)\right)\epsilon^2\right]\,, \qquad (12)$$

$$\mathcal{K}_2 = \left\{\vec{q}^{\,2}\left[\frac{11}{6}\ln\left(\frac{\vec{q}_1^{\,2}\vec{q}_2^{\,2}}{\vec{q}^{\,2}\vec{k}^{\,2}}\right)+\frac{1}{4}\ln\left(\frac{\vec{q}_1^{\,2}}{\vec{q}^{\,2}}\right)\ln\left(\frac{\vec{q}_1^{\,\prime 2}}{\vec{q}^{\,2}}\right)+\frac{1}{4}\ln\left(\frac{\vec{q}_2^{\,2}}{\vec{q}^{\,2}}\right)\ln\left(\frac{\vec{q}_2^{\,\prime 2}}{\vec{q}^{\,2}}\right)\right.\right.$$

$$\left.+\frac{1}{4}\ln^2\left(\frac{\vec{q}_1^{\,2}}{\vec{q}_2^{\,2}}\right)\right]-\frac{\vec{q}_1^{\,2}\vec{q}_2^{\,\prime 2}+\vec{q}_2^{\,2}\vec{q}_1^{\,\prime 2}}{2\vec{k}^{\,2}}\ln^2\left(\frac{\vec{q}_1^{\,2}}{\vec{q}_2^{\,2}}\right)$$

$$\left.+\frac{\vec{q}_1^{\,2}\vec{q}_2^{\,\prime 2}-\vec{q}_2^{\,2}\vec{q}_1^{\,\prime 2}}{\vec{k}^{\,2}}\ln\left(\frac{\vec{q}_1^{\,2}}{\vec{q}_2^{\,2}}\right)\left(\frac{11}{6}-\frac{1}{4}\ln\left(\frac{\vec{q}_1^{\,2}\vec{q}_2^{\,2}}{\vec{k}^4}\right)\right)\right\}+\left\{\vec{q}_i\leftrightarrow\vec{q}_i^{\,\prime}\right\}\,, \qquad (13)$$

$$\mathcal{K}_3 = \left\{\frac{1}{2}\left[\vec{q}^{\,2}(\vec{k}^{\,2}-\vec{q}_1^{\,2}-\vec{q}_2^{\,2})+2\vec{q}_1^{\,2}\vec{q}_2^{\,2}-\vec{q}_1^{\,2}\vec{q}_2^{\,\prime 2}-\vec{q}_2^{\,2}\vec{q}_1^{\,\prime 2}\right.\right.$$

$$\left.\left.+\frac{\vec{q}_1^{\,2}\vec{q}_2^{\,\prime 2}-\vec{q}_2^{\,2}\vec{q}_1^{\,\prime 2}}{\vec{k}^{\,2}}(\vec{q}_1^{\,2}-\vec{q}_2^{\,2})\right]\int_0^1\frac{dx}{(\vec{q}_1(1-x)+\vec{q}_2x)^2}\ln\left(\frac{\vec{q}_1^{\,2}(1-x)+\vec{q}_2^{\,2}x}{\vec{k}^{\,2}x(1-x)}\right)\right\}$$

$$+\left\{\vec{q}_i\leftrightarrow\vec{q}_i^{\,\prime}\right\}\,. \qquad (14)$$

The integral in $R^{(1)}(\vec{q}_1,\vec{q})$ is already known with $O(\epsilon^0)$ accuracy (see the Appendix of Ref. [9]):

$$(\vec{q}^{\,2})^{1-\epsilon}\int\frac{d^{D-2}k}{\Gamma(1-\epsilon)\pi^{1+\epsilon}}\frac{\ln(\vec{q}^{\,2}/\vec{k}^{\,2})}{(\vec{k}-\vec{q}_1)^2(\vec{k}+\vec{q}_1^{\,\prime})^2} = -\frac{1}{\epsilon}\ln\left(\frac{\vec{q}_1^{\,2}\vec{q}_1^{\,\prime 2}}{(\vec{q}^{\,2})^2}\right)-\frac{1}{2}\ln^2\left(\frac{\vec{q}_1^{\,2}}{\vec{q}_1^{\,\prime 2}}\right).$$

Using the above result, the I strong bootstrap condition (9) takes the following form, with $O(\epsilon^0)$ accuracy:

$$\frac{(\vec{q}^{\,2})^{-2\epsilon}}{\pi^{1+\epsilon}\Gamma(1-\epsilon)}\int\frac{d^{D-2}q_2}{\vec{q}_2^{\,2}\vec{q}_2^{\,\prime 2}}\left[\mathcal{K}_1+\mathcal{K}_2+\mathcal{K}_3+f_B\left(-\frac{11}{6}\ln\left(\frac{\vec{q}_2^{\,2}\vec{q}_2^{\,\prime 2}}{\vec{q}_1^{\,2}\vec{q}_1^{\,\prime 2}}\right)\right.\right.$$

$$+\frac{1}{2}\ln\left(\frac{\vec{q}_1^{\,2}}{\vec{q}^{\,2}}\right)\ln\left(\frac{\vec{q}_1^{\,\prime\,2}}{\vec{q}^{\,2}}\right)-\frac{1}{2}\ln\left(\frac{\vec{q}_2^{\,2}}{\vec{q}^{\,2}}\right)\ln\left(\frac{\vec{q}_2^{\,\prime\,2}}{\vec{q}^{\,2}}\right)\Bigg)\Bigg]$$

$$=-\frac{1}{\epsilon^2}\left[\frac{11}{3}+\left(2\psi'(1)-\frac{67}{9}\right)\epsilon+\left(\frac{404}{27}+\psi''(1)-\frac{22}{3}\psi'(1)\right)\epsilon^2\right]$$

$$-\frac{2}{\epsilon}\left[\frac{11}{3}+\left(2\psi'(1)-\frac{67}{9}\right)\epsilon\right]\ln\left(\frac{\vec{q}_1^{\,2}\vec{q}_1^{\,\prime\,2}}{(\vec{q}^{\,2})^2}\right)-\frac{22}{3}\left(\ln^2\left(\frac{\vec{q}_1^{\,2}}{\vec{q}^{\,2}}\right)+\ln^2\left(\frac{\vec{q}_1^{\,\prime\,2}}{\vec{q}^{\,2}}\right)\right) \tag{15}$$

For details on the calculation of the remaining integrals, we refer to [10]; here we merely quote the final results. The integral from $\mathcal{K}_1$ is

$$\frac{(\vec{q}^{\,2})^{-2\epsilon}}{\pi^{1+\epsilon}\Gamma(1-\epsilon)}\int\frac{d^{D-2}q_2}{\vec{q}_2^{\,2}\vec{q}_2^{\,\prime\,2}}\left(\frac{\vec{q}_1^{\,2}\vec{q}_2^{\,\prime\,2}+\vec{q}_2^{\,2}\vec{q}_1^{\,\prime\,2}}{(\vec{q}_1-\vec{q}_2)^2}-\vec{q}^{\,2}\right)\frac{[(\vec{q}_1-\vec{q}_2)^2]^\epsilon}{\epsilon}$$

$$=\frac{1}{\epsilon^2}\left[1+2\epsilon\ln\left(\frac{\vec{q}_1^{\,2}\vec{q}_1^{\,\prime\,2}}{(\vec{q}^{\,2})^2}\right)+\epsilon^2\left(2\ln^2\left(\frac{\vec{q}_1^{\,2}}{\vec{q}^{\,2}}\right)+2\ln^2\left(\frac{\vec{q}_1^{\,\prime\,2}}{\vec{q}^{\,2}}\right)\right.\right.$$

$$\left.\left.+\ln\left(\frac{\vec{q}_1^{\,2}}{\vec{q}^{\,2}}\right)\ln\left(\frac{\vec{q}_1^{\,\prime\,2}}{\vec{q}^{\,2}}\right)-\psi'(1)\right)\right]+O(\epsilon)\,. \tag{16}$$

Among the remaining integrals, there are four which can be calculated in a straightforward way using the generalized Feynman parametrization for arbitrary space-time dimension (although we quote here only the result with $O(\epsilon^0)$ accuracy):

$$I_1(\vec{q})=\frac{(\vec{q}^{\,2})^{1-\epsilon}}{\pi^{1+\epsilon}\Gamma(1-\epsilon)}\int d^{D-2}q_2\,\frac{1}{\vec{q}_2^{\,2}(\vec{q}_2-\vec{q})^2}=\frac{2}{\epsilon}+O(\epsilon)\,,$$

$$I_2(\vec{q})=\frac{(\vec{q}^{\,2})^{1-\epsilon}}{\pi^{1+\epsilon}\Gamma(1-\epsilon)}\int d^{D-2}q_2\,\frac{\ln(\vec{q}_2^{\,2}/\vec{q}^{\,2})}{\vec{q}_2^{\,2}(\vec{q}_2-\vec{q})^2}=-\frac{1}{\epsilon^2}+\psi'(1)+O(\epsilon)\,,$$

$$I_3(\vec{q})=\frac{(\vec{q}^{\,2})^{1-\epsilon}}{\pi^{1+\epsilon}\Gamma(1-\epsilon)}\int d^{D-2}q_2\,\frac{\ln^2(\vec{q}_2^{\,2}/\vec{q}^{\,2})}{\vec{q}_2^{\,2}(\vec{q}_2-\vec{q})^2}=\frac{2}{\epsilon^3}-\frac{2\psi'(1)}{\epsilon}-2\psi''(1)+O(\epsilon)\,,$$

$$I_4(\vec{q})=\frac{(\vec{q}^{\,2})^{1-\epsilon}}{\pi^{1+\epsilon}\Gamma(1-\epsilon)}\int d^{D-2}q_2\,\frac{\ln(\vec{q}_2^{\,2}/\vec{q}^{\,2})\ln((\vec{q}_2-\vec{q})^2/\vec{q}^{\,2})}{\vec{q}_2^{\,2}(\vec{q}_2-\vec{q})^2}=-2\psi''(1)+O(\epsilon).$$

Replacing the above expressions and (16) in Eq. (15), the bootstrap condition takes the form

$$\Bigg[\Bigg\{-3\psi''(1)+\frac{1}{24}\ln^3\left(\frac{\vec{q}_1^{\,2}}{\vec{q}^{\,2}}\right)-\frac{1}{4}\ln^2\left(\frac{\vec{q}_1^{\,2}}{\vec{q}^{\,2}}\right)\ln\left(\frac{\vec{q}_1^{\,\prime\,2}}{\vec{q}^{\,2}}\right)-\frac{1}{2}I(\vec{q}_1^{\,\prime\,2},\vec{q}^{\,2};\vec{q}_1^{\,2})$$

$$+\frac{1}{2}I(\vec{q}^{\,2},\vec{q}_1^{\,2};\vec{q}_1^{\,\prime\,2})-\frac{3}{4}J(\vec{q}_1^{\,2},\vec{q}^{\,2};\vec{q}_1^{\,\prime\,2})\Big\}+\Big\{\vec{q}_1\leftrightarrow\vec{q}_1^{\,\prime}\Big\}\Big]+\frac{1}{\pi}\int\frac{d\vec{q}_2}{\vec{q}_2^{\,2}\vec{q}_2^{\,\prime\,2}}\,\mathcal{K}_3=0\,, \tag{17}$$

with

$$I(\vec{p}_1^{\,2},\vec{p}_2^{\,2};(\vec{p}_1-\vec{p}_2)^2)=\frac{(\vec{p}_1-\vec{p}_2)^2}{\pi}\int d\vec{p}\,\frac{\ln(\vec{p}_2^{\,2}/\vec{p}^{\,2})\ln((\vec{p}_1-\vec{p}_2)^2/(\vec{p}-\vec{p}_2)^2)}{(\vec{p}-\vec{p}_1)^2(\vec{p}-\vec{p}_2)^2}\,,$$

$$J(\vec{p}_1^{\,2},\vec{p}_2^{\,2};(\vec{p}_1-\vec{p}_2)^2)=\frac{(\vec{p}_1-\vec{p}_2)^2}{\pi}\int d\vec{p}\,\frac{\ln(\vec{p}_1^{\,2}/\vec{p}^{\,2})\ln(\vec{p}_2^{\,2}/\vec{p}^{\,2})}{(\vec{p}-\vec{p}_1)^2(\vec{p}-\vec{p}_2)^2}\,.$$

The integrals $I(\vec{p}_1^{\,2},\vec{p}_2^{\,2};(\vec{p}_1-\vec{p}_2)^2)$ and $J(\vec{p}_1^{\,2},\vec{p}_2^{\,2};(\vec{p}_1-\vec{p}_2)^2)$ can be reduced to one-dimensional integrals:

$$I(\vec{p}_1^{\,2},\vec{p}_2^{\,2};(\vec{p}_1-\vec{p}_2)^2)=\psi'(1)\ln\vec{k}_1^{\,2}+\frac{1}{2}\int_0^1\frac{dx}{x}\ln^2\left(\frac{D_0}{\vec{k}_2^{\,2}}\right)$$

$$+\frac{1}{2}\int_0^1\frac{dx}{1-x}\ln^2\left(\frac{D_0}{\vec{k}_1^{\,2}}\right)+\ln\vec{k}_2^{\,2}\int_0^1\frac{dx}{x}\ln\left(\frac{D_0}{\vec{k}_2^{\,2}}\right)+\ln\vec{k}_1^{\,2}\int_0^1\frac{dx}{1-x}\ln\left(\frac{D_0}{\vec{k}_1^{\,2}}\right)$$

$$+\int_0^1\frac{dx}{1-x}\ln x\ln D_0-\int_0^1\frac{dx}{x}\ln x\ln\left(\frac{D_0}{\vec{k}_2^{\,2}}\right)-2\int_0^1\frac{dx}{1-x}\ln(1-x)\ln\left(\frac{D_0}{\vec{k}_1^{\,2}}\right)$$

$$-\int_0^1 dx\ln\left(\frac{x}{1-x}\right)\ln\left(\frac{D_1}{D_0}\right)\frac{\vec{k}_1^{\,2}-\vec{k}_2^{\,2}-(1-2x)}{D_0-D_1}\,, \tag{18}$$

$$J(\vec{p}_1^{\,2},\vec{p}_2^{\,2};(\vec{p}_1-\vec{p}_2)^2)=\ln\left(\frac{\vec{k}_1^{\,2}}{\vec{k}_2^{\,2}}\right)\left[\int_0^1\frac{dx}{1-x}\ln\left(\frac{D_0}{\vec{k}_1^{\,2}}\right)-\int_0^1\frac{dx}{x}\ln\left(\frac{D_0}{\vec{k}_2^{\,2}}\right)\right]$$

$$+\int_0^1\frac{dx}{x}\ln^2\left(\frac{D_0}{\vec{k}_2^{\,2}}\right)+\int_0^1\frac{dx}{1-x}\ln^2\left(\frac{D_0}{\vec{k}_1^{\,2}}\right)$$

$$-2\int_0^1\frac{dx}{D_1-D_0}\ln\left(\frac{x}{1-x}\right)\ln\left(\frac{D_1}{D_0}\right)\left[(1-2x)-\frac{D_1(\vec{k}_1^{\,2}-\vec{k}_2^{\,2})}{D_0}\right]\,, \tag{19}$$

with

$$\vec{k}_1\equiv\frac{\vec{p}_1}{|\vec{p}_1-\vec{p}_2|}\,,\quad\vec{k}_2\equiv\frac{\vec{p}_2}{|\vec{p}_1-\vec{p}_2|}\,,\quad D_0\equiv x\vec{k}_1^{\,2}+(1-x)\vec{k}_2^{\,2}\,,\quad D_1\equiv x(1-x)\,.$$

The integral from $\mathcal{K}_3$ can be re-expressed as follows:

$$A\equiv\frac{1}{\pi}\int\frac{d^2q_2}{\vec{q}_2^{\,2}\vec{q}_2^{\,\prime\,2}}\mathcal{K}_3=(A_1+A_2+A_3)+(\vec{q}_1\leftrightarrow\vec{q}_1^{\,\prime})\,, \tag{20}$$

with

$$A_1 = \frac{1}{2}\int_0^1 \frac{dz}{\vec{q}^{\,2}(1-z)+\vec{q}_1^{\,\prime\,2}z-\vec{q}_1^{\,2}z(1-z)}$$

$$\times\left[\left(\frac{2\vec{q}^{\,2}\vec{q}_1^{\,2}z}{\vec{q}^{\,2}(1-z)+\vec{q}_1^{\,\prime\,2}z-\vec{q}_1^{\,2}z}+\vec{q}^{\,2}+\vec{q}_1^{\,2}-\vec{q}_1^{\,\prime\,2}\right)\right.$$

$$\times\ln\left(\frac{\vec{q}^{\,2}(1-z)+\vec{q}_1^{\,\prime\,2}z-\vec{q}_1^{\,2}z(1-z)}{z(\vec{q}^{\,2}(1-z)+\vec{q}_1^{\,\prime\,2}z)}\right)\ln\left(\frac{\vec{q}^{\,2}(1-z)+\vec{q}_1^{\,\prime\,2}z}{\vec{q}_1^{\,2}z}\right)$$

$$\left.+\left(\frac{2\vec{q}^{\,2}}{z}-\vec{q}^{\,2}-\vec{q}_1^{\,2}+\vec{q}_1^{\,\prime\,2}\right)\ln(1-z)\ln\left(\frac{\vec{q}^{\,2}(1-z)+\vec{q}_1^{\,\prime\,2}z-\vec{q}_1^{\,2}z(1-z)}{\vec{q}^{\,2}(1-z)}\right)\right],$$

$$A_2 = A_1(\vec{q}_1^{\,\prime}\leftrightarrow-\vec{q})\,,\qquad A_3=-A_1(\vec{q}_1^{\,\prime}=0)\,.$$

The one-dimensional integrals in Eqs. (18), (19) and (20) can be calculated analytically, but the arising expressions are discouragingly long and cumbersome. Nevertheless, the proof can be greatly simplified for the following reason. The integrals and the explicit logarithms entering the bootstrap condition (17) are functions of the variables $q_1^2\equiv-\vec{q}_1^{\,2}$, $q_1^{\prime\,2}\equiv-\vec{q}_1^{\,\prime\,2}$ and $q^2\equiv-\vec{q}^{\,2}$. At fixed $q_1^2\le0$, $q_1^{\prime\,2}\le0$, these functions are analytical functions of q^2, real for $q^2<0$ and with the cut $0\le q^2<\infty$. Any such function can be determined by its discontinuity on the cut, up to another function of q_1^2 and $q_1^{\prime\,2}$ (but not of q^2). This last function can be simply obtained by evaluating the first function at $\vec{q}^{\,2}=\infty$. Operatively, to calculate discontinuities, we have to make the replacement $\vec{q}^{\,2}\to-q^2-i0$ in the integrals and in the explicit logarithms entering the bootstrap condition, to calculate their imaginary parts on the upper edge of the cut (they give the discontinuities after multiplication by $2i$) and to check the strong bootstrap for the imaginary parts (which is the same as for discontinuities). Then, what remains to be done is to check the bootstrap in the limit $\vec{q}^{\,2}\gg\vec{q}_1^{\,2}$, $\vec{q}^{\,2}\gg\vec{q}_1^{\,\prime\,2}$.

The bootstrap relation for imaginary parts (divided by π) reads

$$\left\{\left[-\frac{1}{8}\ln^2\left(\frac{\vec{q}_1^{\,2}}{q^2}\right)+\frac{5\pi^2}{24}-\frac{1}{2}\ln\left(\frac{\vec{q}_1^{\,2}}{q^2}\right)\ln\left(\frac{\vec{q}_1^{\,\prime\,2}}{q^2}\right)-\frac{1}{2}\frac{1}{\pi}\Im I(\vec{q}_1^{\,2},-q^2-i0;\vec{q}_1^{\,\prime\,2})\right.\right.$$

$$\left.\left.+\frac{1}{2}\frac{1}{\pi}\Im I(-q^2-i0,\vec{q}_1^{\,2};\vec{q}_1^{\,\prime\,2})-\frac{3}{4}\frac{1}{\pi}\Im J(\vec{q}_1^{\,2},-q^2-i0;\vec{q}_1^{\,\prime\,2})\right]+\left[\vec{q}_1^{\,2}\leftrightarrow\vec{q}_1^{\,\prime\,2}\right]\right\}$$

$$+\frac{\Im A}{\pi}=0\,. \tag{21}$$

The imaginary part of the integrals I and J in Eqs. (18) and (19) can be easily calculated:

$$-\frac{1}{2\pi}\Im I(\vec{q}_1^{\,2},-q^2-i0;\vec{q}_1^{\,\prime\,2})+\frac{1}{2\pi}\Im I(-q^2-i0,\vec{q}_1^{\,2};\vec{q}_1^{\,\prime\,2})=-\frac{\psi'(1)}{2}+\frac{1}{4}\ln^2\left(\frac{\vec{q}_1^{\,2}}{q^2}\right), \tag{22}$$

$$\frac{1}{\pi}\Im J(\vec{q}_1^{\,2}, -q^2 - i0; \vec{q}_1^{\,\prime\,2}) + (\vec{q}_1^{\,2} \leftrightarrow \vec{q}_1^{\,\prime\,2})$$

$$= -\ln\left(\frac{\kappa^-}{q^2}\right)\ln\left(\frac{\kappa^+}{q^2}\right) + \frac{1}{2}\ln^2\left(\frac{\vec{q}_1^{\,2}}{q^2}\right) + \psi'(1) + (\vec{q}_1^{\,2} \leftrightarrow \vec{q}_1^{\,\prime\,2})\,. \tag{23}$$

To calculate the discontinuity of the integral A defined in Eq. (20) at $q^2 = -\vec{q}^{\,2} \geq 0$, it is convenient to rewrite the integral over q_2 in Minkowski space and to use the Cutkosky rules for the calculation of the discontinuity. First, we represent the integral over x appearing in $\mathcal{K}_3$ (see Eq. (14) as

$$I = \int_0^1 dx \int_0^\infty dz \frac{1}{z - k^2 x(1-x) - i0}\,\frac{1}{z - q_1^2(1-x) - q_2^2 x - i0}\,, \tag{24}$$

where k, q_1, q_2 are considered as vectors in the two-dimensional Minkowski space, i.e. $k^2 = -\vec{k}^{\,2}$, $q_1^2 = -\vec{q}_1^{\,2}$, $q_2^2 = -\vec{q}_2^{\,2}$. This representation can be used for arbitrary values of k^2, q_1^2, q_2^2. Analogous representation can be written for $I(q_i \leftrightarrow q_i')$. It permits to rewrite the integral with $\mathcal{K}_3$ in the bootstrap relation in the form

$$A = \frac{1}{i\pi}\int \frac{d^2 q_2}{(q_2^2 + i0)((q - q_2)^2 + i0)}\mathcal{K}_3\,, \tag{25}$$

where now

$$d^2 q_2 = dq_2^{(0)} dq_2^{(1)}, \quad q_2^2 = (q_2^{(0)})^2 - (q_2^{(1)})^2\,, \tag{26}$$

etc., which determines A as function of q_1^2, $q_1^{\prime\,2}$ and q^2 for arbitrary values of these variables. For $q_1^2 \equiv -\vec{q}_1^{\,2} \leq 0$, $q_1^{\prime\,2} \equiv -\vec{q}_1^{\,\prime\,2} \leq 0$ and $q^2 \equiv -\vec{q}^{\,2} \leq 0$ it is just the function entering (17), that is easily seen by making the Wick rotation of the contour of integration over $q_2^{(0)}$. We are interested in the region $q_1^2 \leq 0$, $q_1^{\prime\,2} \leq 0$ and $q^2 \geq 0$. According to the Cutkosky rules, the discontinuity of A related to the terms with I is determined by the two cuts, with the contributions obtained by the substitutions:

$$\frac{1}{(q_2^2 + i0)((q - q_2)^2 + i0)} \to (-2\pi i)^2 \delta(q_2^2)\delta((q - q_2)^2) \tag{27}$$

and

$$\frac{1}{(z - q_1^2(1-x) - q_2^2 x - i0)((q - q_2)^2 + i0)}$$
$$\to -(-2\pi i)^2 \delta(z - q_1^2(1-x) - q_2^2 x)\delta((q - q_2)^2)\,. \tag{28}$$

Using these rules and removing the δ-functions by the integration over q_2 (the most appropriate system for this is $q^{(1)} = 0$, $q^2 = (q^{(0)})^2$), we obtain

$$\frac{\Im A}{\pi} = -\frac{3}{2}\ln\left(\frac{\kappa^-}{q^2}\right)\ln\left(\frac{\kappa^+}{q^2}\right) + \frac{1}{4}\ln^2\left(\frac{\vec{q}_1^{\,2}\vec{q}_1^{\,\prime\,2}}{(q^2)^2}\right) + \frac{1}{2}\ln\left(\frac{\vec{q}_1^{\,2}}{q^2}\right)\ln\left(\frac{\vec{q}_1^{\,\prime\,2}}{q^2}\right), \tag{29}$$

with

$$\kappa^{\pm} = \frac{1}{2}\left(q^2 + \vec{q}_1^{\,2} + \vec{q}_1^{\,\prime 2} \pm \sqrt{(q^2 + \vec{q}_1^{\,2} + \vec{q}_1^{\,\prime 2})^2 - 4\vec{q}_1^{\,2}\vec{q}_1^{\,\prime 2}}\right) .$$

Using Eqs. (22), (23) and (29), it is easy to see that the imaginary part of the bootstrap relation, Eq. (21), is satisfied. Finally, the bootstrap has to be considered in the limit $\vec{q}^{\,2} \gg \vec{q}_1^{\,2}$, $\vec{q}^{\,2} \gg \vec{q}_1^{\,\prime 2}$. We have in this limit

$$I(\vec{q}_1^{\,2}, \vec{q}^{\,2}; \vec{q}_1^{\,\prime 2}) \simeq -\zeta(2) \ln\left(\frac{\vec{q}^{\,2}}{\vec{q}_1^{\,2}}\right) + 2\zeta(3) ,$$

$$I(\vec{q}^{\,2}, \vec{q}_1^{\,2}; \vec{q}_1^{\,\prime 2}) \simeq -\frac{1}{6}\ln^3\left(\frac{\vec{q}^{\,2}}{\vec{q}_1^{\,2}}\right) - \zeta(2)\ln\left(\frac{\vec{q}^{\,2}}{\vec{q}_1^{\,2}}\right) + 2\zeta(3) ,$$

$$J(\vec{q}_1^{\,2}, \vec{q}^{\,2}; \vec{q}_1^{\,\prime 2}) \simeq -\frac{1}{6}\ln^3\left(\frac{\vec{q}^{\,2}}{\vec{q}_1^{\,2}}\right) - 2\zeta(2)\ln\left(\frac{\vec{q}^{\,2}}{\vec{q}_1^{\,2}}\right) + 4\zeta(3)$$

and

$$\frac{1}{\pi}\int \frac{d\vec{q}_2}{\vec{q}_2^{\,2}\vec{q}_2^{\,\prime 2}} \mathcal{K}_3 \simeq -\frac{1}{4}\ln\left(\frac{\vec{q}^{\,2}}{\vec{q}_1^{\,\prime 2}}\right)\ln\left(\frac{\vec{q}^{\,2}}{\vec{q}_1^{\,2}}\right)\left(\ln\left(\frac{\vec{q}^{\,2}}{\vec{q}_1^{\,2}}\right) + \ln\left(\frac{\vec{q}^{\,2}}{\vec{q}_1^{\,\prime 2}}\right)\right)$$

$$-\frac{3\zeta(2)}{2}\left(\ln\left(\frac{\vec{q}^{\,2}}{\vec{q}_1^{\,2}}\right) + \ln\left(\frac{\vec{q}^{\,2}}{\vec{q}_1^{\,\prime 2}}\right)\right) - 6\zeta(3) .$$

Again, it is easy to see that the bootstrap condition (17) is satisfied in the limit of large $\vec{q}^{\,2}$.

References

1. Fadin, V.S (2002) these proceedings and references therein.
2. Fadin, V.S. and Fiore, R. (1998) The generalized nonforward BFKL equation and the 'bootstrap' condition for the gluon Reggeization in the NLLA, *Phys. Lett.*, **Vol. no. B440**, pp. 359–366
3. Fadin, V.S., Fiore, R. and Papa, A. (1999) The quark part of the nonforward BFKL kernel and the 'bootstrap' for the gluon Reggeization, *Phys. Rev.*, **Vol. no. D60**, 074025 (13 pages)
4. Fadin, V.S., Fiore, R. and Kotsky, M.I. (2000) The compatibility of the gluon Reggeization with the *s* channel unitarity, *Phys. Lett.*, **Vol. no. B494**, pp. 100–108
5. Kotsky, M.I. (2002) these proceedings
6. Fadin, V.S., Fiore, R., Kotsky, M.I. and Papa, A. (2000) Quark impact factors, *Phys. Rev.*, **Vol. no. D61**, 094006 (16 pages)
7. Fadin, V.S., Fiore, R., Kotsky, M.I. and Papa, A. (2000) Gluon impact factors, *Phys. Rev.*, **Vol. no. D61**, 094005 (22 pages)
8. Braun, M. (1999) Comments on the second order bootstrap relation, hep-ph/9901447; Braun, M. and Vacca, G.P. (2000) The bootstrap for impact factors and the gluon wave function, *Phys. Lett.*, **Vol. no. B447**, pp. 156–162
9. Fadin, V.S., Fiore, R., Kotsky, M.I. and Papa, A. (2000) Strong bootstrap conditions, *Phys. Lett.*, **Vol. no. B495**, pp. 329–337
10. Fadin, V.S. and Papa, A. (2002) A proof of fulfillment of the strong bootstrap condition, *Nucl. Phys.*, **Vol. no. B640**, pp. 309–330

REGGEIZED GLUON INTERACTION

M.I. KOTSKY
Budker Institute for Nuclear Physics
630090, Lavrenteva st. 11, Novosibirsk, Russia

Abstract. The NLO interaction of the Reggeized gluon with particles is considered in a relation to the bootstrap conditions for the gluon Reggeization in QCD.

1. Introduction

The gluon Reggeization in QCD, which is the base of the BFKL approach [1], was rigorously proved in the leading order (LO) approximation [2], while in the next-to-leading order (NLO) this property has the status of assumption at the moment. To prove the NLO gluon Reggeization, one has to verify the fulfilment of so-called bootstrap conditions [3], among which there is the condition for colour antisymmetric octet impact factors, which we will be concentrated on in this note. Impact factors describe the interaction of external particles with two Reggeized t- channel gluons and, before the projection on a definite colour state in the t- channel, have the following form in the NLO (see Ref. [4] for the details)

$$\Phi_{AA'}^{cc'}(\vec{q}_1, \vec{q}_1{}')$$
$$= \int \frac{d\tilde{s}d\rho_f}{2\pi}\theta(s_1 - \tilde{s})\Gamma^c_{\{f\}A}(q_1)\left(\Gamma^{c'}_{\{f\}A'}(q_1')\right)^* + \Phi_{AA'}^{cc'(count)}(\vec{q}_1, \vec{q}_1{}') \ . \quad (1)$$

The integration here is performed over the phase space ρ_f of produced system $\{f\}$ and over its squared invariant mass $\tilde{s}$, and the sum over all such systems produced in the NLO as well as over their discrete quantum numbers is assumed. $\Gamma^c_{\{f\}A}(q_1)$ is the effective interaction vertex for the transition $AR \to \{f\}$ with R being the Reggeized gluon with the momentum q_1 and the colour index c and $\Gamma^{c'}_{\{f\}A'}(q_1')$ is the analogous notation for the other effective vertex. Vector notations are used for the transverse to the initial particles momenta plane components of momenta and the counterterm $\Phi_{AA'}^{cc'(count)}$ is to cancel the multi-Regge-kinematics contribution of

R. Fiore et al. (eds.), Diffraction 2002, 257–266.

the first term in the Eq. (1) so that the logarithmic dependence on $s_1 \to \infty$ disappear in their sum. This counterterm has the form

$$\Phi_{AA'}^{cc'(count)}(\vec{q}_1, \vec{q}_1{}') = \frac{-1}{2} \int \frac{d^{D-2}q_2}{\vec{q}_2{}^2 \vec{q}_2{}'^2} \Phi_{AA'}^{c_1 c_1'(0)}(\vec{q}_2, \vec{q}_2{}')$$

$$\times \left(\mathcal{K}_r^{(0)}\right)_{c_1 c}^{c_1' c'} (\vec{q}_2, \vec{q}_1, \vec{q}) \ln\left(\frac{s_1^2}{s_0(\vec{q}_2 - \vec{q}_1)^2}\right) , \quad (2)$$

where $D = 4 + 2\epsilon$ is different from 4 for the regularization, the superscript (0) means the Born approximation and $\mathcal{K}_r$ denotes the part of the BFKL equation kernel related to the real particles production. The effective vertices in the Eq. (1), as well as the impact factor itself depend on the energy scale s_0 and we use in this note the simplified notation $k' = k - q$, where q is the t- channel momentum transfer.

The bootstrap condition for colour antisymmetric octet impact factors looks as follows [3, 5]

$$\Phi_{A'A}^{a}(\vec{q}_1, \vec{q}_1{}') = R(\vec{q}_1, \vec{q}_1{}')\Gamma_{A'A}^{a}(q) , \quad \Phi_{A'A}^{a}(\vec{q}_1, \vec{q}_1{}') = \frac{iT_{cc'}^{a}}{\sqrt{N}} \Phi_{AA'}^{cc'}(\vec{q}_1, \vec{q}_1{}') , \quad (3)$$

where T is the colour $SU(N)$ group generator in the adjoint representation and the coefficient function $R(\vec{q}_1, \vec{q}_1{}')$ is universal, i.e. it does not depend on the transition $A \to A'$. This function is known in massless QCD from Ref. [5] where it was obtained by direct comparison of known results for the NLO impact factors and effective vertices for particular processes of quark and gluon scattering. Here we report the general proof of the condition (3) in the NLO approximation for an arbitrary $A \to A'$ transition.

2. Representation of the effective vertices

There are two commonly known ways to calculate the effective vertices for Reggeon interaction in the NLO. First of them consists of direct calculation of quasielastic amplitude $AB \to A'B'$ in the Regge kinematics

$$|p_A^2| \sim |p_{A'}^2| \sim |p_B^2| \sim |p_{B'}^2| \sim |s\alpha| \sim |s\beta| \sim \vec{q}^{\,2} \simeq -t = -(p_{A'} - p_A)^2 - \text{fixed} ,$$

$$s = 2p_1 p_2 \to \infty , \quad q = p_{A'} - p_A = \beta p_1 + \alpha p_2 + q_\perp , \quad (4)$$

with (p_1, p_2) being the light-cone basis of the (p_A, p_B) plane

$$p_A = p_1 + \frac{p_A^2}{s} p_2 , \quad p_B = p_2 + \frac{p_B^2}{s} p_1 , \quad p_1^2 = p_2^2 = 0 . \quad (5)$$

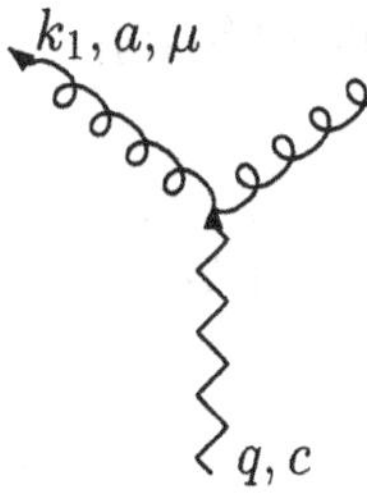

k_1, a, μ k_2, b, ν

q, c

$$= -igT^c_{ab}\left[g^{\mu\nu}\frac{p_2(k_1-k_2)}{s} - \frac{p_2^\mu}{s}(2k_1+k_2)^\nu\right.$$

$$\left. + \frac{p_2^\nu}{s}(2k_2+k_1)^\mu - \frac{p_2^\mu p_2^\nu}{s}\frac{2q^2}{p_2(k_1-k_2)}\right]\theta\left(\left|\frac{p_2(k_1-k_2)}{s}\right| - \beta_0\right) .$$

Figure 1. The Reggeon (zigzag line) elementary interaction with gluons.

This amplitude with gluon colour quantum numbers and negative signature in the t- channel is mediated in the NLO by t- channel Reggeized gluon exchange and has the following form

$$\left(\mathcal{A}^{(8,-)}\right)^{A'B'}_{AB} = \Gamma^c_{B'B}(q)\frac{s}{t}\left[\left(\frac{s}{s_0}\right)^{\omega(t)} + \left(\frac{-s}{s_0}\right)^{\omega(t)}\right]\Gamma^c_{A'A}(q) , \quad (6)$$

where $\omega(t)$ is the Reggeized gluon trajectory. Comparison of the directly calculated one-loop result for this amplitude with the above Reggeized form gives the NLO effective vertices. The disadvantage of such approach is that one deals here not with effective vertices themselves but with complete amplitudes, which is not so convenient.

There is a way to calculate the effective vertices directly [6] without consideration of complete amplitudes. In this approach one calculates an usual one-loop QCD amplitude

$$\tilde{\Gamma}^c_{A'A}(q) = -q^2\langle A'|\left(\frac{p_2}{s}A^c(0)\right)|A\rangle , \quad (7)$$

where $A^c_\mu(x)$ is to denote the gluon field, with replacement of the external virtual gluon with momentum q, colour index c and polarization $-p_2/s$ by the Reggeon whose elementary interaction with quarks remains the ordinary QCD one and the elementary interaction with gluons is shown in the Fig. 1. The Feynman gauge must be used for internal virtual gluons while for external on-mass-shell gluons the gauge

$$ek = ep_2 = 0 \quad (8)$$

is fixed. Since the effective vertex $\Gamma^c_{A'A}(q)$ does not depend on the Sudakov component of q along p_1, it is neglected here so that the momentum transfer becomes $q = \alpha p_2 + q_\perp$ (compare with the Eqs. (4)) and therefore the term of the Fig. 1 with nonlocal interaction produces a "Regge" divergency in the Eq. (7) that is regularized by the θ- function with the regulator

$$\beta_0 \to 0 , \quad s\beta_0 \to \infty . \quad (9)$$

The NLO Reggeon effective vertex is related to the amplitude $\tilde{\Gamma}^c_{A'A}(q)$ as follows [6]

$$\Gamma^c_{A'A}(q) = \left(1 + \frac{\omega^{(1)}(q^2)}{2}\left[\ln\left(\frac{s_0\beta_0^2}{-q^2}\right) + \frac{1}{\epsilon} - \frac{1}{1+2\epsilon} + 2\psi(1+2\epsilon)\right.\right.$$

$$\left.\left. -2\psi(1+\epsilon) + \psi(1-\epsilon) - \psi(1) - P_F(q^2)\right]\right)\tilde{\Gamma}^c_{A'A}(q) \,, \qquad (10)$$

where

$$\omega^{(1)}(q^2) = -g^2 N \frac{\Gamma(1-\epsilon)}{(4\pi)^{2+\epsilon}} \frac{\Gamma^2(\epsilon)}{\Gamma(2\epsilon)} \left(-q^2\right)^\epsilon \qquad (11)$$

is the one-loop Reggeized gluon trajectory and the value $P_F(q^2)$ is defined through one-loop correction to the gluon propagator in the Feynman gauge

$$(D_F)^{\mu\nu}_{ab}(k) = \frac{-i\delta^{ab}}{\left(1 - \tilde{P}_F(k^2)\right)(k^2 + i\delta)} \left[g^{\mu\nu} - \tilde{P}_F(k^2)\frac{k^\mu k^\nu}{k^2}\right] \,,$$

$$\tilde{P}^{(1)}_F(k^2) = \omega^{(1)}(k^2) P_F(k^2) \,. \qquad (12)$$

In the above relations g is the gauge coupling constant; $\Gamma(z)$ and $\psi(z)$ are the Euler gamma-function and its logarithmic derivative respectively.

Here we will use one more representation of the effective vertices, which is very convenient to study general properties of the impact factors. It can be shown that (the details will be given elsewhere [7])

$$\Gamma^c_{A'A}(q) = \left(1 + \frac{\omega^{(1)}(q^2)}{2}\left[\ln\left(\frac{s_0\beta^2}{-q^2}\right) - \frac{1}{1+2\epsilon} + \psi(1-\epsilon) - \psi(1)\right.\right.$$

$$\left.\left. -P_F(q^2) + 2P_L(q^2)\right]\right)\frac{ip_2^\mu}{s}\langle A'|A^c_\mu(0)|A\rangle^{(amp)} \,, \qquad (13)$$

where the matrix element is calculated in QCD in the light-cone gauge for which the second of conditions (8) is put not only for external gluons, but also for internal gluon fields, and the superscript (amp) is to denote the amputated matrix element, i.e. without the external virtual gluon propagator. Analogously to the Eqs. (12), the $P_L(q^2)$ is related to the one-loop correction to the gluon propagator in the light-cone gauge

$$(D_L)^{\mu\nu}_{ab}(k) = \frac{-i\delta^{ab}}{\left(1 - \tilde{P}_L(k^2)\right)(k^2 + i\delta)} \left[g^{\mu\nu} - \frac{k^\mu p_2^\nu + p_2^\mu k^\nu}{kp_2}\right] \,,$$

$$\tilde{P}^{(1)}_L(k^2) = \omega^{(1)}(k^2) P_L(k^2) \,. \qquad (14)$$

Let us note that here we keep a small nonzero β, i.e. the component of the momentum transfer q along p_1, since the matrix element in the Eq. (13) does not exist in the particular kinematics $qp_2 = 0$; of course, the dependence on this regulator is cancelled in the Eq. (13). The representation (13) is valid as for NLO production of "unexcited" states, which we have been mostly concentrated on in above, as well as for "excited" ones (see Ref. [6] for the details). In the later case the expression (13) should be taken at the Born level. So we see, that in the NLO approximation the effective vertex for Reggeon interaction coincides with ordinary QCD vertex-like amplitude calculated in the light-cone gauge up to the properly chosen coefficient (13), that provides more freedom with respect to former representations, in use of the gauge invariance consequences, for instance. In the next Sections we will see how this helps to consider general properties of impact factors.

3. NLO antisymmetric colour octet impact factor

Let us consider the symmetrized with respect to $\vec{q}_1{}' \leftrightarrow -\vec{q}_1$ NLO colour antisymmetric octet impact factor for the $A \to A'$ transition. The reason for such the symmetrization is that the antisymmetric part never plays a role in the NLO approximation and the definition (1) could be given in the symmetric form from the beginning. We first consider the counterterm (2), which for the gluon colour quantum numbers in the t- channel has the form

$$\Phi_{A'A}^{a(count)}(\vec{q}_1, \vec{q}_1{}') = -\frac{g^2 N}{4} \int \frac{d^{D-2} q_2}{(2\pi)^{D-1}} \frac{\Phi_{A'A}^{a(0)}(\vec{q}_2, \vec{q}_2{}')}{\vec{q}_2{}^2 \vec{q}_2{}'^2}$$

$$\times \left(\frac{\vec{q}_1{}^2 \vec{q}_2{}'^2 + \vec{q}_2{}^2 \vec{q}_1{}'^2}{(\vec{q}_2 - \vec{q}_1)^2} - \vec{q}^{\,2} \right) \ln \left(\frac{s_1^2}{s_0 (\vec{q}_2 - \vec{q}_1)^2} \right) . \qquad (15)$$

Using the Born bootstrap condition (3) with the coefficient function $R^{(0)} = -ig\sqrt{N}/2$, one comes to the conclusion of $\vec{q}_2$- independence of the Born impact factor in the above relation so that it can be put out of the integral there. Therefore, to the NLO accuracy, the counterterm of the colour antisymmetric octet impact factor can be written as a factor to the first term of the Eq. (1). Then, using also the representation (13), we get for the impact factor after some simple algebra

$$\sqrt{N} \Phi_{A'A}^{c}(\vec{q}_1, \vec{q}_1{}') = \left(1 - \frac{\omega^{(1)}(-\vec{q}^{\,2})}{2} \left[\ln \left(\frac{s_1^2}{s_0 \vec{q}^{\,2}} \right) + \tilde{K}_1 \right] \right.$$

$$+ \frac{\omega^{(1)}(-\vec{q}_1{}^2)}{2} \left[\ln \left(\frac{s_1 \beta_1}{\vec{q}_1{}^2} \right)^2 - P_F(-\vec{q}_1{}^2) + 2 P_L(-\vec{q}_1{}^2) \right]$$

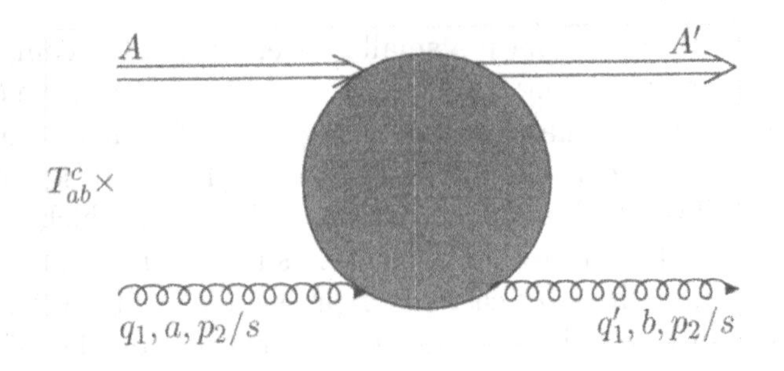

Figure 2. Schematic representation of the amplitude $\mathcal{A}^c$.

$$+\frac{\omega^{(1)}(-\vec{q}_1^{\,\prime 2})}{2}\left[\ln\left(\frac{s_1\beta_1'}{\vec{q}_1^{\,\prime 2}}\right)^2 - P_F(-\vec{q}_1^{\,\prime 2}) + 2P_L(-\vec{q}_1^{\,\prime 2})\right]$$

$$+\frac{\omega^{(1)}(-\vec{q}_1^{\,2})+\omega^{(1)}(-\vec{q}_1^{\,\prime 2})}{2}\left[\frac{1}{2\epsilon(1+2\epsilon)} + 2\psi(1-\epsilon)\right.$$

$$\left.\left. -2\psi(1) + \psi(1+2\epsilon) - \psi(1+\epsilon)\right]\right)$$

$$\times \int \frac{d\tilde{s}d\rho_f}{2\pi}\theta(s_1-\tilde{s})iT^c_{ab}\frac{p_2^\mu p_2^\nu}{s^2}\langle f|A^a_\mu(0)|A\rangle^{(amp)}\left(\langle f|A^b_\nu(0)|A'\rangle^{(amp)}\right)^* , \quad (16)$$

with the integral $\tilde{K}_1$ defined as (see also Ref. [5])

$$\tilde{K}_1 = \frac{(4\pi)^{2+\epsilon}\Gamma(1+2\epsilon)\epsilon\left(\vec{q}^{\,2}\right)^{-\epsilon}}{4\Gamma(1-\epsilon)\Gamma^2(1+\epsilon)}\int\frac{d^{D-2}k}{(2\pi)^{D-1}}\frac{\vec{q}^{\,2}\ln\left(\vec{q}^{\,2}/\vec{k}^{\,2}\right)}{(\vec{k}-\vec{q}_1)^2(\vec{k}-\vec{q}_1^{\,\prime})^2} , \quad (17)$$

and with β_1 and β_1' being the Reggeons momenta fractions along p_1.

Consider the last multiplier of the Eq. (16)

$$\Phi^c = \int \frac{d\tilde{s}d\rho_f}{2\pi}\theta(s_1-\tilde{s})iT^c_{ab}\frac{p_2^\mu p_2^\nu}{s^2}\langle f|A^a_\mu(0)|A\rangle^{(amp)}\left(\langle f|A^b_\nu(0)|A'\rangle^{(amp)}\right)^*$$

$$= \int_0^{s_1}\frac{d\tilde{s}}{2\pi}\Delta_{\tilde{s}}\mathcal{A}^c(\tilde{s}) , \quad (18)$$

where $\Delta_{\tilde{s}}\mathcal{A}^c(\tilde{s})$ is the $\tilde{s}$- channel discontinuity of the amplitude

$$\mathcal{A}^c(\tilde{s}) = -iT^c_{ab}\frac{p_2^\mu p_2^\nu}{s^2}\left[\int d^Dx e^{-iq_1x}\langle A'|TA^a_\mu(x)A^b_\nu(0)|A\rangle\right]^{(amp)} , \quad (19)$$

which is schematically presented by the Fig. 2. This amplitude is defined in

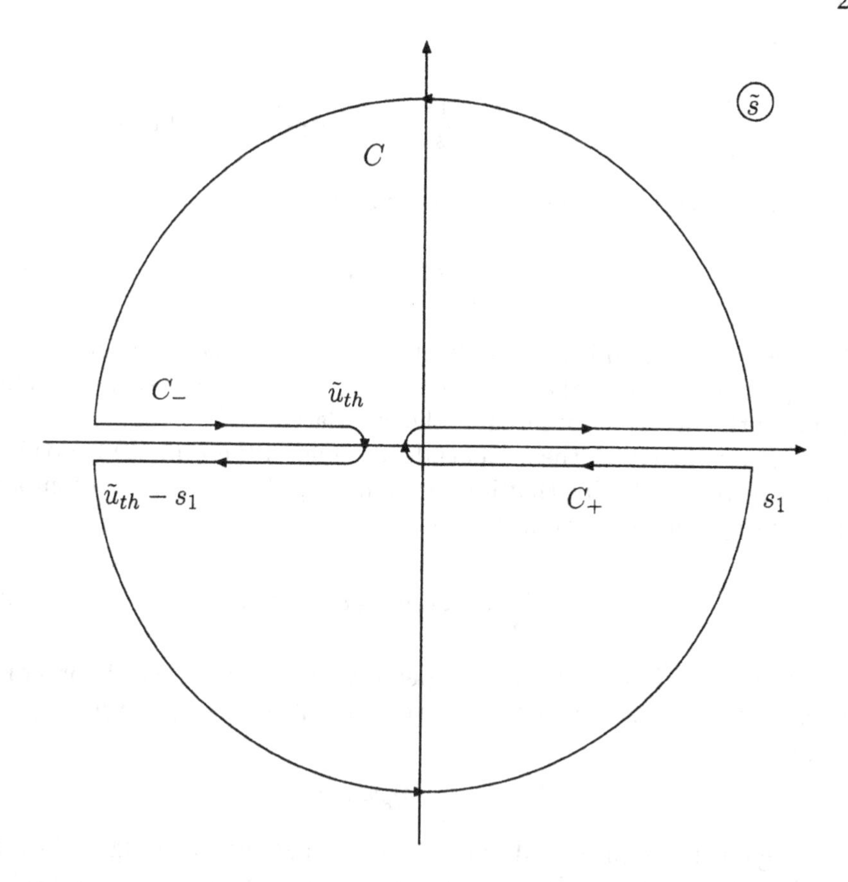

Figure 3. The singularities of the amplitude $\mathcal{A}^c(\tilde{s})$ in the complex $\tilde{s}$ plane.

the light-cone gauge and has to be calculated with the one-loop accuracy; its $\tilde{s}$- dependence enters through the external virtual gluons momenta

$$q_1 \simeq \beta_1 p_1 + \frac{\tilde{s} + \vec{q}_1^{\,2} - p_A^2}{s} p_2 + q_{1\perp} \simeq \beta_1 p_1 - \frac{\tilde{u} + \vec{q}_1^{\,\prime 2} - \vec{q}^{\,2} - p_{A'}^2}{s} p_2 + q_{1\perp} \,,$$

$$q_1' \simeq \beta_1' p_1 - \frac{\tilde{u} + \vec{q}_1^{\,\prime 2} - p_A^2}{s} p_2 + q_{1\perp}' \simeq \beta_1' p_1 + \frac{\tilde{s} + \vec{q}_1^{\,2} - \vec{q}^{\,2} - p_{A'}^2}{s} p_2 + q_{1\perp}' \,,$$

$$\tilde{s} = (p_A + q_1)^2 \,, \quad \tilde{u} = (p_A - q_1')^2 = \tilde{u}_{th} - \tilde{s} \,, \quad \tilde{u}_{th} \simeq \vec{q}^{\,2} + p_{A'}^2 + p_A^2 - \vec{q}_1^{\,\prime 2} - \vec{q}_1^{\,2} \,. \tag{20}$$

The singularities of the amplitude $\mathcal{A}^c(\tilde{s})$ in the complex $\tilde{s}$ plane are depictured at the Fig. 3. Now we notice that for the negative t- channel signature the amplitude $\mathcal{A}^c$ has, the integrals over left and right cuts of the Fig. 3 coincide so that the relation (18) can be rewritten in terms of the integral over the infinite circle C only. Therefore, using the results of such integration

$$\int_C \frac{d\tilde{s}}{2\pi i}\frac{1}{\tilde{s}} = 1 \ , \quad \int_C \frac{d\tilde{s}}{2\pi i}\frac{\ln \tilde{s}}{\tilde{s}} = \int_C \frac{d\tilde{s}}{2\pi i}\frac{\ln(-\tilde{s})}{\tilde{s}} = \ln(s_1) \tag{21}$$

for possible energy dependencies of the amplitude $\mathcal{A}^c(\tilde{s})$, we conclude

$$\Phi^c = \frac{-is_1}{2}(\hat{Re})_{s_1}\mathcal{A}^c(s_1) \ . \tag{22}$$

The $(\hat{Re})_{s_1}$- operation here consists of elimination of the s_1- channel imaginary part, or, that is the same, of the replacement $\ln(-s_1) \to \ln(s_1)$ whenever the $\ln(-s_1)$ appeared at the calculation.

We also notice, that the s- dependence of Φ^c of the Eq. (22) is artificial and can be removed using that it enters only as p_2/s, i.e. there is a complete scaling in p_2. Therefore we introduce

$$\frac{p_2'}{s_1} = \frac{p_2}{s} \ , \quad 2p_1p_2' = s_1 \to \infty \ , \tag{23}$$

that allows to replace all the p_2/s- dependence by the p_2'/s_1 one. In order not to deal with so many new notations we then replace back $p_2' \to p_2$, $s_1 \to s$ and get

$$\Phi^c = \frac{-i}{2}(\hat{Re})_s s\mathcal{A}^c(s) \ , \tag{24}$$

where again the amplitude $\mathcal{A}^c(s)$ is depictured at the same Fig. 2, but the meaning of the virtual gluons momenta there is different with respect to the Eq. (20)

$$q = \beta p_1 + \frac{p_{A'}^2 + \vec{q}^{\,2} - (1+\beta)p_A^2}{s(1+\beta)}p_2 + q_\perp \simeq \beta p_1 + \frac{p_{A'}^2 + \vec{q}^{\,2} - p_A^2}{s}p_2 + q_\perp \ ,$$

$$q_1 = \beta_1 p_1 + p_2 + q_{1\perp} \ , \quad q_1' = q_1 - q \ , \quad s = 2p_1p_2 \to \infty \ , \tag{25}$$

that evidently corresponds to an ordinary Regge kinematics. Formally nothing else is changed with respect to the previous definition of the amplitude: it is to be calculated in the light-cone gauge with the gauge fixing vector p_2 and the replacements made touch neither βs nor the transverse momenta. We therefore finally present the impact factor in the form

$$\sqrt{N}\Phi^c_{A'A}(\vec{q}_1, \vec{q}_1{}') = \frac{-i}{2}\Big(1 - \frac{\omega^{(1)}(-\vec{q}^{\,2})}{2}\left[\ln\left(\frac{s^2}{s_0\vec{q}^{\,2}}\right) + \tilde{K}_1\right]$$

$$+\frac{\omega^{(1)}(-\vec{q}_1{}^2)}{2}\left[\ln\left(\frac{2q_1p_2}{\vec{q}_1{}^2}\right)^2 - P_F(-\vec{q}_1{}^2) + 2P_L(-\vec{q}_1{}^2)\right]$$

$$+\frac{\omega^{(1)}(-\vec{q}_1^{\,\prime 2})}{2}\left[\ln\left(\frac{2q_1'p_2}{\vec{q}_1^{\,\prime 2}}\right)^2 - P_F(-\vec{q}_1^{\,\prime 2}) + 2P_L(-\vec{q}_1^{\,\prime 2})\right]$$

$$+\frac{\omega^{(1)}(-\vec{q}_1^{\,2}) + \omega^{(1)}(-\vec{q}_1^{\,\prime 2})}{2}\left[\frac{1}{2\epsilon(1+2\epsilon)} + 2\psi(1-\epsilon)\right.$$

$$\left.\left.-2\psi(1) + \psi(1+2\epsilon) - \psi(1+\epsilon)\right]\right)\left(\hat{Re}\right)_s s\mathcal{A}^c(s)\ . \tag{26}$$

4. Discussion

We saw that, in order to understand the properties of the NLO colour antisymmetric octet impact factors, we need to study the high-energy asymptotics of the amplitude $\mathcal{A}^c(s)$, defined in the previous Section. The idea how to show the fulfilment of the bootstrap condition (3) is clear from the representation (26) for the impact factor: if the amplitude $\mathcal{A}^c(s)$ has the Regge form like (6) in its high-energy asymptotics, then the desirable proportionality (3) is indeed fulfilled in the process- independent form as necessary. This amplitude has all the necessary to be Reggeized: it has the gluon colour quantum numbers and negative signature in the t- channel and must be taken in the Regge kinematics, but, nevertheless, its Reggeization is not so clear because of the "nonsense" virtual gluons polarization p_2/s. In such the situation, very unusually, the diagrams without gluonic exchanges in the t- channel can also contribute so that the traditional (see Ref. [6], for instance) approaches to show the Reggeization are not directly applicable. We will nevertheless show that the amplitude $\mathcal{A}^c(s)$ does have the Reggeized form.

For this purpose let us decompose

$$p_2 = q_1 - e_1 \simeq q_1' - e_1'\ , \quad e_1 = \beta_1 p_1 + q_{1\perp}\ , \quad e_1' = \beta_1' p_1 + q_{1\perp}'\ ,$$

$$s\mathcal{A}^c(s) = \frac{-iT^c_{ab}}{s}\left(p_2^\mu q_1^{\prime\nu} + q_1^\mu p_2^\nu - q_1^\mu q_1^{\prime\nu} + e_1^\mu e_1^{\prime\nu}\right)$$

$$\times\left[\int d^D x e^{-iq_1 x}\langle A'|TA^a_\mu(x)A^b_\nu(0)|A\rangle\right]^{(amp)}\ . \tag{27}$$

The convenience of this decomposition is provided by the possibility to apply Ward identities (the details will be given in the Ref.[7]) to those terms of the Eq. (27) generated by first three tensor structures there. Instead, in the last term generated by the last tensor structure we have avoided the "nonsense" virtual gluons polarization, so that only "normal" diagrams with gluonic exchanges in the t- channel can contribute there as it is shown in the Fig. 4. The amplitude presented by this figure can be considered by the

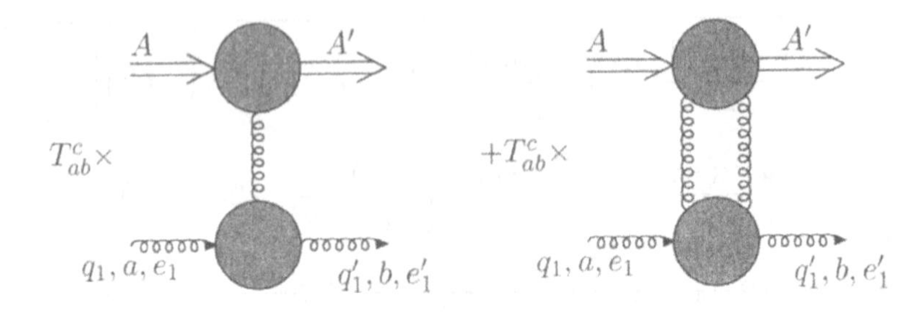

Figure 4. Schematic representation of the amplitude generated by the last tensor structure of the Eq. (27).

traditional methods and has the Reggeized form. Instead, the application of the Ward identities to the remaining contribution to the Eq. (27) gives for it

$$gN\frac{(1-P_L(q_1^2))\,(q_1p_2)(q_1q)+(1-P_L(q_1'^2))\,(q_1'p_2)(q_1'q)}{(1-P_L(q^2))\,q^2(qp_2)}$$
$$\times\frac{ip_2^\mu}{s}\langle A'|A^c_\mu(0)|A\rangle^{(amp)}\,, \tag{28}$$

i.e. the form is again factorized. Therefore we have got the verification of the NLO bootstrap for the impact factors (3). Since we have factorized the coefficient function $R(\vec{q}_1,\vec{q}_1{}')$ in the process- independent way, and we know the bootstrap to be fulfilled in particular cases (see Ref. [5] for instance), we conclude it is fulfilled for the arbitrary transitions as well. This statement is valid in the most general case of QCD with massive quarks and without any ϵ- expansion.

Acknowledgments: Work was supported in part by INTAS and in part by the Russian Fund of Basic Researches.

References

1. Fadin, V.S., Kuraev, E.A. and Lipatov, L.N. (1975), *Phys. Lett.*, **Vol. no. B60**, p. 50; Kuraev, E.A., Lipatov, L.N. and Fadin, V.S. (1976), *Sov. Phys. JETP*, **Vol. no. 44**, p. 443; (1977), *Sov. Phys. JETP*, **Vol. no. 45**, p. 199; Balitskii, Ya.Ya. and Lipatov, L.N. (1978) *Sov. J. Nucl. Phys.*, **Vol. no. 28**, p. 822.
2. Balitskii, Ya.Ya., Lipatov, L.N. and Fadin, V.S. (1979) *in Proceedings of Leningrad Winter School on Physics of Elementary Particles*, p. 109 (in Russian).
3. Fadin, V.S., *the talk in these proceedings.*
4. Fadin, V.S., Fiore, R., Kotsky, M.I. and Papa, A. (2000) *Phys. Rev.*, **Vol. no. D61**, p. 094005; p. 094006.
5. Fadin, V.S., Fiore, R., Kotsky, M.I. and Papa, A. (2000) *Phys. Lett.*, **Vol. no. B495**, p. 329-337.
6. Fadin, V.S., Fiore, R. (2001) *Phys. Rev.*, **Vol. no. D64**, p. 114012.
7. Fadin, V.S., Fiore, R., Kotsky, M.I. and Papa, A., to appear.

JET VERTEX IN THE NEXT-TO-LEADING LOG(S) APPROXIMATION

G. P. VACCA
Dipartimento di Fisica, Università di Bologna and
INFN, Sez. di Bologna, via Irnerio 46, 40126 Bologna, Italy

Abstract. The next-to-leading corrections to the jet vertex which is relevant for the Mueller-Navelet jets production in hadronic collisions and for the forward jet cross section in lepton-hadron collisions are presented in the context of a k_t factorizazion formula which resums the leading and next-to-leading logarithms of the energy.

1. Introduction

Recently, in the study of QCD in the Regge limit, a novel element has been defined and computed at the next-to-leading order (NLO). It is the jet vertex [1], which represents one of the building blocks in the production of Mueller-Navelet jets at hadron hadron colliders and of forward jets [2] in deep inelastic electron proton scattering. Such processes should provide a kinematical environment for which the BFKL Pomeron [3] QCD analysis could apply, provided that the transverse energy of the jet fixes a perturbative scale and the large energy yields a large rapidity interval.

In a strong Regge regime important contributions, or even dominant, come, in the perturbative language, from diagrams beyond NLO and NNLO at fixed order in α_s. This is the main reason for considering a resummation of the leading and next-to-leading logarithmic contributions as computed in the BFKL Pomeron framework. The lackness of unitarity, if the related corrections are not taken into account, forces one to consider an upper bound on the energy to suppress them. It is already known that the leading logarithmic (LL) analysis is not accurate enough [4], being the kinematics selected by experimental cuts far from any asymptotic regime. Moreover at this level of accuracy there is a maximal dependence in the dif-

R. Fiore et al. (eds.), Diffraction 2002, 267–276.

ferent scales involved (renormalization, collinear factorization and energy scales). For the Mueller-Navelet jet production process the only element still not known at the NLO level was the "impact factor", which describes the hadron emitting one inclusive jet when interacting with the reggeized gluon which belongs to the BFKL ladder, accurate up to NLL [5]. The jet vertex, now computed, is the building block of this interaction. For the so called forward jet production in DIS the extra ingredient necessary is the photon impact factor, whose calculation is currently in progress [6, 7]. Let us also remind that NLL BFKL approach has recently gained more theoretical solidity since the bootstrap condition in its strong form, which is the one necessary for the self-consistency of the assumption of Reggeized form of the production amplitudes, has been stated and formally proved [8]. This relation is a very remarkable property of QCD in the high energy limit.

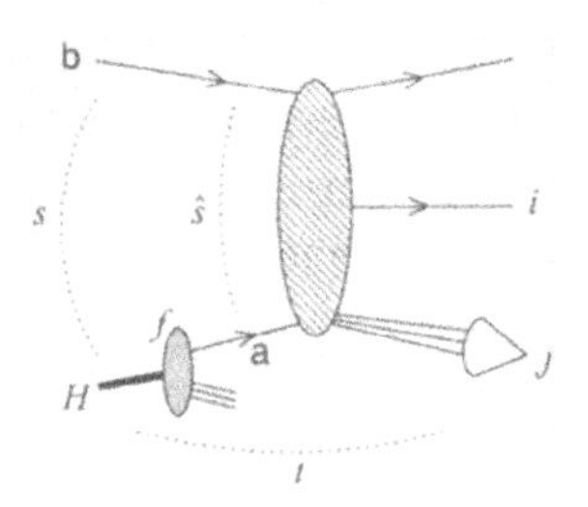

Figure 1. High energy process with jet production. H is the incoming hadron providing a parton a (gluon g/quark q) with distribution density f_a which scatters with the parton b with production of a jet J in the forward direction (w.r.t H) and i is the generic label for outgoing particles.

A theoretical challenge, interesting by itself and appearing in the calculation, is related to the special kinematics. The processes to be analyzed is illustrated in Fig.1: the lower parton emitted from the hadron H scatters with the upper parton q and produces the jet J. Because of the large transverse momentum of the jet the parton is hard and the collinear factorization allows for a partonic scale dependence described by the DGLAP evolution equations [9]. Above the jet, on the other hand, the kinematics chosen requires a large rapidity gap between the jet and the outgoing parton q: such a situation is described by BFKL dynamics. Therefore the jet vertex lies at the interface between DGLAP and BFKL dynamics, a situation which appears for the first time in a non trivial way. As an essential result of our analysis we find that it is possible to separate, inside the jet vertex, the collinear infrared divergences that go into the parton evolution of the incoming gluon/quark from the high energy gluon radiation inside the rapidity gap which belongs to the first rung of the LO BFKL ladder.

2. Jet Vertex: definition and outline of the derivation

Let us consider, for the process illustrated in Fig. 1, the kinematic variables

$$\begin{aligned} p_H &= \left(\sqrt{s/2}, 0, \mathbf{0}\right) \quad , \quad p_{\mathsf{a}} = x\, p_H \quad , \quad p_{\mathsf{b}} = \left(0, \sqrt{s/2}, \mathbf{0}\right) \\ p_i &= E_i \left(\mathrm{e}^{y_i}/\sqrt{2}, \mathrm{e}^{-y_i}/\sqrt{2}, \boldsymbol{\phi}_i\right) \quad , \quad s := (p_H + p_{\mathsf{b}})^2 \,. \end{aligned} \tag{1}$$

We study the partonic subprocess $\mathsf{a} + \mathsf{b} \to X + jet$ in the high energy limit

$$\Lambda^2_{\mathrm{QCD}} \ll E_J^2 \sim -t\,(\text{fixed}) \ll s \to \infty. \tag{2}$$

Starting from the parton model, we assume the physical cross section to be given by the corresponding partonic cross section $\mathrm{d}\hat{\sigma}$ (computable in perturbation theory) convoluted with the parton distribution densities (PDF) f_{a} of the partons a inside the hadron H. A jet distribution S_J, with the usual infrared safe behaviour, selects the final states contributing to the one jet inclusive cross section that we are considering. We choose, as jet variables, rapidity, transverse energy and azimuthal angle and the one jet inclusive cross section initiated by quarks and gluons in hadron H is therefore written as

$$\frac{\mathrm{d}\sigma}{\mathrm{d}J} := \frac{\mathrm{d}\sigma_{\mathsf{b}H}}{\mathrm{d}y_J \mathrm{d}E_J \mathrm{d}\phi_J} = \sum_{\mathsf{a}=\mathsf{q},\mathsf{g}} \int \mathrm{d}x \; \mathrm{d}\hat{\sigma}_{\mathsf{ba}}(x)\, S_J(x) f_{\mathsf{a}}^{(0)}(x) \,. \tag{3}$$

At the *lowest order* the jet cross section, dominated by a t-channel gluon exchange, can be written as [1]

$$\frac{\mathrm{d}\sigma^{(0)}}{\mathrm{d}J} = \sum_{\mathsf{a}=\mathsf{q},\mathsf{g}} \int \mathrm{d}x \int \mathrm{d}\boldsymbol{k}\; h_{\mathsf{b}}^{(0)}(\boldsymbol{k}) V_{\mathsf{a}}^{(0)}(\boldsymbol{k}, x) f_{\mathsf{a}}^{(0)}(x) \tag{4}$$

where $V_{\mathsf{a}}^{(0)}(\boldsymbol{k}, x) = h_{\mathsf{a}}^{(0)}(\boldsymbol{k}) S_J^{(2)}(\boldsymbol{k}, x)$ is the jet vertex induced by parton a, $h_{\mathsf{a}}^{(0)}(\boldsymbol{k})$ is the partonic impact factor (generally expressed in $D = 4 + 2\varepsilon$ dimensions for regularization pourposes at the NLO) and $f_{\mathsf{a}}^{(0)}(x)$ is the parton distribution density (PDF). The jet distribution is in this case trivial

$$S_J^{(2)}(\boldsymbol{k}, x) := S_J^{(2)}(p_1, p_2; p_{\mathsf{g}}, p_{\mathsf{q}}) = \delta\left(1 - x_J/x\right) E_J^{1+2\epsilon} \delta(\boldsymbol{k} - \boldsymbol{k}_J) \tag{5}$$

with $x_J := E_J \mathrm{e}^{y_J}/\sqrt{s}$.

In the *NLO approximation* virtual and real corrections enter in the calculation of the partonic cross section $\mathrm{d}\hat{\sigma}_{\mathsf{ba}}$. The three partons, produced in the real contributions, in the upper, and lower rapidity region are denoted by 2 and 1 respectively, while the third, which can be emitted everywhere in rapidity, by 3. Moreover we shall call $k = p_b - p_2$ and $k' = p_1 - p_a$ and $q = k - k'$. A Sudakov parametrization is chosen

$$k = -\bar{w} p_g + w p_q + k_\perp \quad , \quad k_\perp = (0, 0, \boldsymbol{k}) \tag{6}$$
$$k' = -z p_g + \bar{z} p_q + k'_\perp \quad , \quad k'_\perp = (0, 0, \boldsymbol{k}') \tag{7}$$

with the typical inequalities $\bar{w} \sim \bar{z} \ll w \sim z \ll 1$ valid in the multi Regge kinematics (MRK). Dimensional regularization is used, as usual, to trace the infrared and ultraviolet divergences. On taking into account the infrared (IR) properties of the jet distribution $S_J^{(3)}$, the following structure is matched exactly up to NLO (i.e. α_s^3) [1]

$$\frac{d\sigma}{dJ} = \sum_{a=q,g} \int dx \int d\boldsymbol{k}\, d\boldsymbol{k}'\; h_b(\boldsymbol{k}) G(xs, \boldsymbol{k}, \boldsymbol{k}') V_a(\boldsymbol{k}', x) f_a(x) , \tag{8}$$

where

$$\begin{aligned} h &= h^{(0)} + \alpha_s h^{(1)} + \cdots , \\ V &= V^{(0)} + \alpha_s V^{(1)} + \cdots , \\ f &= f^{(0)} + \alpha_s f^{(1)} + \cdots , \\ G(xs, \boldsymbol{k}, \boldsymbol{k}') &:= \delta(\boldsymbol{k} - \boldsymbol{k}') + \alpha_s K^{(0)}(\boldsymbol{k}, \boldsymbol{k}') \log \frac{xs}{s_0} + \cdots \end{aligned} \tag{9}$$

The partonic impact factor correction in forward direction $h^{(1)}$ is well known [10], the PDF's f_a are the standard ones satisfying DGLAP evolution equations

$$\alpha_s f_a^{(1)}(x, \mu_F^2) := \frac{\alpha_s}{2\pi} \frac{1}{\varepsilon} \left(\frac{\mu_F^2}{\mu^2} \right)^{\varepsilon} \sum_b P_{ab} \otimes f_b^{(0)} \tag{10}$$

with μ_F the collinear factorization scale, and the BFKL Green's function G is defined by the LO BFKL kernel $K^{(0)}$. The new element is the correction to the jet vertex $V^{(1)}$ whose expressions can be found in [1].

Let us note that the previous form in (8) is clearly suggested by the structure of the leading logarithmic part, but at this level it is not a trivial ansatz. Only after a proper treatment of all the IR singular terms one will be able to extract and define the jet verteces initiated by quarks and gluons in the hadron.

The NLO virtual corrections in the high energy limit are relatively simple and they can be summarized in the following expression:

$$\begin{aligned} \frac{d\sigma^{(\text{virt})}}{dJ} &= \alpha_s \sum_a \int dx \int d\boldsymbol{k}\; h_q^{(0)}(\boldsymbol{k}) \left[2\omega^{(1)}(\boldsymbol{k}) \log \frac{xs}{\boldsymbol{k}^2} + \tilde{\Pi}_b(\boldsymbol{k}) + \tilde{\Pi}_a(\boldsymbol{k}) \right] \\ &\quad \times V_a^{(0)}(\boldsymbol{k}, x) f_a^{(0)}(x) , \end{aligned} \tag{11}$$

where $\omega^{(1)}(\boldsymbol{k})$ is the 1-loop reggeized gluon trajectory and the $\widetilde{\Pi}_i(\boldsymbol{k})$ are related to the part constant in energy, after having performed the ultraviolet (UV) ε-pole subtraction occuring in the renormalization of the coupling.

The NLO real corrections, even in the high energy limit, are much more involved since the interplay of different IR soft and collinear singularities follows a more complicated pattern.

One starts from the partonic differential cross sections $d\hat{\sigma}_{\mathsf{ba}}(x)$, appearing in (3), which have been also computed in the high energy regime [10], since in such a case all terms suppressed by powers of s may be neglected. The form of the partonic differential cross section turns out to be quite simple when restricted to one of the two halves of the phase space, which are obtained introducing the rapidity $y' = y + \frac{1}{2}\log\frac{1}{x}$ "measured" in the partonic center of mass frame, and splitting the phase space into two semi-spaces defined by $y'_1, y'_3 > 0$ (*lower half*) and $y'_3 < 0 < y'_1$ (*upper half*).

In the quark initiated case one obtains for both regions a reasonable simple expression, while for the gluon initiated part it is convenient to consider separately the $\mathsf{q}\bar{\mathsf{q}}$ and gg final state cases.

The "upper half region", with $y'_3 < 0$, leads to the rederivation of the partonic b-impact factor, already well known at NLL. The "lower half region", wherein $y'_3 > 0$, which corresponds to $z > z_{\mathrm{cut}} := \frac{E_3}{\sqrt{xs}}$ is precisely the one from which we are able to extract the jet verteces.

A very important ingredient, as previosly mentioned, is given by the jet definition $S_J^{(n)}$. As usual it selects from a generic n-particle final state the configurations contributing to our one jet inclusive observable and it must be IR safe. The last requirement simply corresponds to the fact that emission of a soft particle cannot be distinguished from the analogous state without soft emission. Furthermore, collinear emissions of partons cannot be distinguished from the corresponding state where the collinear partons are replaced by a single parton carrying the sum of their quantum numbers.

The extraction of all the IR singularities from the real contributions and the matching with the virtual ones is done on employing the standard subtraction method. This consists in approximating the amplitudes in any singular region in order to extract the exact analytic singular behavior, and in leaving the remaining finite part in a form suitable for numerical integration, since normally a full analytical computation is tecnically impossible. The analytic singular terms of both virtual and real contributions are therefore combined. Moreover, with respect to standard NLO jet analysis, one has also to deal with the leading logarithmic terms which are entangled to the other IR singular terms. For the process considered it can be shown [1] that finally one is left with the collinear singularities associated to the impact factor of the parton b, to the definition of the distribution function for the parton a in H and to the BFKL kernel term. The finite remaning parts correspond to the impact factors and jet verteces.

We recall that the jet distribution functions become essential in disentangling the collinear singularities, the soft singularities, and the leading $\log s$ pieces when the particle 3 is a gluon. The same basic mechanism may be used for both the quark and the gluon initiated cases.

- When the outgoing parton 1 is in the collinear region of the incoming parton a, i.e., $y_1 \to \infty$, it cannot enter the jet; only gluon 3 can thus be the jet, y_3 is fixed and no logarithm of the energy can arise due to the lack of evolution in the gluon rapidity. No other singular configuration is found when $J = \{3\}$.
- In the composite jet configuration, i.e., $J = \{1, 3\}$, the gluon rapidity is bounded within a small range of values, and also in this case no $\log s$ can arise. There could be a singularity for vanishing gluon 3 momentum: even if the $1 \parallel 3$ collinear singularity is absent, we have seen that, at very low z, a soft singular integrand arises. However, the divergence is prevented by the jet cone boundary, which causes a shrinkage of the domain of integration $\sim z^2$ for $z \to 0$ and thus compensates the growth of the integrand.
- The jet configuration $J = \{1\}$ corresponds to the situation wherein gluon 3 spans the whole phase space, apart, of course, from the jet region itself. The LL term arises from gluon configurations in the central region. Therefore, it is crucial to understand to what extent the differential cross section provides a leading contribution. It turns out that the coherence of QCD radiation suppresses the emission probability for gluon 3 rapidity y_3 being larger than the rapidity y_1 of the parton 1, namely an angular ordering prescription holds. This will provide the final form of the leading term, i.e., the appropriate scale of the energy and, as a consequence, a finite and definite expression for the one-loop jet vertex correction.

It is therefore clear that the energy scale s_0 associated to the BFKL rapidity evolution plays a crucial role. The calculations show a natural choice, due to angular ordered preferred gluon emission and the presence of the jet defining distribution, which is also crucial to obtain the full collinear singularities which factorize into the PDF's. In particular, on considering the term giving the real part of the LL contribution, one can see that outside the angular ordered region

$$\frac{E_3}{z} > \frac{E_1}{1-z} \quad \Longleftrightarrow \quad \theta_3 > \theta_1 \quad \Longleftrightarrow \quad y_3 < y_1 \,, \tag{12}$$

there is a small contribution to the cross section. This ordering sets the natural energy scale which, for example, leads to the expression of the jet vertex for the case $s_0(\boldsymbol{k},\boldsymbol{k}') := (|\boldsymbol{k}'|+|\boldsymbol{q}|)(|\boldsymbol{k}|+|\boldsymbol{q}|)$. A mild modification of such a scale can be performed without introducing extra singularities, contrary to what happens for a generic choice. In any case the use of a different scale requires the introduction of modifing terms. For the useful symmetric Regge type energy scale $s_R = |\boldsymbol{k}||\boldsymbol{k}'|$, one has

$$G(xs,\boldsymbol{k},\boldsymbol{k}') = (\mathbf{1}+\alpha_{\rm s}H_L)\left[\mathbf{1}+\alpha_{\rm s}K^{(0)}\log\frac{xs}{|\boldsymbol{k}||\boldsymbol{k}'|}\right](\mathbf{1}+\alpha_{\rm s}H_R), \qquad (13)$$

where $H_L(\boldsymbol{k},\boldsymbol{k}') = -K^{(0)}(\boldsymbol{k},\boldsymbol{k}')\log\left((|\boldsymbol{k}|+|\boldsymbol{q}|)/|\boldsymbol{k}|\right) = H_R(\boldsymbol{k}',\boldsymbol{k})$.

We do not present the final vertex expressions here, they can be found in [1].

3. Remarks and final formulas

In order to extend the results of the two loop calculations, which have allowed to obtain the full NLL partonic cross sections, to our inclusive jet production case, it is necessary to check the relation between the partonic impact factors and the jet verteces. More precisely one may define the one loop correction to the "jet impact factor" as $[V*f]^{(1)} = V^{(1)}*f^{(0)} + V^{(0)}*f^{(1)}$. This latter object and the corrections to the partonic impact factor $h^{(1)}$ are different in one important aspects, i.e. their collinear singularities. Essentially the IR singularity associated to the LL real term is absent (subtracted) in the partonic impact factors, while this does not happen in the "jet impact factors" due to the constraining presence of the jet. In the "jet impact factors" are therefore hidden the full collinear singularities which should be factorized to obtain the right behavior of the PDF's satisfying the DGLAP evolution equations.

Apart from this important difference there is a full correspondence, as shown in [1], between these two objects. In particular the same angular ordering mechanism, which sets a reference for the energy scale, is present. Thus any tuning of the BFKL kernel energy scale is done precisely in the same way for both impact factors and jet vertices and, thanks to the Regge factorization property valid up to NLL in QCD, one is allowed to use the known form of the NLL BFKL Green's function setting the scale to s_R, as previously discussed.

From a two loop analysis, up to α_s^4 order, one expects to find the contributions

$$\begin{aligned}\frac{1}{\alpha_{\rm s}^4}\frac{{\rm d}\sigma}{{\rm d}J}^{(2)} &= \frac{1}{2}\log^2\frac{xs}{s_R}\,h^{(0)}K^{(0)}K^{(0)}V^{(0)}f^{(0)} + \log\frac{xs}{s_R}\Big[h^{(1)}K^{(0)}V^{(0)}f^{(0)} \\ &\quad + h^{(0)}(H_LK^{(0)}+K^{(0)}H_R+K^{(1)})V^{(0)}f^{(0)} + h^{(0)}K^{(0)}V^{(1)}f^{(0)} \\ &\quad + h^{(0)}K^{(0)}V^{(0)}f^{(1)}\Big], \qquad (14)\end{aligned}$$

where the first term on the RHS is LL while the terms collected in the square brackets are NLL.

Consequently, to obtain the jet cross section with accuracy up to NLL terms, one has to consider the NLL BFKL kernel

$$K = \alpha_s K^{(0)} + \alpha_s^2 K^{(1)} \tag{15}$$

which has been computed with the Regge type scale $s_R = |\boldsymbol{k}_1||\boldsymbol{k}_2'|$ and leads to a Green's function, which resums all terms up to NLL in (8),

$$G(xs, \boldsymbol{k}_1, \boldsymbol{k}_2) = \int \frac{\mathrm{d}\omega}{2\pi\mathrm{i}} \left(\frac{xs}{s_R}\right)^{\omega} \langle \boldsymbol{k}_1 | (\mathbf{1} + \alpha_s H_L)[\omega - K]^{-1}(\mathbf{1} + \alpha_s H_R) | \boldsymbol{k}_2 \rangle \,. \tag{16}$$

The formula for the Mueller-Navelet jets [1] can be easily derived symmetrizing the formula (8) for the two jet case. More precisely one obtains

$$\begin{aligned} \frac{\mathrm{d}^2\sigma}{\mathrm{d}J_1 \mathrm{d}J_2} &= \sum_{\mathsf{a},\mathsf{b}} \int \mathrm{d}x_1 \mathrm{d}x_2 \int \mathrm{d}\boldsymbol{k}_1 \mathrm{d}\boldsymbol{k}_2 \; f_{\mathsf{a}}(x_1) V_{\mathsf{a}}(\boldsymbol{k}_1, x_1) \\ &\quad G(x_1 x_2 s, \boldsymbol{k}_1, \boldsymbol{k}_2) V_{\mathsf{b}}(\boldsymbol{k}_2, x_2) f_{\mathsf{b}}(x_2) \end{aligned} \tag{17}$$

where $\mathsf{a}, \mathsf{b} = \mathsf{q}, \mathsf{g}$, and the subscripts 1 and 2 refer to jet 1 and 2 in hadron 1 and 2, resp. Once the definition of the jets is given, all the elements are known and a definitive computation may be carried on.

Regarding the Green's function, two more remarks should be made. Since the high energy asymptotic regime is far, a few iterations of the kernel instead of a full resummation, as given in eq. (16), may be of some interest in a phenomenological application. The known large corrections of the BFKL kernel are related to the presence of large logs. This problem has been solved using optimal renormalization scale methods [11] or by considering an improved version of the kernel, equivalent at NLL level, but including correcting higher order terms, following a collinear prescription [12].

4. Conclusions

A brief review of the definitions and calculations at the NLO accuracy of the jet vertex relevant for the Mueller-Navelet jets at hadron-hadron colliders and for forward jets in deep inelastic electron-proton scattering has been given. This object, defined for high energies, due to specific kinematics, lies at the interface between DGLAP and BFKL dynamics. Being known all the ingredients, it is now possibile to perform a numerical analysis of the production of Mueller-Navelet jets at NLL. For the forward jet processes further theoretical studies on the NLL photon impact factors are still required.

Acknowledgments

The results discussed have been obtained in collaboration with J. Bartels and D. Colferai [1].

References

1. Bartels J., Colferai D. and Vacca G.P. (2002), The NLO jet vertex for Mueller-Navelet and forward jets: The quark part, *Eur. Phys. J.* **C** 24, p. 83 [hep-ph/0112283];
   Bartels J., Colferai D. and Vacca G.P. (2002), The NLO jet vertex for Mueller-Navelet and forward jets: The gluon part, [hep-ph/0206290].
2. Mueller A.H. and Navelet H. (1987), An Inclusive Minijet Cross-Section And The Bare Pomeron In QCD, *Nucl. Phys.* **B** 282, p. 727
   Mueller A.H. (1991), Parton Distributions At Very Small X Values, *Nucl. Phys.* **B** (Proc. Suppl.) **18C**, p. 125;
   Mueller A.H. (1991), Jets At Lep And Hera, *J. Phys.* **G**17, p. 1443.
3. Lipatov L.N. (1976), Reggeization Of The Vector Meson And The Vacuum Singularity In Nonabelian Gauge Theories, *Sov. J. Nucl. Phys.* **23**, p. 338
   Kuraev E.A., Lipatov L.N. and Fadin V.S. (1976), Multi - Reggeon Processes In The Yang-Mills Theory, *Sov. Phys. JETP* **44**, p. 443
   Kuraev E.A. , Lipatov L.N. and Fadin V.S. (1977), The Pomeranchuk Singularity In Nonabelian Gauge Theories, *Sov. Phys. JETP* **45**, p. 199
   Balitskii Y. and Lipatov L.N. (1978), The Pomeranchuk Singularity In Quantum Chromodynamics, *Sov. J. Nucl. Phys.* **28**, p. 822
4. Bartels J. , De Roeck A. and Loewe M. (1992), Measurement of hot spots inside the proton at HERA and LEP/LHC, *Z. Physik* **C** 54, p. 635
   Kwiecinski J., Martin A.D. and Sutton P.J. (1992), Deep inelastic events containing a measured jet as a probe of QCD behavior at small x, *Phys. Rev.* **D** 46, p. 921
   Tang W.K. (1992), The Structure function nu W(2) of hot spots at HERA, *Phys. Lett.* **B** 278, p. 363
   Bartels J. , De Roeck A. and Lotter H. (1996), The gamma* gamma* total cross section and the BFKL pomeron at e+ e- colliders, *Phys. Lett.* **B** 389, p. 742 [hep-ph/9608401];
   Brodsky S.J. , Hautmann F. and Soper D.E. (1997), Virtual photon scattering at high energies as a probe of the short distance pomeron, *Phys. Rev.* **D** 56, p. 6957 [hep-ph/9706427];
   Andersen J.R., Del Duca V., Frixione S., Schmidt C.R. and Stirling W.J. (2001), Mueller-Navelet jets at hadron colliders, *JHEP* **0102**, p. 007.
5. Fadin V.S. and Lipatov L.N. (1998), BFKL pomeron in the next-to-leading approximation, *Phys. Lett.* **B** 429, p. 127 [hep-ph/9802290];
   Ciafaloni M. and Camici G. (1997), Irreducible part of the next-to-leading BFKL kernel, *Phys. Lett.* **B** 412, p. 396
   Ciafaloni M. and Camici G. (1998), Energy scale(s) and next-to-leading BFKL equation, *Phys. Lett.* **B** 430, p. 349 [hep-ph/9803389].
6. Bartels J., Gieseke S. and Qiao C.-F. (2001), The (gamma* → q anti-q) Reggeon vertex in next-to-leading order QCD, *Phys. Rev.* **D** 63, p. 056014[hep-ph/0009102];
   Bartels J., Gieseke S. and Kyrieleis A. (2002), The process gamma(L)* + q → (q anti-q g) + q: Real corrections to the virtual photon impact factor, *Phys. Rev.* **D** 65, p. 014006 [hep-ph/0107152].
7. Fadin V.S. and Martin A.D. (1999), Infrared safety of impact factors for colorless particle interactions, *Phys. Rev.* **D** 60, p. 114008 [hep-ph/9904505];
   Fadin V.S., Ivanov D.Y. and Kotsky M.I. (2001), Photon Reggeon interaction vertices in the NLA, *Phys. Atom. Nucl.* **65**, p.1513, [hep-ph/0106099].

8. Braun M. and Vacca G.P. (1999), The 2nd order corrections to the interaction of two reggeized gluons from the bootstrap, *Phys. Lett.* **B** 454, p. 319hep-ph/9810454];
Braun M. and Vacca G.P. (2000), The bootstrap for impact factors and the gluon wave function, *Phys. Lett.* **B** 477, p. 156 [hep-ph/9910432];
Fadin V.S. and Papa A. (2002), A proof of fulfillment of the strong bootstrap condition, [hep-ph/0206079].
9. Gribov V.N. and Lipatov L.N.(1972), Deep Inelastic E P Scattering In Perturbation Theory, *Sov. J. Nucl. Phys.* **15**, p. 438
Altarelli G. and Parisi G. (1977), Asymptotic Freedom In Parton Language, *Nucl. Phys.* **B** 126, p. 298
Dokshitzer Y.L. (1977), Calculation Of The Structure Functions For Deep Inelastic Scattering And E+ E- Annihilation By Perturbation Theory In Quantum Chromodynamics. (In Russian), *Sov. Phys. JETP* **46**, p. 641
10. Ciafaloni M. (1998), Energy scale and coherence effects in small-x equations, *Phys. Lett.* **B** 429, p. 363[hep-ph/9801322] ;
Ciafaloni M. and Colferai D. (1999), k-factorization and impact factors at next-to-leading level, *Nucl. Phys.* **B** 538, p. 187 [hep-ph/9806350].
11. Brodsky S.J., Fadin V.S., Kim V.T., Lipatov L.N. and Pivovarov G.B. (2002), High-energy QCD asymptotics of photon photon collisions, [arXiv:hep-ph/0207297].
12. Ciafaloni M., Colferai D. and Salam G.P. (1999), Renormalization group improved small-x equation, *Phys. Rev.* **D** 60, p. 114036 [arXiv:hep-ph/9905566].

THE QCD COUPLING BEHAVIOR IN THE INFRARED REGION

D.V. SHIRKOV
Bogoliubov Lab. of Theor. Phys., JINR,
Dubna, 141980, Russia, shirkovd@thsun1.jinr.ru

Abstract. Short resume of the talk delivered at the "Diffraction-2002" conference (Alushta, Sept 2002). In a more detail the contents of this contribution has been recently published in refs.[1] and [2].

The summary of nonperturbative results for the QCD effective invariant coupling (IC) $\bar{\alpha}_s$ obtained by numerical lattice simulations for the functional integral and by solution of the approximate Dyson–Schwinger equations reveals (see, e.g., refs.[3] — [9]) a remarkable variety of IR behaviors of $\bar{\alpha}_s(Q^2)$ even at the qualitative level. In turn, this rises the question of correspondence between the results obtained so far by different groups of researchers.

We analyze this issue by mass-dependent coupling-constant transformations[10, 11]. The mass dependence plays an important role in the perturbative QCD for analysis of the threshold effects[12, 13]. Now, in our opinion, it reveals itself in the current nonperturbative analysis of the IR properties in QCD. Here, various groups use different definitions of the QCD coupling constant and IC in the IR region. For instance, in the IC defining the Tübingen group uses[3] the gluon–ghost vertex, the Orsay group[4, 5] — three–gluon one, the "Transoceanic team" [6, 7] — the gluon–quark vertex, while the ALPHA collaboration employs quite different approach [8, 9] based on the so–called Schrödinger functional with IC defined as a function of the spatial size of the lattice $\alpha_{\rm SF}(L)$, i.e., in the space–time representation.

We remind the origin of the IC $\bar{\alpha}_s(Q^2)$ notion and note that it is free of any relation with the weak coupling or UV massless limits. Then, we consider the item of non-uniqueness of the $\bar{\alpha}_s(Q^2)$ definition, and the related question of coupling constant and IC transformations. We discuss this issue in terms of coupling transformation similar to the renormalization–scheme one in the weak coupling limit, but more involved due to its mass

R. Fiore et al. (eds.), Diffraction 2002, 277–279.

dependency. Here, the important feature is *the vertex dependence* [14] of the mass-dependent $\bar{\alpha}_s(Q^2, M^2)$ that is typical for QFT models with the gauge symmetry.

Further on, we return to the item of correspondence between different IR behaviors of $\bar{\alpha}_s(Q^2)$. In particular, we give few examples of transformations that are singular in the IR region and, in some cases, result in the $\bar{\alpha}_s(Q^2, M^2)$ behavior similar to the ones observed in refs.[3] — [9].

In particular, we take an example motivated by the *analytic perturbation theory* (=APT), a recently developed construction [15] — [17] that allows cleaning the perturbative QCD of unphysical singularities such as the Landau pole "without bloodshed" — by imposing the Källen–Lehmann analyticity condition. The coupling-constant transformation

$$\alpha_s \to \alpha_{\rm M}(\alpha_s) = \frac{1}{\pi\beta_0} \arccos \frac{1}{\sqrt{1+\pi^2\beta_0^2\alpha_s^2}} = \frac{1}{\pi\beta_0} \arctan(\pi\beta_0\alpha_s)\,, \qquad (1)$$

induces, by integrating the group differential equation, the following IC transformation

$$\bar{\alpha}_s(Q^2) = \frac{1}{\beta_0 L} \to \tilde{\alpha}(Q^2) = \frac{1}{\pi\beta_0} \arccos \left. \frac{L}{\sqrt{L^2+\pi^2}} \right|_{L>0}; \quad L = \ln\left(\frac{Q^2}{\Lambda^2}\right). \qquad (2)$$

with $\tilde{\alpha}$ known in the APT as the Minkowskian effective QCD coupling. The transformation function $\alpha_{\rm M}(\alpha_s)$ obeys important properties:

- **AF:** it tends to α_s in the weak coupling limit, i.e., $\alpha_i \to \alpha_s$ as $\alpha_s \ll 1$.
- **GhF:** it is finite at $\alpha_s = \infty$.
- **IRf:** it is IR-finite, i.e., $\alpha_i \to 1/\beta_0$ in the limit as $\alpha_s \to -0$.

For the weak coupling, property **AF** ensures the correspondence with the asymptotic freedom. Property **GhF** leads to the absence of ghosts, i.e., unphysical singularities. Property **IRf** yields a finite IR limit. Qualitatively, the ghost-free function $\tilde{\alpha}$ with finite IR limiting value $\tilde{\alpha}(0) = 1/\beta_0$ and $\sim (\beta_0 L)^{-1}$ UV behavior resembles the result for $\bar{\alpha}_s(Q^2)$ obtained[3] by the Tübingen group.

Then, we discuss in short the possible origin of extremely steep rise in the IR limit of the QCD effective coupling of ALPHA collaboration obtained by lattice simulation [8, 9] of the Schrödinger functional. In our opinion, the possible resolving of this issue could be related to the transition from the space-time $\alpha_{\rm SF}(L)$ to the energy–momentum representation $\tilde{\alpha}_{\rm SF}(Q)$ by an appropriate Fourier transformation.

More specifically, our analysis performed with the help of the Tauberian theorem (for detail see our paper [2]) revealed some important feature of the quantum–mechanical correspondence rule "$r \to 1/Q$" relating the asymptotic behavior of a function $f(r)$ as $r \to \infty$ and of its Fourier transform

$F(Q)$

$$F(Q) \sim f(1/Q) \quad \text{as} \quad Q \to 0 . \qquad \text{(QMC)}$$

This rule has been used by the ALPHA collaboration for discussing their lattice simulation results in the IR region. We have demonstrated that (QMC), being a reasonable guide for some class of asymptotics (the power and logarithmic type), has its rigid limits of applicability. It is not valid at all for wide class of asymptotic behaviors, violating the so-called Tauberian conditions[18], like the exponential ones.

In particular, the exponentially rising long-range behavior of QCD coupling in the distance representation $\alpha_{\mathrm{SF}}(L) \sim e^{mL}$ observed by the ALPHA collaboration on the basis of the lattice simulation of the Schrödinger functional can correspond in the momentum (transfer) picture to the finite or b) "slightly singular $\sim 1/Q$" IR asymptotics.

Our conclusion is that the question of the IR behavior of the effective QCD coupling $\bar{\alpha}_s(Q^2)$ (and of propagators) is not a well-defined one. Results of different groups, even formulated in the same momentum representation, should not be compared directly with each other. Calculation of hadronic characteristics remains as a reasonable criterion for comparison of different lattice simulation schemes.

References

1. Shirkov D.V., (2002) *Theor.Math.Phys.* **132** No. 3; pp 1307-1317; hep-ph/0208082
2. Shirkov D.V., (2003) "On the Fourier transformation of Renormalization Invariant Coupling", JINR preprint E2-2003-01; to appear in *Theor.Math.Phys.* hep-th/0210013;.
3. Alkofer L. and von Smekal L., (2001) *Phys. Repts.* **353** 281; hep-ph/0007355.
4. Boucaud Ph. *et al.*, (2002) *Nucl. Phys. Proc. Suppl.* **B 106** 266-268; hep-ph/0110171.
5. Boucaud Ph. *et al.*, (2002) *JHEP* 0201 046; hep-ph/0107278.
6. Skullerud J.I., Kizilersu A., Williams A.G., (2002) *Nucl. Phys. Proc. Suppl.* **B 106** 841-843; hep-lat/0109027.
7. Skullerud J., Kizilersu A., Quark-gluon vertex from lattice QCD, *JHEP* 0209 013; hep-ph/0205318.
8. Bode A. *et al.*, (2001)*Phys. Lett.* **B 515** 49-56; hep-lat/0105003.
9. Heitger J., Simma H., Sommer R. and Wolff U., (2002) *Nucl. Phys. Proc. Suppl.* **B 106** 859-861; hep-lat/0110201.
10. Bogoluibov N.N. and Shirkov D.V., (1956) *Nuovo Cim.* **3** 845-863.
11. see chapter "Renormalization group" in monograph by N.N. Bogoliubov and D.V. Shirkov, Introduction to the Theory of Quantized Fields, Wiley & Intersc. N.Y., 1959 and 1980.
12. Shirkov D.V., (1992) *Nucl.Phys.* **B 371** 467.
13. Shirkov D.V., (1992) *Theor.Math.Phys.* **93** pp 466-472.
14. Dokshitzer Yu.L. and Shirkov D.V., (1995) *Zeit. Phys.* **C 67** 449.
15. Shirkov D.V. and Solovtsov I.L., (1997) *Phys. Rev. Lett.* **79** 1209-1212; hep-ph/9704333.
16. Shirkov D.V. and Solovtsov I.L., (1998) *Phys. Lett.* **B 442** 344; hep-ph/9711251.
17. Shirkov D.V., *Eur. Phys. J.* (2001) **22** 331-340 ; hep-ph/0107282.
18. See,*e.g.*, Chapter 3 in the monograph by V.S. Vladimirov, Yu.N. Drozzhinov and B.I. Zavialov, Multidimensional Tauberian Theorems for Generalized Functions, Kluwer, Dordrecht, 1988.

SINGLE-SPIN ASYMMETRY IN PION PRODUCTION IN POLARIZED PROTON-PROTON COLLISIONS AND ODDERON

E. N. ANTONOV
S-Peterburg Institute for Nuclear Physics
Gatchina, Russia

E. BARTOŠ
Joint Institute for Nuclear Research
141980 Dubna, Russia;
Comenius University
84248 Bratislava, Slovakia

AND

A. AHMEDOV, E. A. KURAEV AND E. ZEMLYANAYA
Joint Institute for Nuclear Research
141980 Dubna, Russia

Abstract. Single-spin asymmetry appears due to the interference of single and double gluon exchange between protons. A heavy fermion model is used to describe the jet production in the interaction of gluon with the proton implying the further averaging over its mass. As usually in one-spin correlations, the imaginary part of the double gluon exchange amplitude play the relevant role. The asymmetry in the inclusive set-up with the pion tagged in the fragmentation region of the polarized proton does not depend on the center of mass energy in the limits of its large values. The lowest order radiative corrections to the polarized and unpolarized contributions to the differential cross sections are calculated in the leading logarithmic approximation. In general, a coefficient at logarithm of the ratio of cms energy to the pion mass depends on transversal momentum of the pion. This ratio of the lowest order contribution to the asymmetry may be interpreted as the partial contribution to the odderon intercept. The ratio of the relevant contributions in the unpolarized case can be associated with the partial contribution to the pomeron intercept. The numerical results given for the model describe the jet as a heavy fermion decay fragments.

R. Fiore et al. (eds.), Diffraction 2002, 281–292.

1. Introduction

Let us consider the inclusive process of the pion creation in the fragmentation region of polarized proton at high energy proton-proton collisions

$$P_1(p_1, a) + P_2(p_2) \to \pi(p) + X_1(p_1') + X_2(p_2'), \tag{1}$$

where a is the transversal to beam (implied by cms) axes spin of initial proton

$$a = (0, 0, \mathbf{a}),\ p = (E\beta, E\beta, \mathbf{p}),\ E = \sqrt{s}/2,\ s = (p_1 + p_2)^2 \gg \mathbf{p}^2 \sim m^2,$$

$$p_1 = E(1, \beta_0, 0, 0),\ p_2 = E(1, -\beta_0, 0, 0),\ \beta_0 = \sqrt{1 - \frac{m^2}{E^2}}, \tag{2}$$

where $\beta \sim 1$ is the energy fraction of pion, $M_{1,2} = \sqrt{p_{1,2}'^2}$ are the invariant masses of the jets, which we will assume to be of the order of nucleon mass m. We study the two jet kinematics with jets X_1, X_2 moving along the initial hadron directions. The jet created by the transversely polarized proton is supposed to contain the detected pion. Moreover, we consider the case when its production is not related with the creation of nucleon resonances . In terms of pion transverse components it corresponds to the condition

$$\tilde{s}_1 = (p + p_1')^2 = \frac{1}{\beta\bar{\beta}}\left[\beta M_1^2 + (\mathbf{p} + \beta\mathbf{k_1})^2\right] > M_{res}^2 - m^2,$$

$$\mathbf{p} + \mathbf{k_1} + \mathbf{p_1'} = \mathbf{0}, \tag{3}$$

where $\mathbf{k_1}$ is the transfer momentum between protons. Through this paper we use Sudakov parameterization of 4–momenta of the problem

$$k_i = \alpha_i q_2 + \beta_i q_1 + k_{i\perp},\ k_{i\perp} = (0, 0, \mathbf{k_i}),\ q_{1,2} = E(1, \pm 1, 0, 0). \tag{4}$$

The azimuthal one-spin asymmetry arises from the interference of the amplitudes with one and two gluon exchanges between nucleons (see Fig. 1)

$$A = \frac{2E_\pi \dfrac{d^3\sigma(\mathbf{a}, \mathbf{p})}{d^3p} - 2E_\pi \dfrac{d^3\sigma(-\mathbf{a}, \mathbf{p})}{d^3p}}{2E_\pi \dfrac{d^3\sigma(\mathbf{a}, \mathbf{p})}{d^3p} + 2E_\pi \dfrac{d^3\sigma(-\mathbf{a}, \mathbf{p})}{d^3p}} = \alpha_s R f(\rho, z), \quad R = |\mathbf{a}| \sin\varphi, \tag{5}$$

$$f(\rho, z) = \frac{I_0^{(3)} T_0 + z I_1^{(3)} T_1}{J_0^{(2)} T_2 + z J_1^{(2)} T_3}, \quad \rho = \frac{|\mathbf{p}|}{m}, \quad z = \frac{\alpha_s}{\pi} \ln \frac{s}{m^2}$$

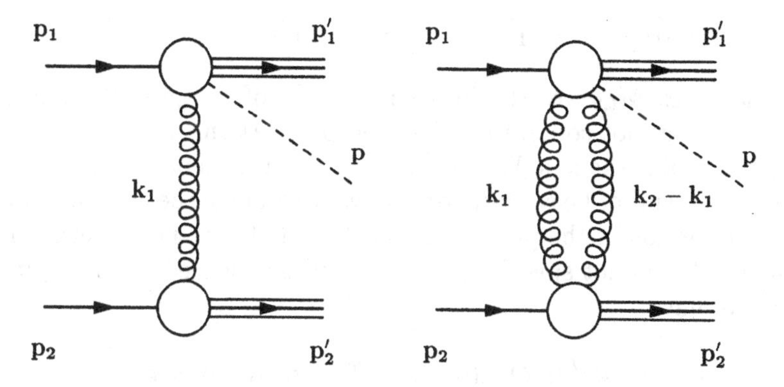

Figure 1. One and two gluon exchange in peripheral collision of protons with π production $P_1(p_1, a) + P_2(p_2) \to \pi(p) + X_1(p_1') + X_2(p_2')$.

where α_s is the strong coupling constant, φ is the azimuthal angle between the 2–vectors $\mathbf{a}$ and $\mathbf{p}$, transverse to the beam axis. The functions I, J as well as the color factors T_i will be specified below. For convenience we put here the alternative form for the phase volume of pion

$$\frac{d^3p}{2E_\pi} = m^2 \rho d\rho d\phi \frac{d\beta}{2\beta}. \tag{6}$$

The one gluon exchange matrix element has a form

$$M_0 = i4\pi\alpha_s \frac{J_\mu^{(1a)} J_\nu^{(2a)} g_{\mu\nu}}{k_1^2}, \tag{7}$$

the currents $J^{(1a)}$, $J^{(2a)}$ are associated with the jets created by particles 1, 2 and index a describes the color state of the jet. Using Gribov representation of the metric tensor

$$g_{\mu\nu} \approx \frac{2}{s} p_2^\mu p_1^\nu, \tag{8}$$

and the gauge condition for the currents

$$k_1^\mu J_\mu^{(1a)} = (\alpha_1 p_2 + k_{1\perp}) J^{(1a)} = 0, \quad k_1^\mu J_\mu^{(2a)} = (\beta_1 p_1 + k_{1\perp}) J^{(2a)} = 0,$$

we express M_0 in the form

$$M_0 = i8s\pi\alpha_s \sum_{a=1}^{N^2-1} \frac{\mathbf{J}^{(1a)} \cdot \mathbf{k}_1 \mathbf{J}^{(2a)} \cdot \mathbf{k}_1}{s_1 s_2 \mathbf{k}_1{}^2},$$

$$s_1 = -s\alpha_1 = \tilde{s}_1 + \mathbf{k_1}^2 - m^2, \quad s_2 = s\beta_1 = M_2^2 - m^2 + \mathbf{k_1}^2. \tag{9}$$

The quantities $M_{1,2}$ are the invariant masses of the jets. We imply that the jet X_1 does not contain the detected pion. At this point we need some model describing the jets. We use the heavy fermion model, i.e., we consider the jet as a result of heavy fermion decay. We assume the coupling constant of the interaction of the pion with a nucleon and with heavy fermion to be the same. We do not specify it as well as it is cancelled in the asymmetry (5). So we have

$$J_\mu^{(1a)} = \bar{u}(p_1')t^a O_\mu u(p_1), \quad J_\mu^{(2a)} = \bar{u}(p_2')t^a \gamma_\mu u(p_2), \tag{10}$$
$$O_\mu = \gamma_5 \frac{\hat{p}_1 - \hat{k}_1 + M_1}{d_1}\gamma_\mu + \gamma_\mu \frac{\hat{p}_1' + \hat{k}_1 + m}{d_2}\gamma_5,$$

t^a are the generators of color group $SU(N)$ and

$$d_1 = \frac{1}{\beta\bar{\beta}}[\beta^2 M_1^2 + (\mathbf{p} + \beta\mathbf{k_1})^2], \quad d_2 = -\frac{m^2}{\beta}[\rho^2 + \beta^2].$$

We use here the spin density matrices of the jets (i.e., heavy fermions)

$$\sum_\lambda u^\lambda(p_1')\bar{u}^\lambda(p_1') = \hat{p}_1' + M_1, \quad \sum_\lambda u^\lambda(p_2')\bar{u}^\lambda(p_2') = \hat{p}_2' + M_2, \tag{11}$$

It is important to note that we impose the gauge invariant form of matrix element and after that we use the heavy fermion model. This operations do not commute as well as the heavy fermion currents do not satisfy the current conservation condition. It is a specific of the considered model.

For the lowest order differential cross section we obtain

$$2E_\pi \frac{d^3\sigma(p)}{d^3p} = \frac{2\alpha_s^2}{\pi^2\bar{\beta}}T_2 \int \frac{d^2\mathbf{k_1}}{\pi} \frac{\Phi_{01}(\mathbf{k_1}, \mathbf{p})\Phi_{02}(\mathbf{k_1})}{(\mathbf{k_1}^2)^2} = \frac{2\alpha_s^2}{\pi^2\bar{\beta}}T_2 J_0^{(2)}, \tag{12}$$

with the explicit expressions for the impact factors Φ_{01}, Φ_{02} given in Appendix A and $T_2 = (N^2 - 1)/4$.

The lowest order spin–dependent contribution to the cross section arises from the interference of imaginary part of 1–loop radiative correction (RC) of Feynman diagram (FD) Fig 1b,c with the Born amplitude Fig 1a. We do not consider the RC from FD Fig. 1, believing that such kind FD contribute to the nucleon resonances formation. Besides it do not contribute in the leading logarithmic approximation (LLA)

$$z \sim 1, \quad \frac{\alpha_s}{\pi} \ll 1. \tag{13}$$

We obtain in the lowest order

$$2E_\pi \frac{d\sigma}{d^3p} = \frac{2\alpha_s^3}{\pi^2\beta} R T_0 I_0^{(3)}, \tag{14}$$

with

$$I_0^{(3)} = -\int \frac{d^2\mathbf{k_1}}{\pi}\frac{d^2\mathbf{k}}{\pi}\frac{\Phi_{11}\Phi_{22}}{\mathbf{k_1}^2\mathbf{k}^2(\mathbf{k_1}-\mathbf{k})^2}, \tag{15}$$

and the color factor

$$T_0 = |Tr(t^a t^b t^c)|^2 = \frac{1}{16}\left[f_{abc}^2 + d_{abc}^2\right] = \frac{(N^2-1)(N^2-2)}{8N}.$$

The impact factors $\Phi_{11,22}$ are given in Appendix B.

Impact factors $\Phi_{11,22}$ contain the new mass parameters $\tilde{M}_{1,2}$ which are intermediate jet state masses.

2. Ladder expansion

We had shown that the lowest order unpolarized and polarized cross sections can be expressed in terms of impact factors of projectiles moving in opposite directions which where introduced first in the papers of H. Cheng and T. T. Wu [1]. The calculation of RC to them can be done following the method developed by J. Balitski and L. N. Lipatov [2]. It was shown by these authors that in the LLA, the cross section has as well the form of conversion of impact factors of colliding particles with some universal kernel. Physically it corresponds to the replacement of exchanged gluons by the reggeized gluons. The reggeization states are taking into account in two factors. First the Regge factor $(s/m^2)^{a(t)}$ must be introduced, where $a(t)$ is the Regge trajectory of gluon with the momentum squared t. The second factor takes into account the contribution of inelastic processes of emission of real gluons. These both contributions suffer from the infrared divergences, however the total sum is free of them. For the RC to the unpolarized cross section we have

$$2E_\pi \frac{d\sigma}{d^3p} = \frac{2\alpha_s^3}{\pi^3\beta} T_3 \ln\frac{s}{m^2} J_1^{(2)}, \tag{16}$$

with color factor

$$T_3 = Tr(t^a t^b) Tr(t^{a'} t^{b'})(-f_{aa'd} f_{bdb'}) = \frac{N}{4}\left(N^2-1\right)$$

and

$$J_1^{(2)} = 2\int \frac{d^2\mathbf{k_1}}{\pi}\frac{d^2\mathbf{k'}}{\pi}\frac{\Phi_1(\mathbf{k_1},\mathbf{p})}{\mathbf{k_1}^2(\mathbf{k_1}-\mathbf{k'})^2}\left[\frac{\Phi_2(\mathbf{k'})}{\mathbf{k'}^2} - \frac{\Phi_2(\mathbf{k_1})}{\mathbf{k'}^2 + (\mathbf{k_1}-\mathbf{k'})^2}\right], \tag{17}$$

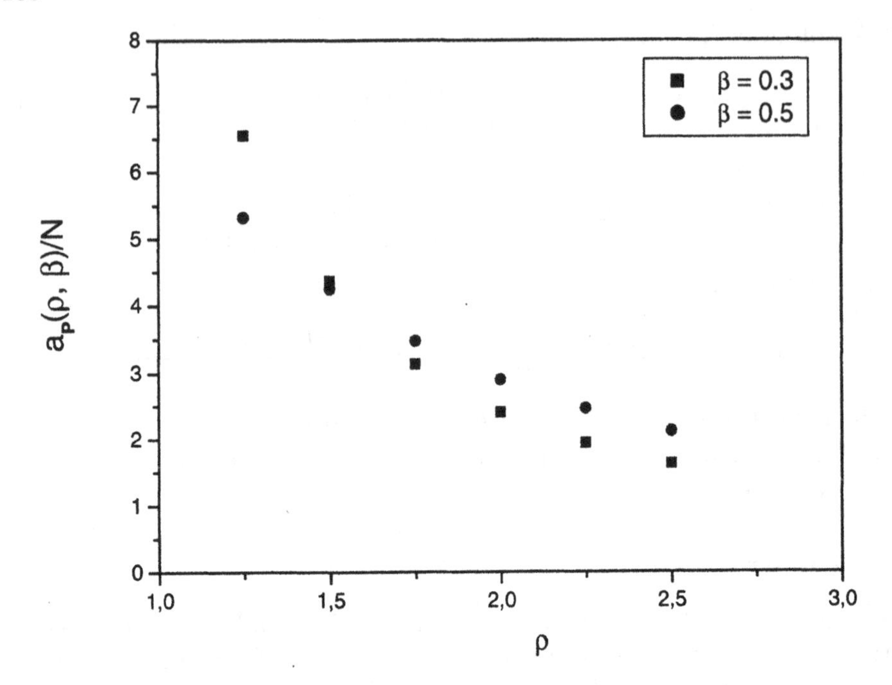

Figure 2. The ρ, β dependence of a partial contribution to the pomeron intercept.

with Φ_1 and Φ_2 given in Appendix C.

This formula can be inferred from the result for the non forward high energy scattering amplitude obtained in the paper [2]

$$I_1 = 3is\frac{\alpha_s}{2}\ln(s/m^2)\int\frac{d^2\mathbf{k}}{\pi}\frac{\Phi^{AA'}(\mathbf{k},\mathbf{q})}{\mathbf{k}^2(\mathbf{q}-\mathbf{k})^2}\Phi^{BB'}*K(\mathbf{k},\mathbf{k}',\mathbf{q}), \tag{18}$$

with definition

$$\Phi^{BB'}*K(\mathbf{k},\mathbf{k}',\mathbf{q}) =$$

$$\int\frac{d^2\mathbf{k}'}{\pi}\left\{\left[-\mathbf{q}^2+\frac{\mathbf{k}^2(\mathbf{q}-\mathbf{k}')^2+\mathbf{k}'^2(\mathbf{q}-\mathbf{k})^2}{(\mathbf{k}-\mathbf{k}')^2}\right]\frac{\Phi^{BB'}(\mathbf{k}',\mathbf{q})}{\mathbf{k}'^2(\mathbf{q}-\mathbf{k}')^2}\right. \tag{19}$$
$$\left.-\frac{\Phi^{BB'}(\mathbf{k},\mathbf{q})}{(\mathbf{k}-\mathbf{k}')^2}\left[\frac{\mathbf{k}^2}{\mathbf{k}'^2+(\mathbf{k}-\mathbf{k}')^2}+\frac{(\mathbf{q}-\mathbf{k})^2}{(\mathbf{q}-\mathbf{k}')^2+(\mathbf{k}-\mathbf{k}')^2}\right]\right\}.$$

The formula (17) can be obtained from the last general one by putting $\mathbf{q}=0$. Note that in the formula obtained in [2] the used color group was $SU(2)$.

Let now consider the LLA RC to the polarized part of the differential cross section. There are presented three types of contributions corresponding to three different choices of two gluons which are involved in the

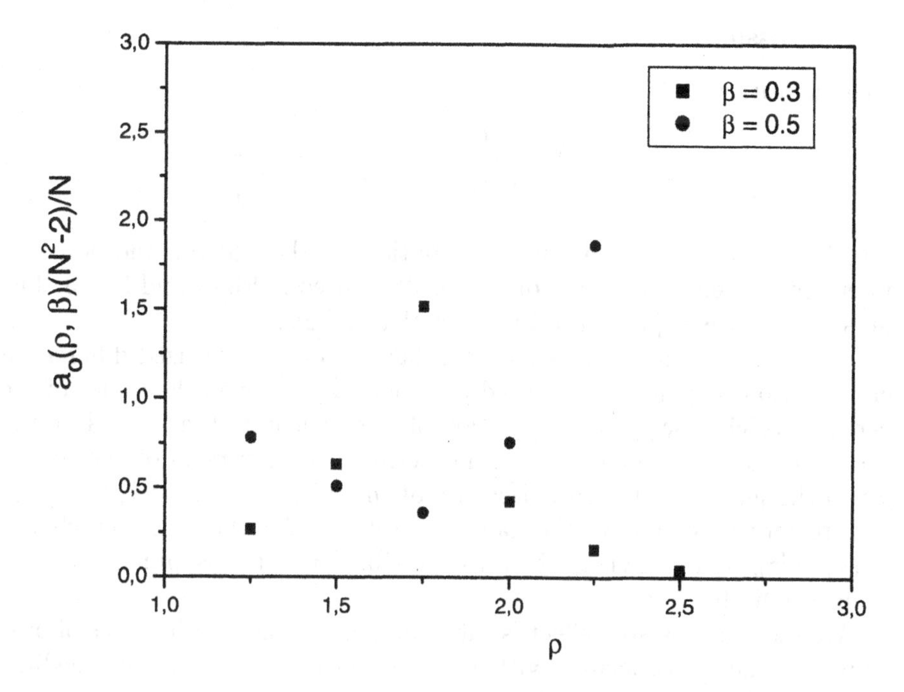

Figure 3. The ρ, β dependence of a partial contribution to the odderon intercept.

reggeization procedure in the lowest order RC

$$2E_\pi \frac{d\sigma}{d^3p} = \frac{\alpha_s^4}{\pi^3} R T_1 I_1^{(3)} \ln \frac{s}{m^2}, \quad I_1^{(3)} = \int \frac{d^2\mathbf{k}}{\pi} \frac{d^2\mathbf{k}'}{\pi} \frac{d^2\mathbf{k_1}}{\pi} \left[I_{12} + I_{13} + I_{23} \right], \tag{20}$$

with the color factor

$$T_1 = Tr t^a t^b t^c Tr t^{b'} t^{a'} t^c f_{aa'd} f_{bdb'} = \frac{(N^2-1)}{8}.$$

Symmetry reasons lead to the conclusion that I_{13} and I_{23} contributions are equal to I_{12} one. For I_{12} we have

$$I_{12} = \frac{\Phi_1^{(12)}(\mathbf{k}, \mathbf{k_1})}{\mathbf{k}^2(\mathbf{k_1}-\mathbf{k})^2} \left\{ \frac{\Phi_2^{(12)}(\mathbf{k}', \mathbf{k_1})}{\mathbf{k}'^2(\mathbf{k_1}-\mathbf{k}')^2} \left[-\mathbf{k_1}^2 + \frac{\mathbf{k}^2(\mathbf{k_1}-\mathbf{k}')^2 + \mathbf{k}'^2(\mathbf{k_1}-\mathbf{k})^2}{(\mathbf{k}-\mathbf{k}')^2} \right] \right.$$

$$\left. - \frac{\Phi_2^{(12)}(\mathbf{k}, \mathbf{k_1})}{(\mathbf{k}-\mathbf{k}')^2} \left[\frac{\mathbf{k}^2}{\mathbf{k}'^2 + (\mathbf{k}-\mathbf{k}')^2} + \frac{(\mathbf{k_1}-\mathbf{k})^2}{(\mathbf{k_1}-\mathbf{k}')^2 + (\mathbf{k}-\mathbf{k}')^2} \right] \right\}. \tag{21}$$

The details of impact factor calculations are given in the Appendices A–D. In the Appendix E we give some details used for performing the integration over the transversal component of the loop momenta.

3. Discussion

The quantities

$$a_{\mathbf{O}}(\rho,\beta)=\frac{N}{(N^2-2)}\frac{I_1^{(3)}}{I_0^{(3)}},\quad a_{\mathbb{P}}(\rho,\beta)=N\frac{J_1^{(2)}}{J_0^{(2)}}, \tag{22}$$

may be interpreted as a partial contributions to the odderon and pomeron intercepts. Their dependence on ρ, β is illustrated in Fig. 2 and Fig. 3. The jet's masses was supposed to be larger than 1 GeV.

The size of contributions to the polarized and unpolarized differential cross sections depends on the used jet model as well as on the choice of the vertices which describe the transition of the nucleon to the jet and on the choice of the vertex function which describes the conversion of one sort of jet to the jet of another sort. Because of the here used choice $V_\mu(k)=\gamma_\mu$, we are forced to put on the gauge conditions. Another possible choice, $V_\mu(k)=[\gamma_\mu,\hat{k}]/M$, leads to the zero contribution to the asymmetry (in the limit of infinite large s).

We see that one-spin effect is rather large. Another mechanisms of one-spin asymmetry associated with the nucleon resonances in intermediate state and final state interaction in $ep\to ep\pi$ process was considered in papers [4, 5] where the effect was of the same order.

Acknowledgments

The work was supported by INTAS-00366. E. Z. is grateful to RFBR for the grant 0001-00617.

Appendix A

The explicit expressions for the lowest order impact factors in the unpolarized case are

$$\Phi_{01}(\mathbf{k_1},\mathbf{p})=\frac{1}{4s^2}Sp(\hat{p}_1'+m)O_\mu(\hat{p}_1+m)\tilde{O}_\nu p_2^\mu p_2^\nu\ ,$$

$$\Phi_{02}(\mathbf{k_1})=\frac{1}{4s_2^2}Sp(\hat{p}_2'+M_2)\gamma_\mu(\hat{p}_2+m)\gamma_\nu k_{1\perp}^\mu k_{1\perp}^\nu. \tag{A.1}$$

To simplify the calculation of the traces we write down here the useful relations

$$p_1'=\frac{a_1}{s}q_2+\bar{\beta}q_1-p_\perp-k_{1\perp},\quad p_1'+k_1=p_1-\frac{\mathbf{p}^2}{s}q_2-\beta q_1-p_\perp\ ,$$

$$p_1-k_1=p_1+\frac{s_1}{s}q_2-k_{1\perp},\quad p_2'=p_2+\frac{s_2}{s}q_1+k_{1\perp}\ ,$$

$$a_1 = (M_1^2 + (\mathbf{p} + \mathbf{k_1})^2)/\bar{\beta}, \tag{A.2}$$

which follow from the on mass shell conditions. Explicit expressions for Φ_{01} and Φ_{02} are

$$\Phi_{01}(\mathbf{k_1}, \mathbf{p}) = \frac{\beta^4 \bar{\beta} \mathbf{k_1}^2}{2\left[\beta^2 m^2 + \mathbf{p}^2\right]\left[\beta^2 m^2 + (\mathbf{p} - \mathbf{k_1}\beta)^2\right]},$$

$$\Phi_{02}(\mathbf{k_1}) = \frac{2\mathbf{k_1}^2\left[\mathbf{k_1}^2 + (M_2 - m)^2\right]}{\left(\mathbf{q}^2 + M_2^2 - m^2\right)^2} \tag{A.3}$$

Appendix B

The lowest order contribution to the impact factor corresponding to the polarized proton with 4-momentum p_1 is

$$\Phi_{11} = (1/\bar{\beta})\Phi_1^b + \Phi_1^c. \tag{B.1}$$

The quantity Φ_1^b has a form

$$\Phi_1^b = \frac{1}{4s^3 R} Sp(\hat{p}_1 + m)(-\gamma_5 \hat{a})\tilde{O}_\mu p_2^\mu (\hat{p}_1' + m)\hat{p}_2 \tag{B.2}$$

$$\times (\hat{p}_1' + \hat{k}_1 - \hat{k} + m) O_{1\nu} p_2^\nu .$$

The Sudakov decomposition of the 4-vectors p_1' is given in (A.2). The exchanged gluon expansions are

$$k = \alpha q_2 + k_\perp, \quad s\alpha = \mathbf{k}^2 - d_3 , \tag{B.3}$$

$$k_1 = \alpha_1 q_2 + k_{1\perp}, \quad s\alpha_1 = \mathbf{k_1}^2 - d_1,$$

the quantity $O_{1\mu}$ is equal to

$$O_{1\mu} = (1/d_2)\gamma_\mu(\hat{p}_1' + \hat{k}_1 + m)\gamma_5 + (1/d_3)\gamma_5(\hat{p}_1 - \hat{p}_1 + m)\gamma_\mu,$$

with $d_{1,2}$ given in (11) and

$$d_3 = \frac{1}{\beta\bar{\beta}}\left[\beta^2 m^2 + (\mathbf{p} + \beta \mathbf{k})^2\right].$$

The quantity Φ_1^c has a form

$$\Phi_1^c = \frac{1}{4s^3 R} Sp(\hat{p}_1 + m)(-\gamma_5 \hat{a})\tilde{O}_\mu p_2^\mu (\hat{p}_1' + m) O_{2\nu} p_2^\nu (\hat{p}_1 - \hat{k} + m)\hat{p}_2 ,$$

$$O_{2\mu} = \gamma_\mu \frac{\hat{p}_1' + \hat{k}_1 - \hat{k} + m}{d_4} \gamma_5 + \gamma_5 \frac{\hat{p}_1 - \hat{k} + m}{d_1} \gamma_\mu, \tag{B.4}$$

with

$$d_4 = -\frac{1}{\beta\bar{\beta}} \left[\beta^2 m^2 + (\mathbf{p} + \beta\mathbf{k})^2\right]$$

and with the same expression for p_1'.

The relevant representation for the exchanged gluon 4–momenta is following

$$k = \alpha q_2 + k_\perp, \quad s\alpha = \mathbf{k}^2 \ ,$$

$$k_1 = \alpha_1 q_2 + k_{1\perp}, \quad s\alpha_1 = \mathbf{k_1}^2 - d_1 \tag{B.5}$$

and the same expression for p_1'.

Impact factor for the unpolarized proton Φ_2 has a form

$$\Phi_{22} = \frac{1}{s\beta s\beta_1 s(\beta - \beta_1)} \frac{1}{4} Sp(\hat{p}_2 + m)\hat{k}_{1\perp}$$

$$\times (\hat{p}_2 + \hat{k}_1 + M_2)(\hat{k}_{1\perp} - \hat{k}_\perp)(\hat{p}_2 + \hat{k} + \tilde{M}_2)\hat{k}_\perp. \tag{B.6}$$

For then calculation of the trace for Φ_2 we used Sudakov representation

$$k = \beta q_1 + k_\perp, \quad s\beta = \tilde{M}_2^2 - m^2 + \mathbf{k}^2,$$
$$k_1 = \beta_1 q_1 + k_{1\perp}, \quad s\beta_1 = M_2^2 - m^2 + \mathbf{k_1}^2 \ . \tag{B.7}$$

Appendix C

The impact factors for the case of unpolarized protons are

$$\Phi_1(\mathbf{k_1}, \mathbf{p}) = \Phi_{01}(\mathbf{k_1}, \mathbf{p}) \ , \tag{C.1}$$

$$\Phi_2(\mathbf{k}') = \frac{1}{4ss_2} Sp(\hat{p}_2 + \hat{k}_\perp' + \beta' q_1 + M_2)\hat{p}_1(\hat{p}_2 + m)\hat{k}_\perp' \ ,$$

where

$$s_2 = M_2^2 - m^2 + \mathbf{k}'^2 \ .$$

Appendix D

Let us now give the expressions for the impact factors in the case of RC to the polarized cross sections. For the $\Phi_1^{(12)}$ in (21) we have

$$\Phi_1^{(12)}(\mathbf{k}, \mathbf{k_1}) = \Phi_{1b}^{(12)} + \Phi_{1c}^{(12)}, \tag{D.1}$$

with

$$\Phi_{1b}^{(12)} = \frac{1}{4sRs_{11}s_{12}} Sp(\hat{p}_1 + m)(-\gamma_5 \hat{a})\tilde{O}_\mu p_2^\mu (\hat{p}_1' + m) \tag{D.2}$$

$$\times (\hat{k}_{1\perp} - \hat{k}_\perp)(\hat{p}_1' + \hat{k}_1 - \hat{k} + m) O_{1\nu} k_\perp^\nu \ ;$$

and

$$\Phi_{1c}^{(12)} = \frac{1}{4sRs_{21}s_{22}} Sp(\hat{p}_1 + m)(-\gamma_5 \hat{a})\tilde{O}_\mu p_2^\mu (\hat{p}_1' + m) \tag{D.3}$$

$$\times O_{2\mu}(k_{1\perp} - k_\perp)^\mu (\hat{p}_1 - \hat{k} + m)\hat{k}_\perp \ ,$$

with the substitutions similar to ones for Φ_1^b from Appendix B for the momenta including $\Phi_{1b}^{(12)}$ and the substitutions similar to ones for Φ_1^c for $\Phi_{1c}^{(12)}$. Besides

$$s_{11} = (\mathbf{k_1} + \mathbf{p})^2 - (\mathbf{k} + \mathbf{p})^2, \quad s_{12} = m^2 - \tilde{s}_1(m^2) - \mathbf{k}^2 \ , \tag{D.4}$$

$$s_{21} = -\mathbf{k}^2, \quad s_{22} = m^2 - \tilde{s}_1 + \mathbf{k}^2 - \mathbf{k_1}^2.$$

The impact factor of unpolarized proton in this case have a form

$$\Phi_2^{(12)}(\mathbf{k}', \mathbf{k_1}) = \frac{1}{4s_2 s_{23} s_{13}} Sp(\hat{p}_2 + m)\hat{k}_{1\perp}(\hat{p}_2 + \hat{k}_{1\perp} + \beta_1 \hat{q}_1 + M_2)$$

$$\times (\hat{k}_{1\perp} - \hat{k}'_\perp)(\hat{p}_2 + \hat{k}'_\perp + \beta' \hat{q}_1 + \tilde{M}_2)\hat{k}'_\perp, \tag{D.5}$$

with

$$s_{13} = \tilde{M}_2^2 - M_2^2 - \mathbf{k_1}^2 + \mathbf{k}'^2, \quad s_{23} = \tilde{M}_2^2 - m^2 + \mathbf{k}'^2, \tag{D.6}$$

and

$$s\beta_1 = M_2^2 - m^2 + \mathbf{k_1}^2, \quad s\beta' = \tilde{M}_2^2 - m^2 + \mathbf{k}'^2. \tag{D.7}$$

Appendix E

Here we give some details used in performing the loop momenta integration.

When calculating the relevant trace we use the Shouthen identity

$$(p_1 p_2 p_3 p_4) Q_\mu = (\mu p_2 p_3 p_4) Q p_1 + (p_1 \mu p_3 p_4) Q p_2 + (p_1 p_2 \mu p_4) Q p_3 + (p_1 p_2 p_3 \mu) Q p_4.$$

This identity permits to express all conversions with Levi-Chivita tensor in the standard form. For instance

$$(p_1 p_2 k_1 p_\perp)(a k_2) = (p_1 p_2 a p_\perp)(k_1 k_2) + (p_1 p_2 k_1 a)(p_\perp k_2).$$

The second term in the right side of this equation can be expressed through the

$$E = (p_1 p_2 a p) = \frac{s}{2} m \rho R, \tag{E.1}$$

using the relation

$$\int \frac{d^2\mathbf{k}}{\pi}\frac{d^2\mathbf{k_1}}{\pi}\frac{d^2\mathbf{k'}}{\pi}F(\mathbf{k},\mathbf{k_1},\mathbf{k'},\mathbf{p})\mathbf{k_i} = \frac{\mathbf{p_i}}{\mathbf{p}^2}\int \frac{d^2\mathbf{k}}{\pi}\frac{d^2\mathbf{k_1}}{\pi}\frac{d^2\mathbf{k'}}{\pi}F(\mathbf{k},\mathbf{k_1},\mathbf{k'},\mathbf{p})(\mathbf{p}\cdot\mathbf{k}). \tag{E.2}$$

References

1. Cheng, H. and Wu, T.T (1969),"High Energy colision processes in Quantum Electrodynamics. I",*Phys. Rev.*, **182**, p. 1852-1867; Frolov, G. and Lipatov, L. (1971),"Some processes in quantum electrodynamics at high energies", *Yad. Fiz.*, **13**, p. 588-599.
2. Balitsky, I. and Lipatov, L. (1978),"Pomeranchuk singularity in quantum chromodynamic.", *Yad. Fiz.*, **28**, p. 1596-1611.
3. Fadin, V.S., Lipatov, L.N. and Kuraev, E.A. "Multireggeon processes in the Yang-Mils theory",(1976), *ZhETP*, **71**, p.840-855; "Pomeranchuk singularity in non-abelian gauge theories",(1976) *ZhETP*, **72**, p. 377-390.
4. Arbuzov, A.B., Vasendina, V.A., Kuraev, E.A. and Nikitin, V.A.,"A possibility of mesuring proton transverse polarization in diffractive pion production in *pp* and *ee* collision at high energies.", (1994), *Phys. Atom. Nucl.*, **57**, p. 1065-1070.
5. Ahmedov, A., Akushevich, I.V., Kuraev, E.A. and Ratcliffe, P.G. ,"Single-spin asymmetries for small-angle pion production in high-energy hadron collisions",(1999), *Eur. Phys. J. C*, **11**, p. 703-708.

LEADING PARTICLES AND DIFFRACTIVE SPECTRA IN THE INTERACTING GLUON MODEL

F.O.DURÃES AND F.S.NAVARRA
Instituto de Fisica, Universidade de São Paulo
C.P.66318, 05389-970 São Paulo, SP, Brazil

AND

G.WILK
The Andrzej Soltan Institute for Nuclear Studies,
Nuclear Theory Deaprtment (Zd-PVIII)
ul. Hoża 69, 00-681 Warsaw, Poland

Abstract. We discuss the leading particle spectra and diffractive mass spectra from the novel point of view, namely by treating them as particular examples of the general energy flow phenomena taking place in the multiparticle production processes. We argue that they show a high degree of universality what allows for their simple description in terms of the Interacting Gluon Model developed by us some time ago.

1. Introduction

The multiparticle production processes are the most complicated phenomena as far as the number of finally involved degrees of freedom is concerned. They are also comprising the bulk of all inelastic collisions and therefore are very important - if not *per se* than as a possible background to some other, more specialized reactions measured at high energy collisions of different kinds. The high number of degrees of freedom calls inevitably for some kind of statistical descrition when addressing such processes. However, all corresponding models have to be supplemented by information on the fraction of the initial energy deposited in the initial object(s) (like fireball(s)) being then the subject of further investigations.

Some time ago we have developed a model describing such energy deposit (known sometimes as *inelasticity*) and connecting it with the apparent

R. Fiore et al. (eds.), Diffraction 2002, 293–303.

dominance of multiparticle production processes by the gluonic content of the impinging hadrons, hence its name: *Interacting Gluon Model* (IGM) (Fowler *et al.*, 1989). Its classical application to description of inelasticity (Duraes *et al.*, 1993) and multiparticle production processes in hydrodynamical model approach (Duraes *et al.*, 1994) was soon followed by more refined applications to the leading charm production (Duraes *et al.*, 1995) and to the (single) diffraction dissociation, both in hadronic reactions (Duraes *et al.*, 1997a) and in reactions originated by photons (Duraes *et al.*, 1997b). These works allowed for providing the systematic description of the leading particle spectra (which turned out to be very sensitive to the presence of diffractive component in the calculations, not accounted for before) (Duraes *et al.*, 1998a) and clearly demonstrated that they are very sensitive to the amount of gluonic component in the diffracted hadron (Duraes *et al.*, 1998b) and (Duraes *et al.*, 1998a). We have found it amusing that all the results above were obtained using the same set of basic parameters with differences arising essentially only because of different kinematical limits present in each particular application (i.e., in different allowed phase space). All this points towards the kind of *universality* of energy flow patterns in all the above mentioned reactions.

Two recent developments prompted us to return again to the IGM ideas of energy flow: one was connected with the new, more refined data on the leading proton spectra in $ep \rightarrow e'pX$ obtained recently by ZEUS collaboration [1] (which are apparently different from what has been used by us before in (Duraes *et al.*, 1998a), (Duraes *et al.*, 1998b) and (Duraes *et al.*, 1998a)). The other was recent work on the central mass production in Double Pomeron Exchange (DPE) process reported in (Brandt *et al.*, 2002) allowing in principle for deduction of the Pomeron-Pomeron total cross section σ_{IP-IP}. In what follows we shall therefore provide a brief description of IGM, stressing the universality of energy flow it provides and illustrating it by some selected examples from our previous works. The new results of ZEUS will be then shown again and commented. Finally, we shall present our recent application of the IGM to the DPE processes as well (Duraes *et al.*, 2002).

2. IGM and some of its earlier applications

The main idea of the model is that nucleon-nucleon collisions (or any hadronic collisions in general) at high energies can be treated as an incoherent sum of multiple gluon-gluon collisions, the valence quarks playing a secondary role in particle production. While this idea is well accepted for

[1] Private information from A.Garfagnini, see also ZEUS Collab. presentation in these proceedings.

large momentum transfer between the colliding partons, being on the basis of some models of minijet and jet production (for example HIJING (Wang *et al.*, 92)), in the IGM its validity is extended down to low momentum transfers, only slightly larger than Λ_{QCD}. At first sight this is not justified because at lower scales there are no independent gluons, but rather a highly correlated configuration of color fields. There are, however, some indications coming from lattice QCD calculations, that these soft gluon modes are not so strongly correlated. One of them is the result obtained in (Giacomo *et al.*, 1992), namely that the typical correlation length of the soft gluon fields is close to 0.3 fm. Since this length is still much smaller than the typical hadron size, the gluon fields can, in a first approximation, be treated as uncorrelated. Another independent result concerns the determination of the typical instanton size in the QCD vaccum, which turns out to be of the order of 0.3 fm (Shaefer *et al.*, 1998). As it is well known (and has been recently applied to high energy nucleon-nucleon and nucleus-nucleus collisions) instantons are very important as mediators of soft gluon interactions (Shaefer, 1998). The small size of the active instantons lead to short distance interactions between soft gluons, which can be treated as independent.

These two results taken together suggest that a collision between the two gluon clouds (surrounding the valence quarks) may be viewed as a sum of independent binary gluon-gluon collisions, which is the basic idea of our model. The interaction follows then the usual IGM picture (Fowler *et al.*, 1989) and (Duraes *et al.*, 1998a), namely: the valence quarks fly through essentially undisturbed whereas the gluonic clouds of both projectiles interact strongly with each other (by gluonic clouds we understand a sort of "effective gluons" which include also their fluctuations seen as $\bar{q}q$ sea pairs) forming a kind of central fireball (CF) of mass M. The two impinging projectiles (usually protons/antiprotons and mesons) loose fractions x and y of their original momenta and get excited forming what we call *leading jets* (LJ's) carrying $x_p = 1 - x$ and $x_{\bar{p}} = 1 - y$ fractions of the initial momenta. Depending on the type of the process under consideration one encounters different situation depicted in Fig. 1. In non-diffractive (ND) processes one is mainly interested only in CF of mass M, in single diffractive (SD) ones in masses M_X or M_Y (comprising also the mass of CF) whereas in double Pomeron exchanges (DPE) in a special kind of CF of mass M_{XY}. The only difference between ND and SD or DPE processes is that in the later ones the energy deposition is done by a restricted bunch of gluons which in our language are forming what is regarded as a kind of "kinematical" Pomeron ($I\!P$), the name which we shall use in what follows.

The central quantity in IGM is then the probability to form a CF carrying momentum fractions x and y of two colliding hadrons (Fowler *et al.*,

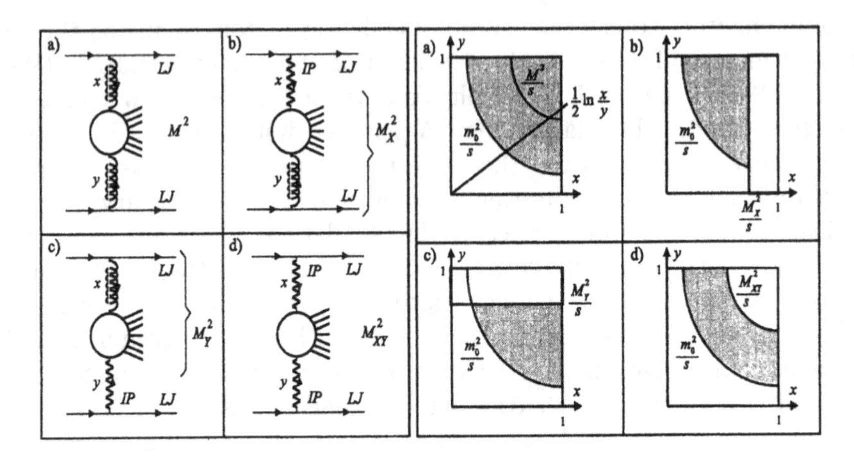

Figure 1. Left panel: schematic IGM pictures for (*a*) non-diffractive (ND), (*b*) and (*c*) single diffractive (SD) and (*d*) double Pomeron exchange (DPE) processes. Their corresponding phase space limits are displayed on the right panel. The $\frac{1}{2}\ln\frac{x}{y}$ line in the right (*a*) panel indicates the rapidity Y of the produced mass M.

1989) and (Duraes *et al.*, 1998a) which is given by:

$$\chi(x,y) = \frac{\chi_0}{2\pi\sqrt{D_{xy}}}\cdot\exp\left\{-\frac{1}{2D_{xy}}\left[\langle y^2\rangle(x-\langle x\rangle)^2 + \langle x^2\rangle(y-\langle y\rangle)^2\right]\right\}$$
$$\times\quad \exp\left\{-\frac{1}{D_{xy}}\langle xy\rangle(x-\langle x\rangle)(y-\langle y\rangle)\right\}, \qquad (1)$$

where $D_{xy} = \langle x^2\rangle\langle y^2\rangle - \langle xy\rangle^2$ and

$$\langle x^n\, y^m\rangle = \int_0^{x_{max}} dx'\, x'^n \int_0^{y_{max}} dy'\, y'^m\, \omega(x',y'), \qquad (2)$$

with χ_0 defined by the normalization condition, $\int_0^1 dx \int_0^1 dy\, \chi(x,y)\theta(xy - K^2_{min}) = 1$, where $K_{min} = \frac{m_0}{\sqrt{s}}$ is the minimal inelasticity defined by the mass m_0 of the lightest possible CF. The spectral function, $\omega(x',y')$, contains all the dynamical input of the IGM. Their soft and semihard components are given by (cf. (Duraes *et al.*, 1993)):

$$\omega(x',y') = \omega^{(S)}(x',y') + \omega^{(H)}(x',y') \qquad (3)$$

with

$$\omega^{(i)}(x',y') = \frac{\hat{\sigma}^{(i)}_{gg}(x'y's)}{\sigma(s)}\, G(x')\, G(y')\, \theta\left(x'y' - \left[K^{(i)}\right]^2_{min}\right), \qquad (4)$$

where $i = S, H$, and $K_{min}^{(S)} = K_{min}$ whereas $K_{min}^{(H)} = 2p_{T_{min}}/\sqrt{s}$. The values of x_{max} and y_{max} depend on the type of the process under consideration (cf. Fig. 2). For ND processes $x_{max} = y_{max} = 1$ (all phase space above the minimal one is allowed) whereas for SD and DPE there are limitations seen in Fig. 2 and discussed in more detail in the appropriate sections below. Here G's denote the effective number of gluons from the corresponding projectiles (approximated in all our works by the respective gluonic structure functions), $\hat{\sigma}_{gg}^{S}$ and $\hat{\sigma}_{gg}^{H}$ are the soft and semihard gluonic cross sections, $p_{T_{min}}$ is the minimum transverse momentum for minijet production and σ denotes the impinging projectiles cross section.

Lets us close this section by mentioning that, as has been shown in (Fowler *et al.*, 1989), (Duraes *et al.*, 1994) and (Duraes *et al.*, 1995), IGM can describe both the hadronic and nuclear collision data (providing initial conditions for the Landau Hydrodynamical Model of hadronization (Duraes *et al.*, 1994)) as well as some peculiar, apparently unexpected features in the leading charm production (Duraes *et al.*, 1995) (mainly via its strict energy-conservation introducing strong correlations between production from the CF and LJ). It was done with the same form of the gluonic structure function in the nucleon used: $G(x) = p(m+1)(1-x)^m/x$ with defold value of $m = 5$ and with the fraction of the energy-momentum allocated to gluons equal to $p = 0.5$ and with $\sigma_{gg}^{(i)}(xys) = const$ (notice that results are sensitive only to the combination of $p^2\sigma_{gg}/\sigma$).

3. Single Diffractive processes in the IGM

In the last years, diffractice scattering processes received increasing attention mainly because of their potential ability to provide information about the most important object in the Regge theory, namely the Pomeron (*IP*), its quark-gluonic structure and cross sections. Not entering the whole discussion (Goulianos, 1983) [2] we would like to show here the possible approach, based on the IGM, towards the mass(M_X) distributions provided by different experiments. In Figs. 1b and 1c the understanding of what such mass means in terms if the IGM is clearly shown: it contains both the central fireball and the LJ from the initial particle which got excited. The only difference between it and, say, the corresponding object which could be formed also in Fig. 1a is that the energy transfer from the diffracted projectile is now done by the highly correlated bunch of gluons dennoted *IP* which are supposed to be in the colour singlet state. The other feature also seen in Fig. 1 (right panel) is that now only the limited part of

[2]See also talk by K.Goulianos in these proceedings.

the phase space supporting the $\chi(x, y)$ distribution is allowed and that the limits depend on the mass M_X we are going to produce (and observe).

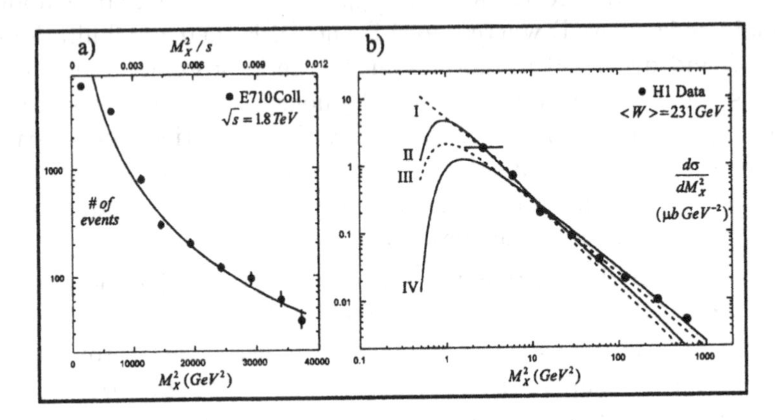

Figure 2. Examples of diffractive mass spectra for (a) $p\bar{p}$ collisions compared with Tevatron data (Abe *et al.*, 1994) (Fig. 4 from (Duraes *et al.*, 1997a)) and for (b) γp collisions compared with H1 data (Adloff *et al.*, 1997) (Fig. 2b from (Duraes *et al.*, 1997b), Vector Dominance Model has been used here with $G^{\rho}(x) = G^{\pi}(x) = 2p(1-x)/x$; the different curves correspond to different choices of (m_0 GeV, σ mb): I= $(0.31, 2.7)$, II= $(0.35, 2.7)$, III= $(0.31, 5.4)$ and IV= $(0.35, 5.4)$).

In technical terms it means that in comparison to the previous applications of the IGM we are free to change both the possible shape of the function $G_{IP}(x)$ (telling us the number of gluons participating in the process) and the cross section σ in the spectral function ω in eq. (4) above. Actually we have found that we can keep the shape of $G(x)$ the same as before and the only change necessary (and sufficient) to reach agreement with data is the amount of energy-momentum $p = p_{IP}$ allocated to the impinging hadron and which will find its way in the object we call *IP*. It turns out that $p_{IP} \simeq 0.05$ (to be compared with $p \simeq 0.5$ for all gluons encountered so far). In Fig. 3 we provide a sample of results taken from (Duraes *et al.*, 1997a) and (Duraes *et al.*, 1997b). They all have been obtained by putting $x_{max} = 1$ and $y_{max} = M_X^2/s$ in eq. (2) above and by writing

$$\frac{dN}{dM_X^2} = \int_0^1 dx \int_0^1 dy\, \chi(x, y) \delta\left(M_X^2 - sy\right) \Theta\left(xy - K_{min}^2\right). \tag{5}$$

As can be explicitely shown the characteristic $1/M_X^2$ behaviour of diffractive mass spectra are due to the $G(x) \sim 1/x$ behaviour of the gluonic structure functions for small x. The full formula results in small deviations following precisely the trend provided by experimental data (and usually attributed in the Regge model approach to the presence of additional Reggeons (Goulianos, 1983)).

4. Leading Particle spectra in the IGM

With the above development of the IGM one can now think about the systematic survey of the leading particle spectra, both in hadronic and in γp collisions (Duraes *et al.*, 1998a) (cf. Figs. 4 and 5). The specific prediction of the IGM connected with the amount of gluons in the hadron available for interactions (i.e., for the slowing down of the original quark content of the projectile) has been discussed in (Duraes *et al.*, 1997a), (Duraes *et al.*, 1997b) and (Duraes *et al.*, 1998b), (Duraes *et al.*, 1998a), and shown here in Fig. 6. The leading particle can emerge from different regions of the phase space (cf. Figs. 1) and distribution of its momentum fraction x_L is given by (Duraes *et al.*, 1997a):

$$\begin{aligned} F(x_L) = \quad & (1-\alpha)\int_{x_{min}}^{1} dx\, \chi^{(nd)}\,(x; y = 1 - x_L) + \\ & + \sum_{j=1,2} \alpha_j \int_{x_{min}}^{1} dx\, \chi^{(d)}\,(x; y = 1 - x_L)\,, \end{aligned} \tag{6}$$

where $\alpha = \alpha_1 + \alpha_2$ is the total fraction of single diffractive (d) events (from the upper and lower legs in Fig. 1, respectively, both double DD and DPE events are neglected here) and where

$$x_{min} = \mathrm{Max}\left[\frac{m_0^2}{(1-x_L)s}; \frac{(M_{LP} + m_0))^2}{s}\right] \tag{7}$$

with M_{LP} being the mass of the LP under consideration. Notice that the α is essentially a new parameter here, which should be of the order of the ratio between the total diffractive and total inelastic cross sections (Duraes *et al.*, 1997a). All other parameters leading to results in Fig. 4 are the same as established before [3].

We want to stress here the fact that the fair agreement with data observed in the examples shown in Figs. 4 and 5 is possible only because the diffraction processes have been properly incorporated in calculating the LP spectra (Duraes *et al.*, 1997a). As far as the energy flow is concerned the IGM works extremely well (including the pionic LP not shown here but discussed in (Duraes *et al.*, 1997a)) with essentially two parameters only: the nonperturbative gluon-gluon cross section and the fraction of diffractive events. At the same time, assuming the Vector Dominance Model and

[3]In what concerns comparison with ZEUS data see also presentation of ZEUS Collab. in these proceedings. Our present results differ from Fig. 4 in (Duraes *et al.*, 1997a) where the preliminary ZEUS data were used instead. The only difference between the two fits is that whereas in (Duraes *et al.*, 1997a) we were assuming that 30% of the LP comes from diffraction, now it is only 10%.

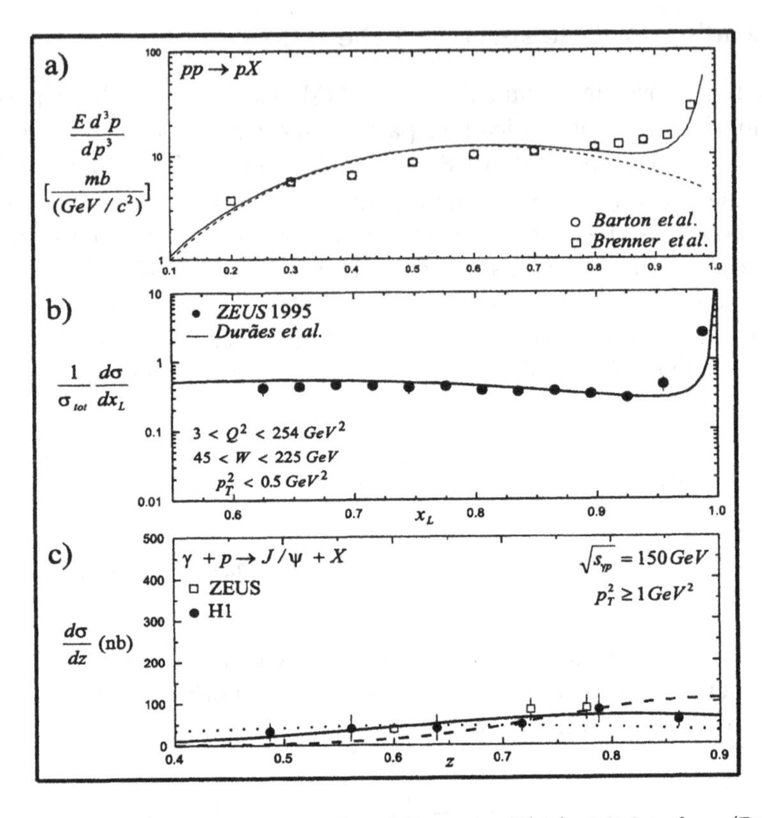

Figure 3. (*a*) Example of comparison of our LP spectra $F(x_L)$ with data from (Barton *et al.*, 1983) and (Brenner *et al.*, 2002) (Fig. 2a from (Duraes *et al.*, 1997a)). (*b*) Comparison between our calculation and and the new data on the leading proton spectrum measured at HERA by the ZEUS Collab. (*c*) Fits to leading J/Ψ spectra as given by ZEUS and H1 groups (Aid *et al.*, 1996), (Derrick *et al.*, 1995) and (Breitweg *et al.*, 1997) (cf. (Duraes *et al.*, 1998b) and (Duraes *et al.*, 1998a); the fixed value of $\sigma^{inel}_{J/\Psi-p} = 9$ mb has been used and results for three different choices of $p^{J/\Psi}$ are displayed: 0.066 - dashed line, 0.033 - solid line and 0.016 -dotted line).

replacing impinging photon by its hadronic component (as in Fig. 3b), we are able to describe also the leading proton spectra observed in $e-p$ reactions. Also here the inclusion of diffractive component turns out to be crucial factor to get agreement with data. We can also describe fairly well pionic LP (not shown here, cf. (Duraes *et al.*, 1997a)) and the observed differences turns out to be due to their different gluonic distributions. The crucial role played by the parameter p representing the energy-momentum fraction of a given hadron allocated to gluons is best seen in the Fig. 3c example showing fit to data for leading J/Ψ photoproduction (Aid *et al.*, 1996), (Derrick *et al.*, 1995) and (Breitweg *et al.*, 1997). It turns out that (after accounting for the proper kinematics in (7) and presence of diffraction processes as discussed above) the only parameter to which results are really sensitive is $p = p^{J/\Psi}$ which, as shown in Fig. 3c, has to be astonishingly small, $p^{J/\Psi} = 0.033$. However, closer scrutiny shows us that this is exactly what could be expected from the fact that charmonium is a non-relativistic system and almost all its mass comes from the quark masses leaving therefore only small fraction,

$$p^{J/\Psi} = \frac{M_{J/\Psi} - 2m_c}{M_{J/\Psi}} \simeq 0.033, \tag{8}$$

for gluons (here $m_c = 1.5$ GeV and $M_{J/\Psi} = 3.1$ GeV).

5. Double Pomeron Exchange in the IGM

Our latest application of the IGM discussed here will be for the DPE processes seen as a specific energy flow (cf. Fig. 1 d) taking place from both colliding particles and directed into the central region. The difference between it and the "normal" energy flow as represented by Fig. 1a is that now the gluons involved in this process must be confined to what is usu-

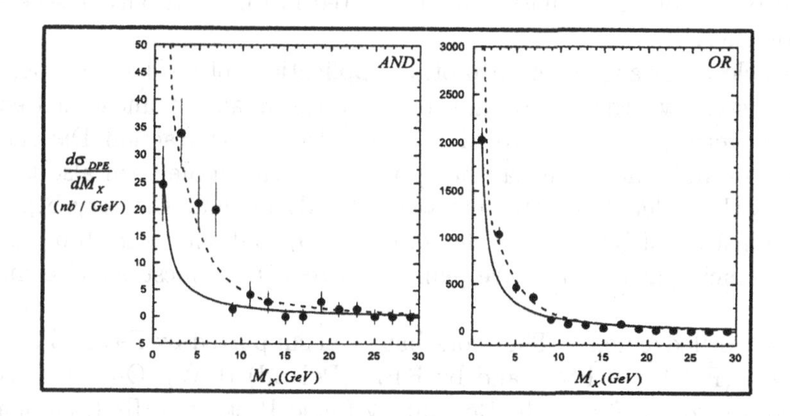

Figure 4. Our fits to two types of DPE diffractive mass distribution given by (Brandt *et al.*, 2002) with $\sigma_{IP-IP} = 1$ mb (solid lines) and 0.5 mb (dashed lines).

ally refered to as Pomeron (*IP*). Such process war recently measured by UA8 (Brandt *et al.*, 2002) and used (using the normal Reggeon calculus arguments) to deduce the *IP* – *IP* cross section, $\sigma_{IP\text{–}IP}$. It turned out that using this method one gets $\sigma_{IP\text{–}IP}$ which apparently depends on the produced massM_{XY}. This fact was tentatively interpreted as signal of glueball formation (Brandt *et al.*, 2002). However, when seen from the IGM point of view mentioned above, where

$$\frac{1}{\sigma}\frac{d\sigma}{dM_{XY}} = \frac{2M_{XY}}{s}\int_{\frac{M_{XY}^2}{s}}^{1}\frac{dx}{x}\chi\left(x, y = \frac{M_{XY}^2}{xs}\right)\Theta\left(M_{XY}^2 - m_0^2\right), \quad (9)$$

the (Brandt *et al.*, 2002) data can be fitted (see Fig. 4) with the same set of parameters as used previously to describe the SD processes (Duraes *et al.*, 1997a) and (Duraes *et al.*, 1997b) and with constant value of $\sigma_{IP-IP} = 0.5$ mb (which is new parameter here). No glueball concept is needed here.

6. Summary and conclusions

The picture which is emerging from the above discussion is that the energy flows, which are present in all multiparticle production reactions, are apparently a kind of universal phenomenon in the following sense: they are all sensitive mainly to the gluonic content of the colliding projectiles (i.e., both to the number of gluons as given by the form of the function $G(x)$ and to the amount of energy-momentum of the hadron, p, they carry and to the gluonic cross section which defines the actual effectiveness of the gluonic component in the energy transfer phenomenon). Their sensitivity to other aspects of the production process (except of kinematical limits provided by the observed masses, as illustrated in Fig. 1) is only of secondary importance.

To close our arguments two other applications of IGM should be, however, tested: whether it can also describe the final, yet unchecked energy flow pattern as the one provided by the Double Diffraction Dissociation processes and whether it can be applied in such simple form also to reactions with nuclei (first attempts were already done at the very beginning of the history of IGM in (Fowler *et al.*, 1989), but they were too crude to be convincing at present). We plan to address these questions elsewhere.

Acknowledgements: This work has been supported by FAPESP, CNPQ (Brazil) (FOD and FSN) and by KBN (Poland) (GW). One of us (GW) would like to thank also the Bogolubow-Infeld Program (JINR, Dubna) for financial help in attending the Diffraction 2002 conference where the above material has been presented.

References

Abe, F. et al. (CDF Collab.) (1994) Measurement of $p\bar{p}$ single difraction dissociation at $\sqrt{s} = 546$ and 1800 GeV, *Physical Review D*, **Vol. no. 50**, pp. 5535-5590

Adloff, C. et al. (1997) Difraction dissociation in photoproduction at HERA, *Zeitschrift für Physik C*, **Vo. no. 74**, 221-235

Aid, S. et al. (H1 Collab.) (1996) Elastic and inelastic photoproduction of J/Ψ mesons at HERA, *Nuclear Physics B*, **Vo. no. 472**, 3-31

Barton, D.S. et al. (1983) Experimental studies of the A dependence of inclusive hadron fragmentation, *Physical Review D*, **Vol. no. 27**, 2580-2599

Brandt A., *et al.* (2002) A Study of Inclusive Double-Pomeron-Exchange in $p\bar{p} \rightarrow pX\bar{p}$ at $\sqrt{s} = 630\,GeV$, *The European Pysical Journal C*, **Vol. no. 25**, pp. 361-377

Breitweg, J. et al. (ZEUS Collab.) (1997) Measurements of inelastic J/Ψ photoproduction at HERA, *Zeitschrift für Physik C*, **Vo. no. 76**, 599-612

Brenner, A.E. et al. (1982) Experimental study of single-particle inclusive hadron scattering and associated multiplicities, *Physical Revie D*, **Vol. no. 26**, 1497-1553

Derrick, M. et al. (ZEUS Collab.) (1995) Measurement of the cross section for the reaction $\gamma p \rightarrow J/\Psi p$ with the ZEUS Detector at HERA, *Physics Letters B*, **Vo. no. 350**, 120-134

Durães, F.O., Navarra, F.S. and Wilk, (391 Minijets and the behavior of inelasticity at high energies, *Physical Review D*, **Vol. no. 47**, pp. 3049-3052

Durães, F.O., Navarra, F.S. and Wilk, G. (1994) Hadronization and inelasticities, *Physical Review D*, **Vol. no. 50**, pp. 6804-6810

Durães, F.O., Navarra, F.S. and Wilk, G. (1995) Hadronization and inelasticities, *Physical Review D*, **Vol. no. 53**, pp. 6136-6143

Durães, F.O., Navarra, F.S. and Wilk, G. (1997) Diffractive dissociation in the interacting gluon model, *Physical Review D*, **Vol. no. 55**, pp. 2708-2717

Durães, F.O., Navarra, F.S. and Wilk, G. (1997) Diffractive mass spectra at DESY HERA in the interacting gluon model, *Physical Review D*, **Vol. no. 56**, pp. R2499-R2503

Durães, F.O., Navarra, F.S. and Wilk, G. (1998) Systematics of leading particle production, *Physical Review D*, **Vol. no. 58**, pp. 094034-1 - 094034-12

Durães, F.O., Navarra, F.S. and Wilk, G. (1998) J/Ψ elasticity distribution in the vector dominance approach, *Modern Physics Letters A*, **Vol. no. 13**, pp. 2873-2885

Durães, F.O., Navarra, F.S. and Wilk, G. (1998) Leading particle effect in the J/Ψ elasticity distribution, *Nrazilian Journal of Physics*, **Vol. no. 28**, pp. 505-509

Durães, F.O., Navarra, F.S., G. Wilk, G. (2002) Extracting the Pomeron-Pomeron Cross Section from Diffractive Mass Spectra, ArXiv:hep-ph/0209140

Fowler, G.N., Navarra, F.S., Plümer, M., Vourdas, A., Weiner, R.M. and Wilk, G. (1989) Interacting gluon model for hadron-nucleus and nucleus-nucleus collisions in the central rapidity region, *Physical Review C*, **Vol. no. 40**, pp. 1219-1233

di Giacomo, A. and Panagopoulos, H. (1992) Field strength correlations in the QCD vacum, *Physics Letters B*, **Vol. no. 285**, pp. 133136

Goulianos, K. (1983) Diffractive interactions of hadrons at high energies, *Physics Reports*, **Vol. no. 101**, pp. 169-219

Schaefer, T. and Shuryak, E. (1998) Instantons in QCD, *Review of Modern Physics*, **Vo. no. 70**, pp. 323-425

Shuryak, E.V. (2000)Towards the non-perturbative description of high energy processes, *Physics Letters B*, **Vol. no. 486**, pp. 378-384

Wang, X.N. and Gyulassy, M. (1992) Systematic study of particle production in $p + (\bar{p})$ collisions via the HIJING model, *Physical Review D* **Vol. no. 45**, pp. 844-856

THE K_T FACTORIZATION PHENOMENOLOGY OF INELASTIC J/Ψ PRODUCTION AT HERA

A.V. LIPATOV
Department of Physics, Moscow State University
119992 Moscow, Russia

AND

N.P. ZOTOV
SINP, Moscow State University
119992 Moscow, Russia

In the framework of k_T-factorization QCD approach and the colour singlet model we consider J/ψ deep inelastic leptoproduction processes at HERA. We investigate the dependences of the single differential cross section on different forms of the unintegrated gluon distribution. The z and $\mathbf{p}_T$ dependences of the spin aligment parameter α are presented also. Our theoretical predictions agree well with recent data taken by the H1 and ZEUS collaborations at HERA. It is shown that experimental study of the polarization J/ψ mesons at low $Q^2 < 1\,\mathrm{GeV}^2$ is an additional test of BFKL gluon dynamics.

1. Introduction

Recently the new experimental data on the deep inelastic J/ψ leptoproduction at HERA were obtained by the H1 collaboration [1]. For the analysis of these data here we will use the so called k_T-factorization approach [2–5].

The k_T-factorization QCD approach is based on Balitsky, Fadin, Kuraev, Lipatov (BFKL) [6] evolution equations. The resummation of the terms $\alpha_S^n \ln^n(\mu^2/\Lambda_{\mathrm{QCD}}^2)$, $\alpha_S^n \ln^n(\mu^2/\Lambda_{\mathrm{QCD}}^2) \ln^n(1/x)$ and $\alpha_S^n \ln^n(1/x)$ in the k_T-factorization approach leads to the unintegrated (dependent from $\mathbf{q}_T$) gluon distribution $\Phi(x, \mathbf{q}_T^2, \mu^2)$ which determine the probability to find a gluon carrying the longitudinal momentum fraction x and transverse momentum $\mathbf{q}_T$ at probing scale μ^2.

To calculate the cross section of a physical process the unintegrated gluon distributions have to be convoluted with off mass shell matrix ele-

R. Fiore et al. (eds.), Diffraction 2002, 305–315.

ments corresponding to the relevant partonic subprocesses [2–5]. In the off mass shell matrix element the virtual gluon polarization tensor is taken in the BFKL form [2–5]:

$$L^{\mu\nu}(q) = \frac{q_T^\mu q_T^\nu}{\mathbf{q}_T^2}. \tag{1}$$

Nowadays, the significance of the k_T-factorization QCD approach becomes more and more commonly recognized [7]. It was already used for the description of a wide class heavy quark and quarkonium production processes [8–23]. It is notable that calculations in k_T-factorization approach provide results which are absent in other approaches, such as fast growth of total cross sections in comparison with the standard parton model (SPM), a broadening of the p_T spectra due to extra the transverse momentum of the colliding partons and other polarization properties of final particles in comparison with the SPM.

We point out that heavy quark and quarkonium cross section calculations within the SPM in the fixed order of pQCD have some problems. For example, the very large discrepancy (by more than an order of magnitude) [24, 25] between the pQCD predictions for hadroproduction J/ψ and Υ mesons and experimental data at Tevatron was found. This fact has resulted in intensive theoretical investigations of such processes. In particular, it was required to use additional transition mechanism from $c\bar{c}$-pair to the J/ψ mesons, so-called the colour octet (CO) model [26], where $c\bar{c}$-pair is produced in the color octet state and transforms into final colour singlet (CS) state by help soft gluon radiation. The CO model was supposed to be applicable to heavy quarkonium hadro- and leptoproduction processes. However, the contributions from the CO mechanism to the J/ψ meson photoproduction contradict the experimental data at HERA for z distribution [27–30].

Another difficulty of the CO model are the J/ψ polarization properties in $p\bar{p}$-interactions at the Tevatron. In the framework of the CO model, the J/ψ mesons should be transverse polarized at the large transverse momenta $\mathbf{p}_T$. However, this is in contradiction with the experimental data, too. The CO model has been applied earlier [31, 32] in an analysis of the J/ψ inelastic production experimental data at HERA [33]. However, the results do not agree with each other [32]. We note that the shapes of the Q^2, $\mathbf{p}_T^2$ and y^* distributions are not reproduced by the calculation [31], and the z distributions [32] contradict the HERA experimental data too. Results obtained within the usual collinear approach and CS model underestimate experimental data by factor about 2.

The deep inelastic J/ψ production at HERA in the CS model with k_T-factorization also was considered in [15, 16]. The results [16] agree with the H1 experimental data [33] both in normalization and shape only at quite small charmed quark mass $m_c = 1.4\,\mathrm{GeV}$. The theoretical prediction [15] are stimulated the experimental analysis of J/ψ polarization properties at HERA conditions.

Based on the above mentioned results here we use the CS model and the k_T-factorization approach for the analysis of the H1 data [1]. We investigate the dependences of the single differential J/ψ production cross section on different forms of the unintegrated gluon distribution. Special attention is drawn to the unintegrated gluon distributions obtained from BFKL evolution equation which has been applied earlier in our previous papers [12–16]. For studying J/ψ meson polarization properties we calculate the $\mathbf{p}_T$ dependences of the spin aligment parameter α.

The outline of this paper is as follows. In Section 2 we present, in analytic form, the differential cross section for the deep inelastic J/ψ leptoproduction in the CS model with k_T-factorization. Section 3 contains the numerical results of our calculations and the comparisons them with the H1 [1] and ZEUS [34] data. Finally, in Section 4, we give some conclusions.

2. Details of the calculations

In the k_T-factorization approach deep inelastic J/ψ meson production is determined by the contribution of six photon-gluon fusion diagrams with "unusual" properties of the gluons in proton. These gluons are off mass shell with the virtuality $q^2 = q_T^2 = -\mathbf{q}_T^2$, and their distribution in x and $\mathbf{q}_T^2$ in a proton is given by the unintegrated gluon structure function $\Phi(x, \mathbf{q}_T^2, \mu^2)$. The differential cross section for the deep inelastic J/ψ production differential cross section in the k_T-factorization approach has the following form [16]:

$$\begin{aligned} d\sigma(e\,p \to e'\,J/\psi\,X) = {} & \frac{1}{128\pi^3}\,\frac{\Phi(x_2,\,\mathbf{q}_{2T}^2,\,\mu^2)}{(x_2\,s)^2\,(1-x_1)}\,\frac{dz}{z\,(1-z)}\,dy_\psi \times \\ & \times \sum |M|^2_{\mathrm{SHA}}(e\,g^* \to e'\,J/\psi\,g')\,d\mathbf{p}_{\psi T}^2\,dQ^2\,d\mathbf{q}_{2T}^2\,\frac{d\phi_1}{2\pi}\,\frac{d\phi_2}{2\pi}\,\frac{d\phi_\psi}{2\pi}, \end{aligned} \tag{2}$$

where $\mathbf{p}_{\psi T}$ and $\mathbf{q}_{2T}$ are transverse 4-momenta of the J/ψ meson and initial BFKL gluon, y_ψ is the rapidity of J/ψ meson (in the *ep* c.m. frame), ϕ_1, ϕ_2 and ϕ_ψ are azimuthal angles of the incoming virtual photon, BFKL gluon and outgoing J/ψ meson, x_1 and x_2 are the photon and gluon momentum fraction.

The matrix element $\sum |M|^2_{\mathrm{SHA}}(e\,g^* \to e'\,J/\psi\,g')$ depends on the BFKL gluon virtuality $\mathbf{q}_{2T}^2$ and differs from the one of the SPM. We used here the results obtained in [16]. For studying J/ψ polarized production we used the 4-vector of the longitudinal polarization ε_L^μ as in [35].

There are several theoretical approximations for $\Phi(x, \mathbf{q}_T^2, \mu^2)$, which are based on solution of the BFKL evolution equations. As in our previous papers [9–12], we used here the so called JB [36] and KMS [37] parametrizations (see also refs. [7, 23] for more details).

3. Numerical results

In this section we present the theoretical results in comparison with recent experimental data taken by the H1 [1] and ZEUS [34] collaborations at HERA.

There are three parameters which determine the common normalization factor of the cross section under consideration: J/ψ meson wave function at the origin $\psi(0)$, charmed quark mass m_c and factorization scale μ. The value of the J/ψ meson wave function at the origin may be calculated in a potential model or obtained from the well known experimental decay width $\Gamma(J/\psi \to \mu^+ \mu^-)$. In our calculation we used $|\psi(0)|^2 = 0.0876\,\mathrm{GeV}^3$ as in [38].

Concerning a charmed quark mass, the situation is not clear: on the one hand, in the nonrelativistic approximation one has $m_c = m_\psi/2 = 1.55\,\mathrm{GeV}$, but on the other hand there are examples when smaller value of a charm mass $m_c = 1.4\,\mathrm{GeV}$ is used [32, 43]. However, in our previous paper [16] we analyzed in detail the influence of charm quark mass on the theoretical results. We found that the main effect of change of the charm quark mass connects with final phase space of J/ψ meson, and in the subprocess matrix elements this effect is neglectable. Taking into account that the value of $m_c = 1.4\,\mathrm{GeV}$ corresponds to the unphysical phase space of J/ψ state, in the present paper we will use value of a charm mass $m_c = 1.55\,\mathrm{GeV}$ only. Also the most significant theoretical uncertanties come from the choice of the factorization scale μ_F and renormalization one μ_R. One of them is related to the evolution of the gluon distributions $\Phi(x, \mathbf{q}_T^2, \mu_F^2)$, the other is responsible for strong coupling constant $\alpha_S(\mu_R^2)$. As often done in literature, we set $\mu_F = \mu_R = \mu$. In the present paper we used the following choice $\mu^2 = \mathbf{q}_{2T}^2$ as in [5, 16].

3.1. INELASTIC J/ψ LEPTOPRODUCTION AT HERA

The integration limits in (2) are taken as given by kinematical conditions of the H1 experimental data [1]. One kinematical region[1] is $2 < Q^2 < 100\,\mathrm{GeV}^2$, $50 < W < 225\,\mathrm{GeV}$, $0.3 < z < 0.9$, $\mathbf{p}_{\psi T}^{*2} > 1\,\mathrm{GeV}^2$ and other kinematical region is $12 < Q^2 < 100\,\mathrm{GeV}^2$, $50 < W < 180\,\mathrm{GeV}$, $\mathbf{p}_{\psi T}^2 > 6.4\,\mathrm{GeV}^2$, $0.3 < z < 0.9$ and $\mathbf{p}_{\psi T}^{*2} > 1\,\mathrm{GeV}^2$. Here and in the following, we used $\Lambda_{\mathrm{QCD}} = 250\,\mathrm{MeV}$.

[1]Here we denote the J/ψ meson transverse momentum and rapidity in the $\gamma^* p$ c.m. frame by $\mathbf{p}_{\psi T}^*$ and y_ψ^*, respectively.

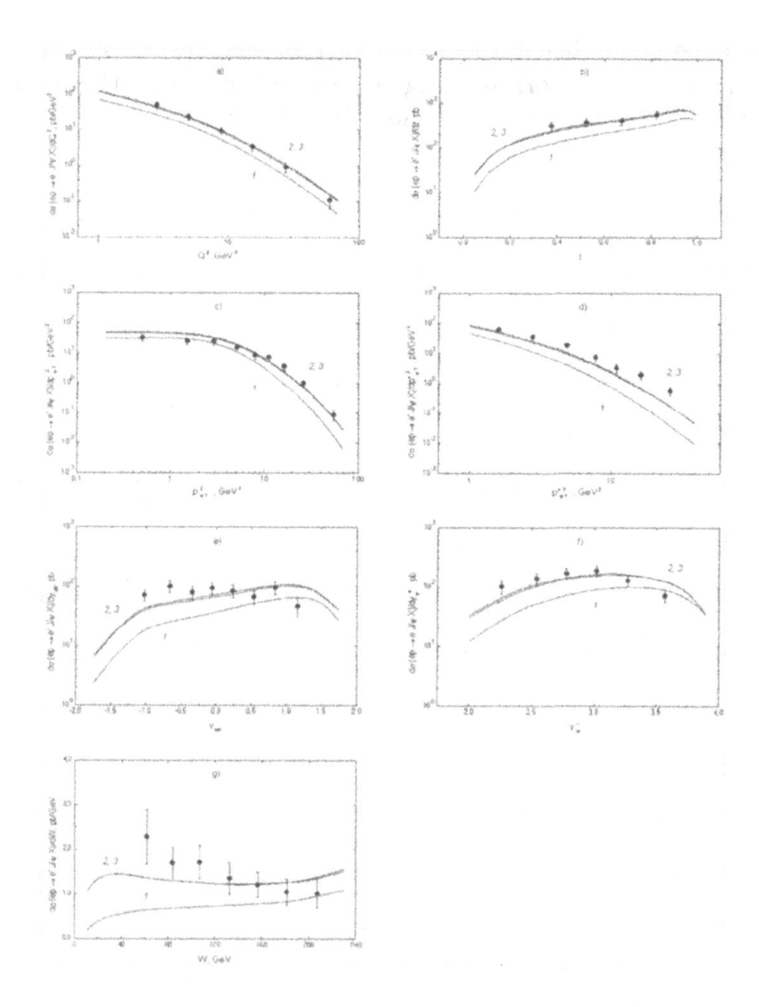

Figure 1. The single differential cross sections of the deep inelastic J/ψ leptoproduction obtained in the kinematical region $2 < Q^2 < 100\,\text{GeV}^2$, $50 < W < 225\,\text{GeV}$, $0.3 < z < 0.9$ and $\mathbf{p}_{\psi\,T}^{*\,2} > 1\,\text{GeV}^2$ at $\sqrt{s} = 314\,\text{GeV}$ in comparison with the H1 data [1]. Curve 1 corresponds to the SPM calculations at the leading order approximation with GRV (LO) gluon density, curves 2 and 3 correspond to the k_T-factorization QCD calculations with the JB and KMS unintegrated gluon distributions.

The results of our calculations are shown in Fig. 1 and 2. Fig. 1 shows the single differential cross sections of the deep inelastic J/ψ meson leptoproduction obtained in the first kinematical region at $\sqrt{s} = 314\,\text{GeV}$. Curve *1* corresponds to the SPM calculations at the leading order approximation with the GRV (LO) gluon density, curves *2* and *3* correspond to the k_T-factorization results with the JB (at $\Delta = 0.35$ [17, 23]) and the KMS unintegrated gluon distributions. One can see that results obtained in the CS model with k_T-factorization agree very well with the H1 experimental data. The SPM calculations are lower than the data by a factor 2 - 3.

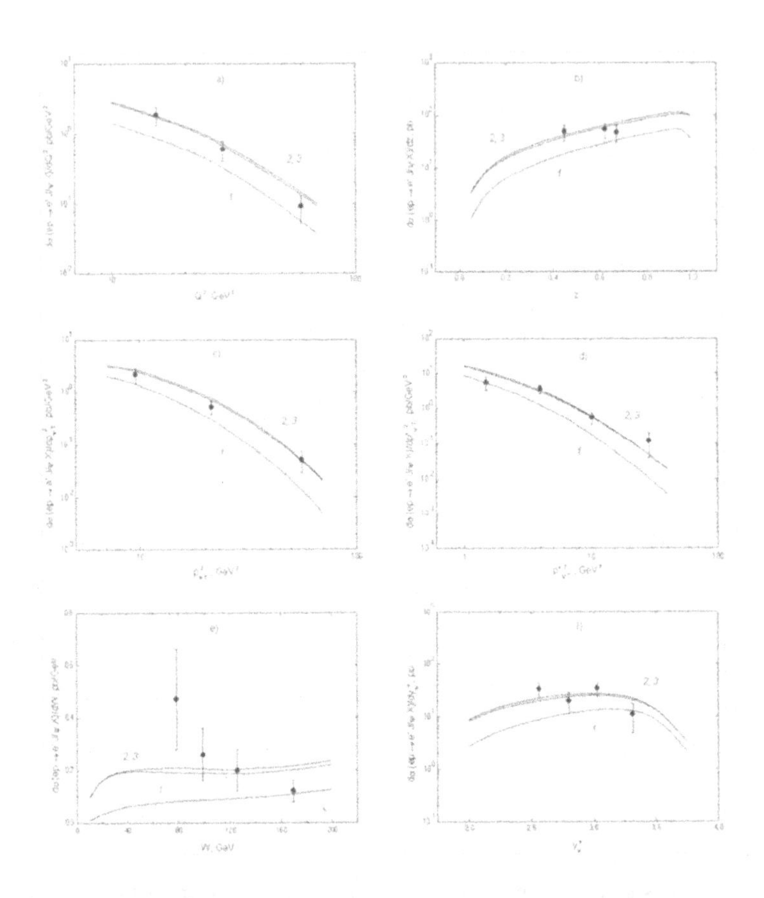

Figure 2. The single differential cross sections of the deep inelastic J/ψ leptoproduction obtained in the kinematical region $12 < Q^2 < 100\,\mathrm{GeV}^2$, $50 < W < 225\,\mathrm{GeV}$, $\mathbf{p}_{\psi\,T}^2 > 6.4\,\mathrm{GeV}^2$, $0.3 < z < 0.9$ and $\mathbf{p}_{\psi\,T}^{*\,2} > 1\,\mathrm{GeV}^2$ at $\sqrt{s} = 314\,\mathrm{GeV}$ in comparison with the H1 data [1]. Curves $1 - 3$ are the same as in Fig. 3.

We would like to note the difference in the transverse momentum distribution shapes between curves obtained using the k_T-factorization approach and the SPM. This difference manifests the p_T broadening effect mentioned in early. It is visible also that only the k_T-factorization approach gives a correct description of the $\mathbf{p}_{\psi\,T}^2$ spectrum. However, we note that the $\mathbf{p}_{\psi\,T}^{*\,2}$ distribution somewhat less well described (in contrast with the $\mathbf{p}_{\psi\,T}^2$ spectrum) at high values of the J/ψ transverse momentum (see Fig. 1d).

Also we point out the good description of the z distributions which obtained in the k_T-factorization approach in contrast with CO model results [32], except for the region $z < 0.3$, where the contribution of the resolved photon process may be important [40].

Fig. 2 shows the single differential cross sections of the inelastic J/ψ meson production obtained in the second kinematical region at $\sqrt{s} = 314\,\mathrm{GeV}$. Curves *1 - 3* are the same as in Fig. 1. We find also good agreement between the results obtained in the CS model with k_T-factorization and the H1 data. It is notable that in this kinematical region in contrast with first one the both $\mathbf{p}_{\psi T}^2$ and $\mathbf{p}_{\psi T}^{*2}$ distributions agree with the data. It is interesting to note that results obtained with the JB unintegrated gluon distribution at $\Delta = 0.35$ and the KMS one, which effectively included about 70% of the full NLO corrections to the value of Δ [37], coincide practically in a wide kinematical region.

Fig. 1 and 2 show that the k_T-factorization results for deep inelastic J/ψ leptoproduction with realistic value of a charm mass $m_c = 1.55\,\mathrm{GeV}$ agree well with the H1 experimental data without any additional $c\bar{c} \to J/\psi$ fragmentation mechanisms, such as the CO contributions.

3.2. POLARIZATION PROPERTIES OF THE J/ψ MESON AT HERA

As it was mentioned above, one of differences between the k_T-factorization approach and the SPM is connected with polarization properties of the final particles. In the present paper for studying J/ψ meson polarization properties we calculate the $\mathbf{p}_T$ dependence of the spin aligment parameter α [14–16]:

$$\alpha(\mathbf{p}_{\psi T}) = \frac{d\sigma/d\mathbf{p}_{\psi T} - 3\,d\sigma_L/d\mathbf{p}_{\psi T}}{d\sigma/d\mathbf{p}_{\psi T} + d\sigma_L/d\mathbf{p}_{\psi T}}, \tag{3}$$

where σ_L is the production cross section for the longitudinally polarized J/ψ mesons. The parameter α controls the angular distribution for leptons in the decay $J/\psi \to \mu^+\,\mu^-$ (in the J/ψ meson rest frame):

$$\frac{d\Gamma(J/\psi \to \mu^+\,\mu^-)}{d\cos\theta} \sim 1 + \alpha\cos^2\theta. \tag{4}$$

The cases $\alpha = 1$ and $\alpha = -1$ correspond to transverse and longitudinal polarization of the J/ψ meson, respectively.

In our previous paper [16] we analyzed in detail the Q^2 and $\mathbf{p}_{\psi T}^2$ dependences of the spin parameter α in leptoproduction case. We found that it is impossible to make of exact conclusions about a BFKL gluon contribution to the polarized J/ψ production cross section because of large additional contribution from initial longitudinal polarization of virtual photons. However at low Q^2 and in photoproduction limit these contributions are negligible. This fact should result in observable spin effects of final J/ψ mesons, connected with the k_T-factorization effects. In this paper we have performed such calculations for the inelastic J/ψ photoproduction process.

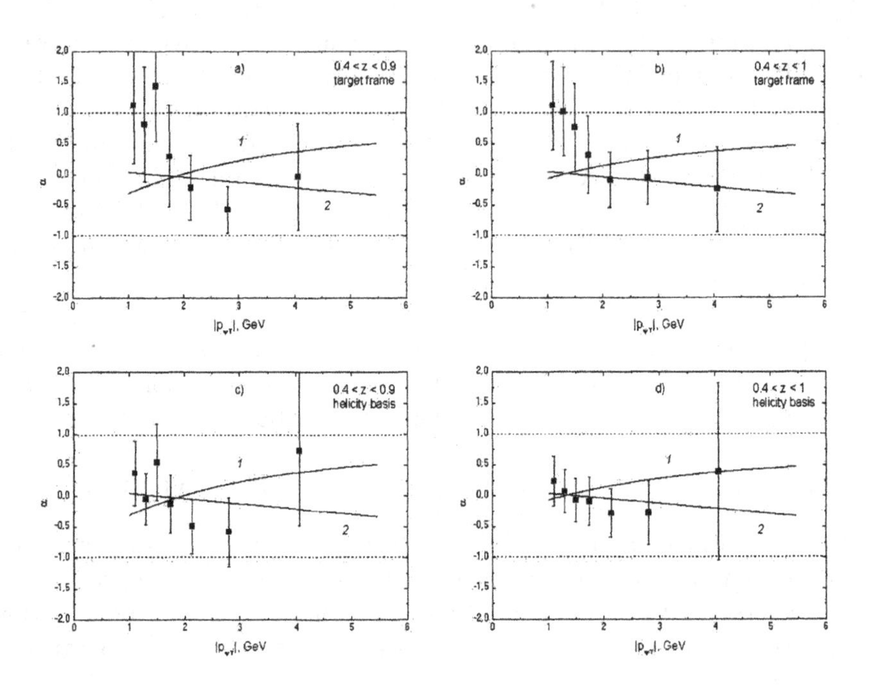

Figure 3. The parameter α as a function $\mathbf{p}_{\psi T}$ for the inelastic J/ψ photoproduction process which obtained in the kinematical region $50 < W < 180\,\text{GeV}$, $0.4 < z < 0.9$ (Fig. 3a, c), $0.4 < z < 1$ (Fig. 3b, d) and $1 < \mathbf{p}_{\psi T}^2 < 60\,\text{GeV}^2$ in comparison with the ZEUS [34] data. Curve 1 corresponds to the SPM calculations at the leading order approximation with GRV (LO) gluon density, curve 2 corresponds to the k_T-factorization QCD calculations with the JB unintegrated gluon distribution.

Fig. 3 shows the $\mathbf{p}_{\psi T}$ dependence of the spin parameter α in comparison with the ZEUS experimental data obtained in the kinematical region $50 < W < 180\,\text{GeV}$, $0.4 < z < 0.9$ (Fig. 3a and 3c) and $0.4 < z < 1$ (Fig. 3b and 3d). We note that in Fig. 3a and 3b the quantisation axis is chosen to be opposite of the incoming proton direction in the J/ψ rest frame, θ is the opening angle between the quantisation axis and the μ^+ direction of flight in the J/ψ rest frame. This frame is known as the "target frame" [34]. In Fig. 3c and 3d, the quantisation axis was defined as the J/ψ direction of flight in the ZEUS coordinate system. This frame is known as the "helicity basis" [34]. Curve *1* corresponds to the SPM calculations in the leading order approximation with the GRV (LO) gluon density, curve *2* corresponds to the k_T-factorization results obtained with the JB (at $\Delta = 0.35$ [17, 23]) unintegrated gluon distribution.

It is visible that only the k_T-factorization approach gives a correct description of the ZEUS data, although the experimental points have large errors. We also have large difference between predictions of the leading order of the SPM and the k_T-factorization approach. The SPM predictions lies somewhat below the data at low $\mathbf{p}_{\psi T}$ and somewhat above at high $\mathbf{p}_{\psi T}$. Therefore experimental measurement of polarization properties of the J/ψ mesons will be an additional test of BFKL gluon dynamics.

4. Conclusions

In this paper we considered the deep inelastic J/ψ meson leptoproduction at HERA in the colour singlet model using the standard parton model in leading order in α_S and the k_T-factorization QCD approach. We investigated the single differential cross sections of inelastic J/ψ production on different forms of the unintegrated gluon distribution. The $\mathbf{p}_T$ dependence of the spin aligment parameter α was done also. We compared the theoretical results with recent experimental data taken by the H1 and ZEUS collaboration at HERA. We have found that the k_T-factorization results (in contrast with the SPM ones) with the JB and KMS unintegrated gluon distributions agree well with the experimental data at realistic value of a charm mass $m_c = 1.55\,\mathrm{GeV}$, $|\psi(0)|^2 = 0.0876\,\mathrm{GeV}^3$ and $\Lambda_{\mathrm{QCD}} = 250\,\mathrm{MeV}$ without any additional transition mechanism from $c\bar{c}$-pair to the J/ψ mesons (such as given by the CO model). We also found that results obtained with the JB unintegrated gluon density at $\Delta = 0.35$ and KMS one, which effectively included about 70% of the full NLO corrections to the Pomeron intercept Δ, practically coincide in a wide kinematical region for J/ψ production processes at HERA conditions. Finally, it is shown that experimental study of a polarization of J/ψ meson at HERA should be additional test of BFKL gluon dynamics.

5. Acknowledgments

The authors would like to thank S. Baranov for encouraging interest and useful discussions. A.L. thanks also V. Saleev for the help on the initial stage of work. The study was supported in part by RFBR grant 02–02–17513 and INTAS grant YS 2002 N399.

References

1. Adloff C. *et al.* (H1 Collab.) (2002) Inelastic Leptoproduction of J/Ψ Mesons at HERA, *Eur. Phys. J.* **25, 1**, 41.
2. Gribov L., Levin E. and Ryskin M. (1983) Semihard Processes in QCD, *Phys. Rep.* **100**, 1.
3. Catani S., Ciafoloni M. and Hautmann F. (1991) High-Energy Factorization and Small x Heavy Flavor Production, *Nucl. Phys.* **B 366**, 135.
4. Collins J. and Ellis R. (1991) Heavy Quark Production in Very High-Energy Hadron Collisions, *Nucl. Phys.* **B 360**, 3.
5. Levin E., Ryskin M., Shabelsky Yu. and Shuvaev A. (1991) Heavy Quark Production in Semihard Nucleon Interactions, *Sov. J. Nucl. Phys.* **53**, 657.
6. Kuraev E., Lipatov L. and Fadin V. (1976) Multi-Reggeon Processes in the Yang-Mills Theory, *Sov. Phys. JETP* **44**, (1976) 443; (1977) The Pomeranchuk Singularity in Nonabelian Gauge Theories, *Sov. Phys. JETP* **45**, 199;
   Balitsky Yu. and Lipatov L. (1978) The Pomeranchuk Singularity in Quantum Chromodynamics, *Sov. J. Nucl. Phys.* **28**, 822.
7. Andersson B. *et al.* (Small x Collab.) (2002) Small x Phenomenology: Summary and Status, *Eur. Phys. J.* **C 25**, 77, hep-ph/0204115.
8. Ryskin M., Shabelski Yu. and Shuvaev A. (1996) Heavy Quark Production in Parton Model and in QCD, *Z. Phys.* **C 69**, 269.
9. Ryskin M., Shabelski Yu. and Shuvaev A. (2001) Comparision of k_T Factorization Approach and QCD Parton Model for Charm and Beauty Hadroproduction, *Phys. Atom. Nucl.* **64**, 1995.
10. Saleev V. and Zotov N. (1994) Heavy Quarkonium Photoproduction at High Energies in Semihard Approach, *Mod. Phys. Lett.* **A 9**, 151.
11. Saleev V. and Zotov N. (1996) Heavy Quark Photoproduction in the Semihard Approach at HERA and beyond, *Mod. Phys. Lett.* **A 11**, 25.
12. Lipatov A. and Zotov N. (2000) Study of BFKL Gluon Dynamics in Heavy Quarkonium Photoproduction at HERA, *Mod. Phys. Lett.* **A 15**, 695.
13. Lipatov A., Saleev V. and Zotov N. (2000) Heavy Quark Photoproduction in the Semihard QCD Approach and the Unintegrated Gluon Distribution, *Mod. Phys. Lett.* **A 15**, 1727.
14. Baranov S., Lipatov A. and Zotov N. (2001) Heavy Quark Production in the Semihard QCD Approach at HERA and beyond, *Proceedings of the 9th International Workshop on DIS and QCD (DIS'2001), Bologna, Italy*; hep-ph/0106229.
15. Baranov S. (1998) Probing the BFKL gluons with J/Ψ leptoproduction, *Phys. Lett.* **B 428**, 377.
16. Lipatov A. and Zotov N. (2002) Deep Inelastic J/Ψ Production at HERA in the Color Singlet Model with k_T Factorization, hep-ph/0208237; to be published in *Yad. Fiz.*.
17. Baranov S. and Zotov N. (1999) Pomeron Intercept and BFKL Gluon Dynamics in the $p_T(D^*)$ Spectra at HERA, *Phys. Lett.* **B 458**, 389; (2000) BFKL Gluon Dynamics and Resolved Photon Processes in D^* and Dijet Associated Photoproduction at HERA, *Phys. Lett.* **B 491**, 111.
18. Baranov S. and Smizanska M. (2000) Semihard B Quark Production at High Energies versus Data and Other Approaches, *Phys. Rev.* **D 62**, 014012.
19. Hagler P., Kirschner R., Schefer A. *et al.* (2000) Heavy Quark Production as Sensetive Test for an Improved Description of High-Energy Hadron Collisions, *Phys. Rev.* **D 62**, 071502.
20. Jung H., (2002) Heavy Quark Production at the TEVATRON and HERA Using k_T Factorization with CCFM Evolution, *Phys. Rev.* **D 65**, 034015; hep-ph/0110034.
21. Hagler P., Kirschner R., Schefer A. *et al.* (2001) Direct J/Ψ Hadroproduction in k_T Factorization and the Color Octet Mechanism, *Phys. Rev.* **D 63**, 077501;
    Yuan F. and Chao K.-T. (2001) Color Singlet Direct J/Ψ and J/Ψ' Production at

TEVATRON in the k_T Factorization Approach, *Phys. Rev.* **D 63**, 034006.
22. Jung H. (2001) Heavy Quark Production at HERA in k_T Factorization Supplemented with CCFM Evolution, Proceedings of Photon'2001, Ascona, Switzerland; hep-ph/0110345.
23. Baranov S., Jung H., Jönsson L. and Zotov N.P. (2002) A Phenomenological Interpretation of Open Charm Production at HERA in Terms of the Semi-Hard Approach, *Eur. Phys. J.* **C 24**, 425; hep-ph/0203025.
24. Braaten E. and Fleming S. (1995) Color Octet Fragmentation and the Ψ' Surplus at the TEVATRON, *Phys. Rev. Lett.* **74**, 3327;
Braaten E. and Yuan T. (1995) Gluon Fragmentation into Spin Triplet S Wave Quarkonium, *Phys. Rev.* **D 52**, 6627.
25. Cho P. and Leibovich A. (1996) Color Octet Quarkonia Production, *Phys. Rev.* **D 53**, 150; Color Octet Quarkonia Production. 2, *Phys. Rev.* **D 53**, 6203.
26. Bodwin G., Braaten E. and Lepage G. (1995) Rigorous QCD Analysis of Inclusive Annihilation and Production of Heavy Quarkonium, *Phys. Rev.* **D 51**, 1125; (1997) *Phys. Rev.* **D 55**, 5853(E).
27. Cacciari M. and Kramer M. (1996) Color Octet Contributions to J/Ψ Photoproduction, *Phys. Rev. Lett.* **76**, 4128.
28. Ko P., Lee J. and Song H. (1996) Color Octet Mechanism in $\gamma + p \to J/\Psi + X$, *Phys. Rev.* **D 54**, 4312.
29. Aid S. *et al.* (H1 Collab.) (1996) A Measurement and QCD Analysis of the Proton Structure Function $F_2(x, Q^2)$ at HERA, *Nucl. Phys.* **B 472**, 3.
30. Breitweg J. *et al.* (ZEUS Collab.) (1997) Measurement of Inelastic J/Ψ Photoproduction at HERA, *Z. Phys.* **C 76**, 599.
31. Fleming S. and Mehen T. (1998) Leptoproduction of J/Ψ, *Phys. Rev.* **D 57**, 1846.
32. Kniehl B. and Zwirner L. (2002) J/Ψ Inclusive Production in *ep* Deep Inelastic Scattering at DESY HERA, *Nucl. Phys.* **B 621**, 337; hep-ph/0112199.
33. Adloff C. *et al.* (H1 Collab.) (1999) Charmonium Production in Deep Inelastic Scattering at HERA, *Eur. Phys. J.* **C 10**, 373.
34. Chekanov S. *et al.* (ZEUS Collab.) (2002) Measurement of Inelastic J/Ψ and J/Ψ' Photoproduction at HERA, DESY 02-163.
35. Beneke M. and Kramer M. (1997) Direct J/Ψ and J/Ψ' Polarization and Cross-Sections at the TEVATRON, *Phys. Rev.* **D 55**, 5269.
36. Blumlein J. (1995) On the k_T Dependent Gluon Density of the Parton, DESY 95-121.
37. Kwiecinski J., Martin A. and Stasto A. (1997) A Unified BFKL and GLAP Description of F_2 Data, *Phys. Rev.* **D 56**, 3991.
38. Kramer M. (1996) QCD Corrections to Inelastic J/Ψ Photoproduction, *Nucl. Phys.* **B 459**, 3.
39. Ball P., Beneke M. and Braun V. (1995) Resummation of Running Coupling Effects in Semileptonic B Meson Decays and Extractionof $|V_{cb}|$, *Phys. Rev.* **D 52**, 3929.
40. Jung H, Schuler G.A. and Terron J. (1992) J/Ψ-Production Mechanisms and Determination of the Gluon Density at HERA, DESY-92-028.

SUBJECT INDEX

AUTHOR INDEX

LIST OF CONTRIBUTORS

M.G. Albrow
Fermilab
MS 122, Wilson Road
Batavia, IL 60510
USA

A. Borissov
Department of Physics and Astronomy
University of Glasgow
Glasgow, G128QQ
United Kingdom

L.P. Csernai
Department of Physics,
University of Bergen
Allegeten 55, N-5007 Bergen
NORWAY

J.R. Cudell
Institut de Physique, Bat. B5
Universite' de Liege
Sart Tilman, B4000,
BELGIUM

L.G. Dakhno
Petersburg Nuclear Physics Institute
188300 Gatchina, St. Petersburg district
RUSSIA

V.V. Davydovsky
Institute for Nuclear Research
Nauki Av. 47, 03680 Kiev
UKRAINE

V.V. Ezhela
COMPAS group, IHEP
142284 Protvino
RUSSIA

V.S. Fadin
Budker Institute for Nuclear Physics
630090, Lavrentieva str. 11
Novosibirsk
RUSSIA

K. Goulianos
The Rockefeller University
1230 York Avenue,
New York, NY 10021
USA

L.L. Jenkovszky
N.N. Bogolyubov Institute for Theoretical Physics
Metrolohichna 14 b, 03143 Kiev
UKRAINE

M.I. Kotsky
Budker Institute for Nuclear Physics
630090 Lavrenteva str. 11, Novosibirsk
RUSSIA

V. Kundrat
Institute of Physics, AS CR
182 21 Praha 8
Czech Republik

E.A. Kuraev
Bogolyubov Lab. of Theor. Physics
Joint Institute for Nuclear Research
141980 Dubna, Moscow reg.
RUSSIA

V.I. Kuvshinov
B.I. Stepanov Institute of Physics
F. Skaryna av. 68
220072 Minsk
BELARUS

A.I. Lengyel
Insitute of Electron Physics
Universitetska 21
UA-88016 Uzhgorod
UKRAINE

L.N. Lipatov
N.N. Bogolyubov Theoretical Physics Department
Petersburg Nuclear Physics Institute
Orlava Roshcha, Gatchina,
1883000, St. Petersburg
RUSSIA

V.K. Magas
CFIF, Instituto Superior Tecnico
Physics Department
Av. Rovisco Pais, 1049-001 Lisbon
PORTUGAL

A.R. Malecki
Instytut Fizyki AP
Ul Podchoraz'ych, 2
Krakow, PL 30-084
POLAND

E.S. Martynov
Institut de Physique, Bat. B5
Universite' de Liege
Sart Tilman, B4000,
BELGIUM

N.N. Nikolaev
Institut f. Kernphysik,
Forschungszentrum Juelich
D-52425 Juelich,
GERMANY

A. Papa
Dipartimento di Fisica
Universita della Calabria
Porte Bucci, cubo 31c
87036 Arcavacata di Rende, Cosenza,
ITALIA

A.Prokudin
Dipartimento di Fisica Teorica
Universita' degli Studi di Torino
Via P.Giuria, 1, 10125 Torino
ITALY

A.A. Savin
University of Wisconsin-Madison
1150 University Av.
Madison, WI 53706-1390
USA

O.V. Selyugin
Institut de Physique, Bt. B5
Universite' de Liege
Sart Tilman, B4000
BELGIUM

D.V. Shirkov
N.N. Bogolyubov Lab. Theor. Physics
Joint Institute for Nuclear Research
Dubna, Moscow reg.
141980 RUSSIA

S.M. Troshin
Theory Department
Institute for High Energy Physics
142281 Protvino, Moscow reg.
RUSSIA

G.P. Vacca
Dipartimento di Fisica,
Universita' di Bologna
Via Irnerio 46,
40126 Bologna
ITALY

G. Wilk
The Andrzej Soltan Institute for Nuclear Studies
Nuclear Physics Dept. (Zd-PVIII)
Ul. Hoza 69, 00-681 Warsaw
POLAND

N.P. Zotov
SINP, Moscow State University
119992 Moscow
RUSSIA